THE PHILADELPHIA NAVY YARD

THE PHILADELF

A BARRA FOUNDATION BOOK

HIA NAVY YARD

From the Birth of the U.S. Navy to the Nuclear Age

JEFFERY M. DORWART

with Jean K. Wolf

UNIVERSITY OF PENNSYLVANIA PRESS

Philadelphia

10 9 8 7 6 5 4 3 2 1

Published by
University of Pennsylvania Press
Philadelphia, Pennsylvania 19104-4011

Library of Congress Cataloging-in-Publication Data

Dorwart, Jeffery M.
 The Philadelphia Navy Yard : from the birth of the U.S. Navy to the nuclear age / Jeffery
M. Dorwart with Jean K. Wolf.
     p.   cm.
 "A Barra Foundation book"
 Includes bibliographical records and index
 ISBN: 0-8122-3575-4 (alk. paper)
 1. Philadelphia Navy Yard—History. 2. Philadelphia (Pa.)—buildings, structures, etc.   I.
Title   II. Wolf, Jean K.
VA70.P5D67   2000
359.7/09748/11—dc21   00-060736

Frontispiece: Panoramic view of League Island Navy Yard, c. 1911. NARAM

Designed by Carl Gross

CONTENTS

ACKNOWLEDGMENTS

CLOSING THE Philadelphia Naval Shipyard marked the final scene in a long history. For 200 years there had been an American navy yard on the Delaware River in Philadelphia. The emotional response to this historic event by so many Delaware Valley residents, civilian and naval, who worked or served at the Philadelphia Navy Yard in the decades before closure reflected an awareness that a part of the region's history had disappeared forever. This realization led to enthusiastic support for researching and writing a history from earliest colonial roots to final deactivation of the Philadelphia Navy Yard. No one proved more supportive than Robert L. McNeil, Jr., president of The Barra Foundation, Inc., an organization dedicated to preserving Philadelphia's cultural, institutional, and historical heritage. Though moving away from book production, McNeil thought it so important to record the complete history of the navy yard that he convinced the Barra Foundation to support one more book in its program of Philadelphia histories. The University of Pennsylvania Press shared McNeil's enthusiasm and agreed to publish this book on the Philadelphia Navy Yard for both scholarly community of naval historians and wider general public who wanted to remember their local navy yard.

Fred Cassady, Jr., Technical Information Specialist for Business and Industry of the NAVSEA Shipbuilding Support Office of the Norfolk Naval Shipyard Detachment located in Philadelphia, procured vital documents, site plans, maps, photographs, and descriptions of historic buildings on League Island for this study. Randy Giancaterino, Michael Mally, and the staff of the Public Affairs Office, Philadelphia Naval Shipyard, allowed me access to the historical files and records even as they prepared for closure by boxing the material for transfer to the National Archives and Records Administration, Mid Atlantic Region in Philadelphia. The staff of the Mid Atlantic Archives, including Regional Administrator Robert Plowman, Assistant Director Kellee Blake, Archivist Shawn Aubitz, and Rebecca Warlow helped guide primary researcher Jean K. Wolf through the early navy yard records already deposited at the Archives.

At the Historical Society of Pennsylvania, Linda Stanley, Laura Beardsley, Dan Rolph, and Max Moeller helped locate the papers of Joshua Humphreys and other rare collections with which to reconstruct the navy yard's earliest years. Bruce Shearer assisted with prompt reproduction of rare images. At the same time, Joseph Benford, head of the Print and Photograph Collection, Free Library of Philadelphia, provided easy access to the collection. Barbara Wright introduced the collection of photographs at the Urban Archives at Temple University, Philadelphia, to Wolf and assistant researcher Christine Taniguchi of the National Park Service.

Special thanks must go to Michael Angelo, Librarian at the Independence Seaport Museum and Todd Bauders, photographer of Bauders Biomedical Photography of Lansdowne, Pennsylvania, for locating and photographing many rare and fragile images from the Lenthall Collection of the Franklin Institute, on loan to the Independence Seaport Museum. The photographic staff of the U.S. Naval Institute

Annapolis copied images of the Philadelphia Navy yard from their collection, while Jack Green facilitated research and rapid reproduction of photographs from the Naval Historical Center, Washington Navy Yard.

I reserve the last and warmest thanks for the home team of my naval history students, colleagues, and former Philadelphia Navy Yard employees, too numerous to recognize individually, with whom I spent hours discussing preparation of this book. My former student Joseph-James Ahern of the Atwater Kent Museum helped organize a day-long symposium at the National Archives on the history of the Philadelphia Navy Yard. The session brought together the last shipyard commander Captain Jon C. Bergner and former staff and employees to discuss recollections and the meaning of the institution to U.S. Navy and Delaware Valley history. Elaine Navarra, Libby Hart, Judy Odom, and the entire staff of the Paul Robeson Library, Rutgers University, Camden, New Jersey, offered constant assistance. Rodney P. Carlisle of Rutgers University and Russell F. Weigley of Temple University offered continual expert advice and support for this project. Loretta Carlisle of the History Department laboriously entered and proofed every chapter of the work on her computer. And last, my special thanks go to the two most important people in making this book possible. Historic Preservation Consultant Jean K. Wolf located and meticulously researched the primary sources, and my wife Nel typed the rough draft and put up with my constant ill temper during the writing of the book.

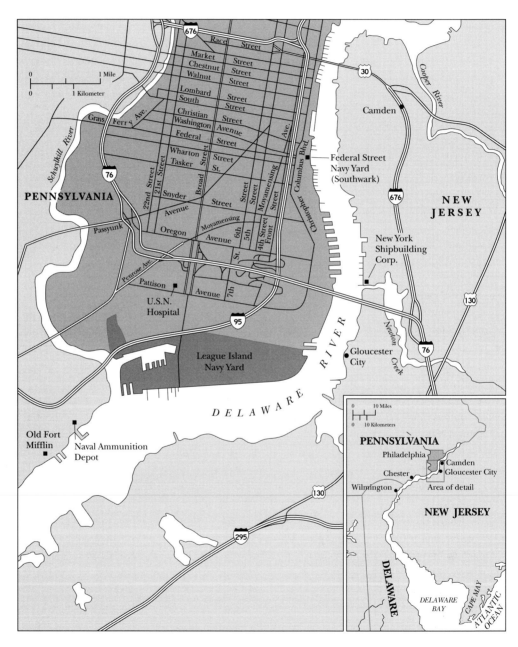

Located 100 miles up the Delaware River from the Atlantic Ocean, the Philadelphia Navy Yard existed from 1801 to 1875 on a 17-acre site in the Southwark district of the city and between 1876 and 1996 on the 900-acre League Island site.

INTRODUCTION

LOCAL NAVY YARDS made the United States into the world's greatest sea power. From ice-choked New England harbors, down the Atlantic seaboard, around the Gulf of Mexico, and up the Pacific coast, U.S. navy yards maintained and often constructed ships for the fleet. But naval officers who commanded these ships took navy yards for granted. "The fleet is the thing [and] the base exists for the fleet," Captain Wat Tyler Cluverius insisted in 1912. "The reference, then, of all work performed at a navy yard is to the ships—a unit of the fleet." Captain Alfred Thayer Mahan, father of the concept of sea power, agreed: "The highest function of a navy yard is to maintain the fleet in efficiency in war, [and] no utility in peace will compensate for the want of this in war."[1]

Naval historians as well often disregarded the navy yard, concentrating on the fighting admirals and ships in combat. Such neglect became more prevalent in the 1960s, when the government began closing navy yards and sending all new ship construction and most repair and conversion work to private shipyards. Gradually, historic U.S. government naval shipyards in New England, New York, Pennsylvania, South Carolina, Florida, and California became institutional relics of the American past, to be recalled through official in-house histories written by enthusiastic advocates eager to promote the unique role of their particular local institution. Some of these histories used their navy yard to illuminate larger organizational, technological, and socioeconomic change over an extended period of time. Most however, were nar-

rowly conceived and lacked the careful historical context and analysis necessary to assess institutional change. Such limited perspective dominated in-house histories of the Philadelphia Navy Yard, an institution that over two hundred years was known variously as the Continental, United States, Southwark, Federal Street, League Island, or Philadelphia Navy Yard, and after 1945 as the Philadelphia Naval Base and Naval Shipyard.[2]

One of the oldest and most historic American institutions, the Philadelphia Navy Yard received hagiographic treatment as America's first navy yard, original naval shipbuilding facility, and birthplace of the U.S. Navy and Marine Corps. In later years, local histories called it "Uncle Sam's Greatest Navy Yard," and the "finest and most extensive dockyard in the world." These superlatives, while satisfying to regional pride, distorted the historical meaning of the Philadelphia Navy Yard and obscured the larger importance of this evolving national institution that stood for two centuries near the center of some of the most significant changes in American history.

In reality, the Philadelphia Navy Yard suffered major limitations as a base to maintain the fleet for naval operations. "The selection of particular sites to serve this end should be governed by this one consideration, of usefulness in war," Captain Mahan announced. But the Philadelphia Navy Yard lay nearly one hundred miles from the Atlantic Ocean, up a Delaware Bay and River that until recently was often closed to navigation by ice during the winter. Shifting sand bars, dangerous shoals, and strong tidal currents annually altered the course of the river. Before the U.S. Army Corps of Engineers dredged a deep ship channel, there were times that no warship with a draft greater then eighteen feet could cross the bar below the Philadelphia Navy Yard. Larger ships built upriver had to be floated over the bar and armed or outfitted below at some other river port. Consequently, generations of naval officers disparaged Philadelphia. "As a home base its lack of strategic 'position' and its distance from the sea up a long narrow channel, puts it out of consideration," contended war planners on the Navy General Board. Mahan thought Philadelphia one of the poorest sites for a navy yard, arguing that New York and Norfolk provided the best location as permanent bases. "Boston, Philadelphia, Charleston, and others may serve for momentary utility, disseminating provision and preparation, but the protection given them as commercial ports will suffice for the inferior use made of them for supplying the fleet."[3]

Despite such continuing criticism, an American naval shore establishment existed in Philadelphia for well over two centuries, and a U.S. Navy Yard operated on the Delaware River near Philadelphia without interruption from early 1801 until late 1996. During its long history the navy yard evolved from an informal, private craft organization the government rented in the Southwark neighborhood into a large industrial naval manufacturing complex on League Island. This history makes the yard one of the most important institutions with which to understand organizational change in American naval history. From the American Revolution through the creation of a federal government, Philadelphia was the focal point of national institutional growth and thus a laboratory for earliest experiments in forming and building structures of government, including an American naval establishment.[4]

Philadelphia and surrounding Delaware River communities emerged during the early nineteenth century as the premier region for the design and development of wooden sailing ships and later iron steam-powered warships. The Navy Yard became the nexus of larger technological and organizational change in naval shipbuilding, particularly in the application of screw propeller technology to steam-powered engines on warships. During the Civil War, Philadelphia stood as the first line of naval defense for the Union after federal naval facilities to the south fell to Confederate forces. Philadelphia built, converted, and outfitted more than one hundred warships during the war, including ironclads. The Navy Yard outgrew its original location in the overcrowded Southwark waterfront district of the city, and in 1876 moved permanently downriver to League Island. At that time, Assistant Secretary of the Navy Gustavus Vasa Fox predicted that the place would become a leading government naval facility because it could lay up the entire ironclad fleet in a safe and secure freshwater basin. "The whole of League Island, embracing an extent of land that exceeds in area the six navy yards on the Atlantic Coast, will be ample for all the requirements of a naval station at Philadelphia, forever."[5]

Instead, the naval station on League Island barely survived, even as private industry and shipyards in the surrounding Delaware Valley built most of the iron and steel steam-powered warships for the late nineteenth-century American navy. Several times flood tides put the entire navy yard underwater and destroyed the floating dry dock. Barely fifteen years after the establishment of League Island, a violent wind storm blew down the only covered shipbuilding ways, and the Navy Department prepared to close its yard. Instead, the government spared the muddy island installation and gradually improved it, opening permanent dry docks in 1891 and 1907 and building research and testing facilities for fuel oil burning engines and boilers, propellers, and wireless telegraphy. In 1911, the Navy transferred the Atlantic Marine Corps Advance Base headquarters and school to League Island. During World War I, the Navy Department established shipbuilding ways, a naval aircraft factory, and a submarine base, and started building a third dry dock that promised to make League Island a first-class naval base and shipbuilding yard. "Philadelphia should be always retained as our leading subsidiary base," argued naval war planner John Hood, "and the navy yard there developed along the lines indicated as a building and manufacturing yard, and as a home station for the reserve fleet and auxiliaries and the advance base outfit."[6]

The Philadelphia Navy Yard constructed few ships during World War I, and soon scrapped two large battle cruisers laid down on the new shipways after the war. Rumors of closure circulated in the postwar era of naval arms limitations and budgetary cutback brought on by the Great Depression, but industrial recovery under the New Deal and naval rearmament for World War II gave Philadelphia its era of greatest importance. The Navy Yard added two 1,000-foot dry docks and dozens of heavy machine and other industrial shops and manufactured fifty-three warships, including three battleships, two heavy cruisers, and three aircraft carriers. The League Island facility expanded research and development laboratories

for turbine engines and boilers, propellers, aircraft, rockets, submarines, and for a moment even the atomic bomb.[7]

Postwar cutbacks led once again to rumors of closure, but the Cold War revived the yard, particularly during the Korean and Vietnam conflicts. The Philadelphia naval shore establishment became a center for assisting friendly European and Latin American navies, and although the naval aviation factory and material center closed, League Island continued research and development in aviation, including technology for space exploration, electronic warfare, and nuclear propulsion. But Philadelphia stopped building warships in the late 1960s, and the naval shipyard never received a license to assemble or overhaul the nuclear-powered submarines and surface vessels that became the backbone of the

Philadelphia Navy Yard Reserve Basin, aerial view, 1919. This view, looking toward the Girard Point grain elevator, shows the role of League Island as a home base for the Reserve Fleet. NARAM.

Dry Dock No. 1 on League Island was built in 1891. This 1919 view shows the Delaware riverfront piers and industrial buildings of the older section of the Navy Yard. NARAM.

Cold War fleet. Gradually a "culture of closure" permeated every aspect of Navy Yard operations in Philadelphia.[8]

Remarkably, the Philadelphia Navy Yard stayed alive for twenty-five years after its last ship launching in January 1969, displaying a resiliency unparalleled in American institutional history. In 1990, the Commission on Base Closure and Realignment recommended permanent closure, but even then political and legal maneuvering delayed closure and provided hope that once more the Navy Yard might remain active. Last-minute efforts to save the place created false expectations throughout a Delaware

Valley that had come to view the Philadelphia Navy Yard as a cherished and permanent institution. Consequently, final closure created intense emotional responses and disbelief that the government could so arbitrarily and abruptly close this historic and seemingly indispensable naval shore establishment.[9]

Many asked why the United States had closed the Philadelphia Navy Yard. But, as the following history demonstrates, the question more appropriately should have been why it had taken so long to do so. The answer lies in tracing the long, complex, and often tumultuous struggle to survive of an important American naval and leading Philadelphia institution.

Battleship *New Jersey* in dry dock on League Island during World War II. The Navy Yard built fifty-three warships during World War II, including the *New Jersey*. NARAM.

Below Launched on 7 December 1943, *Wisconsin* (seen here in the Delaware River off League Island) was the second *Iowa*-class battleship built at the Navy Yard during World War II. NARAM.

Stern of troopship *Henderson* at its 1916 launching. *Henderson* was League Island's first new ship contract. Launching days always drew enthusiastic crowds, establishing over time an emotional bond between the Delaware Valley and the Navy Yard. NARAM.

PART ONE

SOUTHWARK

1 ★ ORIGIN OF A NAVY YARD

LOCATED ON the Delaware River nearly one hundred miles from the open sea, the city of Philadelphia seemed an unlikely place for a navy yard. Founded in 1682 by Quaker entrepreneur William Penn, the Philadelphia river port shunned military and naval developments. Every other colony armed. The New England, Chesapeake Bay, and Carolina colonies constructed small warships for the British Lords of the Admiralty or local colonial governments and converted merchant vessels into heavily armed privateers to attack the commerce of Britain's enemies France and Spain. Virginia maintained a small naval shipyard for the British at Gosport.[1]

But the Quaker peace testimony spoke against such warlike preparation and the Quaker city on the Delaware River ignored maritime defense. Philadelphia, the largest merchant shipbuilding center in British North America, refused to raise a militia, build fortifications, or launch ships-of-war for its own protection. When Spanish and French maritime raiders came up the Delaware Bay to plunder New Castle and threatened the larger Pennsylvania port upriver, a Quaker-dominated assembly opposed construction of guard boats or conversion of privateers. During the French and Indian War (1756–63), Quaker legislators ignored Captain John Sibbald's appeal for armed vessels to protect Philadelphia's maritime trade routes or to raid those of the enemy. With an eye on the counting house as well as the meeting house, Philadelphia Quaker merchants declared that privateers created too much risk with higher insurance rates, prohibitive construction costs in convert-

ing merchant vessels, and inflated prices for naval supplies to outfit them. War raised the wage demands of local ship carpenters, chandlers, and common laborers. Worse, privateers could bring reprisals against the merchants' dominant trade position. "The Uncertain Posture of affairs makes Business very Dull," merchant shipowner Thomas Willing explained, "but [I] am in hopes a Peace will take place, & Commerce go on without Interruption."[2]

Willing's antiwar sentiment masked the Delaware Valley's mid-eighteenth-century emergence as a center for naval and military business. The South Carolina Commission on Fortifications ordered 1,000 iron cannonballs, guns, and military stores in 1757, "the whole purchased in Philadelphia," and further encouraged Delaware River arms traders. "If you think proper to make the Shipment in a Vessel of Your own, or one belonging to Your friends," the commissioners instructed Philadelphia merchant William Allen, "we will do our utmost to promote the Interest of such Vessel." Allen's friends in the growing local weapons trade included Willing, James Logan, Benjamin Franklin, John Wharton, James Penrose, Joseph Turner, and Robert Morris. They provided British army and colonial militia during the French and Indian War with clothing, boots, supply wagons, gun carriages, and fire locks for muskets. Local mechanics, organized in part by Franklin, forged iron and brass guns, manufactured gunpowder, and assembled flintlocks.[3]

Meanwhile, Delaware River ship carpenters constructed Philadelphia's first warship. In April 1762

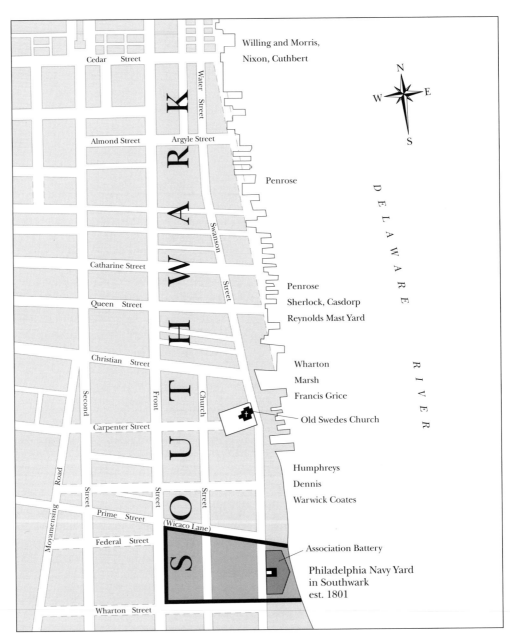

Cedar   Street

Willing and Morris,
Nixon, Cuthbert

Water Street

Almond Street   Argyle Street

Penrose

S O U T H W A R K

N
W   E
S

D E L A W A R E

Swanson

Catharine Street

Penrose

Street

Sherlock, Casdorp

Reynolds Mast Yard

Queen   Street

R I V E R

Christian   Street

Wharton

Marsh

Second

Front

Church

Francis Grice

Old Swedes Church

Carpenter Street

Humphreys

Dennis

Warwick Coates

Road

Street

Street

Street

Prime   Street

(Wicaco Lane)

S O U T H

Association Battery

Philadelphia Navy Yard
in Southwark
est. 1801

Moyamensing

Federal   Street

Wharton   Street

1762–1801. The Southwark waterfront between Cedar (South) Street and the Association Battery at the foot of Wicaco Lane (Prime Street) served between 1762 and 1801 as the informal American naval shore establishment in Philadelphia.

master shipwrights James and Thomas Penrose laid the keel for a 24-gun ship on their riverfront landing above Old Swedes Church in the "hamlet of Wicaco" just below the city. Named *Hero*, this ship became a model for frigates in the first American naval construction program during the Revolutionary War. In December 1775 the Continental Marine Committee (a forerunner of the U.S. Navy Department) authorized construction of two 32-gun and two 24-gun frigates, and "That the 24 Gun Ships be of the Same dimensions as the *Hero* Privateer built in the City of Philadelphia in the last War."[4]

Shortly after *Hero*'s launching, Wicaco shipbuilders petitioned the Pennsylvania provincial legislature to rename their neighborhood Southwark after the London shipbuilding region. Wicaco, the oldest community in the Philadelphia area, had been settled by the Swedish several decades before the British led by William Penn founded a port town immediately upriver. The Swedish settlers of Wicaco built a fortified blockhouse around 1669, a log church in 1677, and in 1700 Gloria Dei (Old Swedes) Church. The Swanson family became the largest riverfront landholders, with Christopher Swanson accumulating at least 150 acres of waterfront properties between Old Swedes Church and a landing place to the north. He subdivided his tract among sons-in-law John Parham, Joseph Knowles, and Anthony Duché, the latter a pottery maker and investor in Southwark's shipbuilding and maritime enterprise, and laid out Christian, Queen, and Catharine Streets along their property divisions above the church lands. These rough dirt tracks, extending slightly northwest from riverfront landing places, became sites for America's first naval shipbuilding community.[5]

The immensely profitable Philadelphia waterfront development delayed exploitation of the Wicaco area. William Penn encouraged maritime enterprise and shipbuilding along his port town's riverfront from Vine Street down to Cedar (later South) Street, the latter providing Wicaco's northern boundary. Before 1700 Penn invested in Charles West's Vine Street shipyard and Bartholomew Penrose's shipyard at the foot of Market Street. Both constructed large merchant vessels. Robert Turner's yard near Mulberry (later Arch) Street Landing also "built ships of considerable burthen." By the first decades of the eighteenth century, Philadelphia shipwrights and carpenters had constructed several hundred smaller vessels. Ropewalks, sawpits, and timber yards grew up around shipbuilding sites. Wharves, docks, storehouses and counting houses, brew and malt houses, and inns filled Philadelphia's compact waterfront. Crowding along the riverfront landings prompted Philadelphia shipbuilding and mercantile firms to seek properties south of Cedar Street.[6]

Philadelphia and Chester County, Pennsylvania, entrepreneurs developed Wicaco during King George's War (1740–48), when Spanish and French privateers on the Delaware threatened Philadelphia. Constant alarms of impending attack sounded, but the Quaker and Proprietary factions failed to raise a defense. Consequently, Provincial Land Office Secretary Richard Peters, civic leader Benjamin Franklin, and militant Quaker James Logan organized voluntary defense associations to arm Philadelphia. They used a lottery to fund the construction of a 13-gun battery on Anthony Atwood's wharf in the city above Cedar Street and a massive 27-gun earth and timber Association Battery in Wicaco. "The grand Battery at Wicacoa below this Town is also nearly finished," Franklin reported in April 1748, "and tis thought will be ready the Beginning of next Week to receive the heavy Cannon borrowed from N. York." Fifty-three years later, the original United States Navy Yard in Philadelphia arose on the exact spot where Franklin had erected this Association Battery.[7]

"Tradesmen for Building and Fitting out a Vessel of Defence" came to the little inlets and river landings between the Association Battery and Old Swedes Church to convert merchant ships into privateers or to build armed vessels. British and German carpenters, blacksmiths, rope and sail makers, and shipwrights left the overcrowded Philadelphia waterfront or migrated from Chester County to settle along the still unpaved Catharine, Queen, and Christian Streets and down the winding lanes (soon called Swanson and Front Streets) that ran north to south along the riverfront. "Southwark is getting

The Philadelphia and Southwark waterfront as seen from the New Jersey shore c. 1752 reveals the informal nature of the eighteenth-century shipbuilding business on the Delaware River, with shipways simply thrown up near landings. The inset shows the Association Battery at the foot of Wicaco Lane (later Prime Street), site of the first Philadelphia Navy Yard. HSP, section one of "An East Perspective View of the City of Philadelphia; taken by George Heap from the Jersey Shore, under the Direction of Nicholas Scull, Surveyor General of the Province of Pennsylvania, 1752" (Bc 864 H 4342).

greatly disfigured by erecting irregular and mean houses," Peters observed, "thereby so marring its beauty that, when [Pennsylvania Proprietor Thomas Penn] shall return, he will lose his usual pretty walk to Wicaco."[8]

While working folk and craftsmen settled Wicaco's narrow dirt lanes, shipwrights raised tiny wooden ship-launching ways and platforms along the open Southwark waterfront. Warwick Coates and Richard Dennis set up shipways near the foot of Wicaco Lane (later Prime Street) near the Association Battery. Between these ways and Swedes Landing (Christian Street) and Queen Street Landing, above Old Swedes Church, shipwrights Joseph Marsh, Thomas Casdorp, Simon Sherlock, Francis Grice, James Doughty, and the Penroses constructed ship and boat building facilities. Captains Henry Dougherty, Joseph Blewer, John Hazelwood, and John Conyngham settled nearby. These Southwark shipwrights, sea captains, and craftsmen formed the core of America's first naval shore establishment.[9]

Farther inland, wealthy Philadelphia merchant and iron forge owner Joseph Turner, merchant entrepreneurs Luke and Anthony Morris, and Thomas and Joseph Wharton, who moved to Southwark from the city for political and economic opportunity and bought riverfront properties, built mansion houses in the Wicaco neighborhood. Thomas Wharton observed that by 1762 the district had changed dramatically. "Knowing thy turn to Politicks," he wrote friend and fellow Wicaco investor William Fisher, I "cannot ommitt mentioning that a few days past our Assembly rose, having present the [Governor] with a Law for Payving the Streets, & for Regulating the Southern side of this City and Erecting it into a Borough, called it Southwark."[10]

In the midst of Southwark's emergence as an important maritime community, Joseph Wharton's nephew John, son of Chester County coroner and saddlemaker John Wharton and Mary Dobbins, invested in a shipbuilding enterprise with shipwright James Penrose, son of shipwright Thomas Penrose the younger and brother-in-law of Southwark shipwright Warwick Coates. Their property transactions revealed how eighteenth-century Philadelphia family and business connections provided the institutional roots for the first American naval shore establishment. With financial backing from Uncles Joseph and Thomas Wharton, Anthony Morris, and the Cedar Street mercantile house of Willing, Morris and Company, John Wharton accumulated Southwark riverfront properties. He purchased a lot adjoining Swanson's son-in-law John Parham's land near Queen Street, cousin James Reynolds's mast yard at the Queen Street Landing, and brother-in-law Joseph Dobbins's shares in the former Joseph Cox estate, which included Casdorp's boatyard.[11]

Wharton's partner James Penrose accumulated waterfront properties below the Penrose family wharves and storehouses on Queen Street and the Joseph and Samuel Wharton wharves and warehouses on the north side of Christian Street. Penrose leased a wharf and stores from uncle Isaac Penrose, a wealthy dry goods merchant, rented Jacob Lewis's lot above Christian that included Simon Sherlock's boatyard, and purchased a three-acre piece of land between Christian and Catharine, part of step-uncle Anthony Duché's original holdings and adjoining uncle Joseph Wharton's landing and docks. Finally, Penrose bought John Lawson's old boatyard between Christian and Queen Streets with an access lane that ran from the Delaware River through Swanson to Front Street.[12]

The Wharton and Penrose shipbuilding enterprise lay within paces of the future U.S. Navy Yard and was the immediate institutional antecedent for the Philadelphia naval shore establishment during the American Revolution. But it took association with John Wharton's cousin Joshua Humphreys, Jr., to complete the family linkage to an American naval establishment.

Business and family connections in the Philadelphia merchant shipbuilding community brought Humphreys to Penrose and Wharton. The Southwark shipbuilders started to construct larger seagoing merchant vessels of up to 300 tons for British and American mercantile interests, particularly the Free Society of Traders and (Society Hill) merchant shipowners Robert Morris, John Nixon, John Maxwell Nesbitt, and Thomas Wharton, Jr.—all intimately involved later in founding an American navy. To build larger ships, Penrose and Wharton placed contracts with the iron workers at the Durham Forge in Chester County, ship chandler James Wharton (John's older brother), and Society Hill timber merchant Joshua Humphreys, the elder.[13]

The senior Humphreys, son of one of the original Quaker settlers of Haverford Township in the "Welsh Tract" along the Schuylkill River west of the city, operated a farm, grist mill, and large sawmill in Humphreysville (later Bryn Mawr). He built a bridge over Haverford Mill Race in 1739 to connect his mills to the "Road Leading toward Phila." and in 1758 bought a timber yard between Front and Second Streets below Spruce, the property of his aunt Mary's father Peregrine Hogg, a wealthy London merchant. This purchase brought Humphreys close to uncle Benjamin, who operated a waterfront inn nearby on Dock Street, and older brother Clement, a sawyer who owned property on the Pine Street wharf. From this base, Humphreys established a prosperous timber enterprise in the postwar Philadelphia house building boom of the 1760s. He supplied sap board, scantlings, lath pine, and cedar board to his Dock Ward neighbor Benjamin Loxley, the wealthiest house carpenter in the city; master carpenters Robert Allison and David Evans; and Society Hill merchants Nixon, Nesbitt, Thomas Wharton, Samuel Rhoads, and cousin Levi Hollingsworth. Humphreys also sold building materials to stonemason John Palmer for the "Fortification of Deep Water Island" (later Fort Mifflin). His most important business was carried on with Southwark shipbuilders Penrose and Wharton. Humphreys opened Carpenter's Yard to supply them with "½ price boards," and more significantly in 1765 apprenticed son Joshua to Penrose "to learn the art of shipbuilding."[14]

With the younger Humphreys on board as an apprentice, the Wharton and Penrose shipbuilding enterprise (which now extended along Swanson Street past Queen, Catharine, and Christian) became the informal center of Philadelphia's ship construction, conversion, and repair industry. In the middle of this neighborhood resided Welsh, Scottish, Irish, and English shipwrights and their Southwark "Negroes," along with Swedish and German ship carpenters, joiners, caulkers, coopers, and common laborers—many destined to convert or construct the first ships of the Pennsylvania State, Continental, and United States navies.[15]

Joshua Humphreys, Jr., directed most of this work throughout the formative process of building the first American naval shore establishment. James Penrose's sudden death in 1771 catapulted his twenty-year-old apprentice to the center of the business. According to family tradition, Humphreys single-handedly completed the 300-ton ship *Sally*, then on the stocks. For a time, master shipwright Francis Grice, Jr. (whose grandson Isaac became a chief naval constructor for the U.S. Navy) replaced Penrose as John Wharton's partner. Grice and Wharton had learned the shipbuilding trade from Richard Dennis, whose wharf lay between the crumbling Association Battery and Old Swedes Church. In late 1773 Wharton bought out Grice's share of the business and gave it to the younger Humphreys, his cousin, in return for all the timber in Carpenter's Yard.[16]

Wharton and Humphreys began their shipbuilding partnership on the eve of a growing American crisis with Great Britain over colonial economic regulations, taxation, and trade restrictions. John Wharton had been active for several years in local nonimportation associations and protest movements against British mercantile policies. His older brother Thomas Wharton, Jr., became a leader of the Secret Committee of Correspondence and the Committee of Safety and Defense and served later as president of the Supreme Executive Council of Pennsylvania. Younger brother Charles Wharton became a member of the Continental Congress. The Whartons undoubtedly welcomed the gathering in Philadelphia in 1774 where they joined likeminded intercolonial delegates at a Continental Congress to discuss common resistance measures against growing British interference with American maritime commerce.[17]

By hosting the Continental Congress, Philadelphia became the base for organizing national institutions. The Second Congress convened in Philadelphia in

May 1775 as news arrived that the British army had clashed with Massachusetts minutemen along the road between the towns of Lexington and Concord and Boston, and that fighting near Boston harbor had caused heavy casualties. The second Congress resolved to take up arms in defense of American rights, form a Continental Army, and issue a Continental currency to support a military establishment. Massachusetts delegate John Adams urged that the delegates also consider building "a Fleet of our own." Representatives from other New England maritime areas joined Adams in calling for a national navy, while South Carolina delegate Christopher Gadsden, a former Royal Navy purser, encouraged such sentiments, assuring Adams that the British navy "is not so formidable to America, as we fear."[18]

Perhaps not, but at the moment the Congress showed little interest in creating a navy. Many delegates still were not committed to a complete break with Britain and viewed a navy as an expression of independence. Even Adams worried that, whereas forming an army could be interpreted as the traditional English right to raise a militia for self-defense, launching a navy implied offensive warfare. The most the Congress would do at this point was to instruct the colonies each to protect local coasts, harbors, and trade "at [their] own expense" by arming boats to discourage British naval patrols. Virginia delegate Thomas Jefferson noted approvingly in July 1775 that each colony would construct its own naval defense. "The New Englanders are fitting out light vessels of war," he told brother-in-law Francis Eppes, "by which it is hoped as shall not only clear the seas and bays here of everything below the size of a ship of war, but that they will visit the coasts of Europe and distress the British trade in every part of the world."[19]

Such thinking troubled Connecticut delegate Silas Deane, advocate of a large naval establishment to protect American overseas trade. Deane expressed his disappointment with the Congress's timidity in founding a Continental fleet to Pennsylvania delegate Benjamin Franklin. Franklin had confronted similar reluctance from the Quaker-dominated Pennsylvania Assembly when he tried to arm ships in

Philadelphia during the recent wars. He sympathized with Deane: "I lament with you the Want of a naval Force," and promised that "the next Winter will be employ'd in forming one."[20]

The traditionally antimilitaristic and parsimonious Pennsylvania government nevertheless handed Franklin the means to arm Philadelphia, passing a pathbreaking defense bill on 30 June 1775 that called for raising a militia, fortifying Philadelphia and the Delaware River, manufacturing arms and gunpowder, and issuing £35,000 in bills of credit for military preparations. To organize the military establishment proposed by the law, the Pennsylvania Assembly created a 25-member Committee of Safety and Defense that elected Franklin president and merchant shipowner Robert Morris vice president. The Committee held extra-legal powers under the military emergency to recruit "Minutemen" in each county, provide arms, and build forts and machines of war (river obstructions of iron-tipped logs called *chevaux-de-frise*). It could manufacture or purchase gunpowder, cannon, and small arms and contract for war supplies. It enforced law and order, coerced local compliance with the boycott of British products, and demanded sympathy for Whig resistance to British ministerial economic policy.[21]

The sixty-year-old Franklin, who attended forty-three of the first forty-six Defense Committee meetings, and Vice President Morris built a bureaucracy of committees and offices to coordinate the business of war. Though a somewhat informal gathering (sometimes only Franklin, Morris, and two or three other members attended), each committee provided an institutional base for a semiformal war department to manage a substantial Pennsylvania military and naval establishment. Franklin mobilized the city's scientists, mechanics, and inventors, particularly Philosophical Society members Owen Biddle, David Rittenhouse, Robert Smith, and Lewis Nicola, to organize war preparations. He rebuilt the craft militia associations from the previous war to produce arms. Franklin confided that his object was to manufacture "200 Muskets per Annum for Ten years, at a good Price, which I doubt not will in that time establish the Manufacture among us: and an

Arsenal with 20,000 good Fire locks in it, will be no bad thing for the Colony."[22]

Franklin advocated a system of fortifications and sunken river obstructions rather than a costly naval force to defend this emerging riverfront arsenal. During earlier wars, he had directed construction of redoubts around the city waterfront, including the Association Battery, and forts in Northampton County, and had helped plan Fort Pitt in western Pennsylvania. He seemed more comfortable with land than sea defenses, and thus dispatched shipbuilders Thomas Penrose, Jr., and John Wharton and naval architect Lewis Nicola to survey sites on the Delaware River for possible land fortification. The fortification committeemen suggested defensive works at Red Bank and Billingsport on the Jersey side of the river and on Mud Island at the confluence of the Schuylkill and Delaware Rivers, but they asked Franklin to consider a naval defense as well. Nicola urged construction of a floating battery to be anchored in the river; the others suggested construction of a Pennsylvania State Navy.[23]

As a consequence, Franklin ordered John Wharton and other local shipbuilders to submit construction plans for a flotilla of armed boats to guard the river. The Defense Committee accepted plans for an armed galley from Wharton and Kensington shipwright Emanuel Eyre, and resolved that Wharton immediately procure material, make preparation to lay the keel for "a Boat or Calevat," and exhibit a model of this boat the following morning. Both gunboat plans called for a double-ended craft with 50-foot keel, 13-foot beam, and drawing less than five feet of water so as to navigate the treacherous shoals and shallows of the Delaware River. Each galley showed a bank of oars, lateen sail, and long rudder sweep for steering. Plans provided that the bows mount heavy iron cannon, an arrangement that made the tiny gunboats unseaworthy.[24]

The Defense Committee accepted two identical flawed designs for armed boats to please the rival shipbuilding neighborhoods in Kensington to the north and Southwark below the city, which after the end of the French and Indian War had filled with unemployed shipwrights, carpenters, laborers, and common sailors. Wharton represented the interests of Southwark and Eyre those of his Kensington community. Both had been active members of the Committee of Sixty-Six that had convinced the Assembly to pass the June 1775 defense bill; thus it was good politics and business to bind these rival riverfront maritime communities to the defense mobilization effort through shipbuilding contracts. Franklin ordered shipbuilders "to immediately employ all the Carpenters & other Workmen that they think necessary, for collecting materials for building 12 Boats," and dispatched Wharton and Chester County militia Captain Anthony Wayne "to go to the Carpenters down Town" and Defense Committeemen Owen Biddle and Captain Henry Dougherty "to go to the Carpenters up Town."[25]

The Defense Committee gave the first two lucrative naval shipbuilding contracts (worth more than 500 pounds each) to the Wharton and Humphreys shipbuilding business. Here Humphreys supervised forty-two ship carpenters, sawyers, caulkers, and laborers on construction of the first Pennsylvania armed boat, named appropriately *Experiment*. Humphreys's crew next built the galley *Washington*. Construction took sixteen days for each boat; then Southwark shipwright Casdorp laid the keel for galley *Chatham*, Coates for *Burke*, Marsh for *Warren*, and Sherlock for *Camden*. Most likely, all six Southwark galleys were constructed on the Wharton and Humphreys shipbuilding grounds between Christian and Queen Streets. Meanwhile in Kensington, Emanuel Eyre built row galleys *Franklin*, *Congress*, and *Bull Dog*, shipbuilding partner John Rice *Dickinson*, and William Williams *Hancock*.[26]

In September 1775 Franklin submitted to the Pennsylvania Assembly "an Estimate of Moneys already expended" for the construction of "13 Armed Boats or Gondolos Built, armed and Equipped." The first American naval program cost 7,500 pounds Pennsylvania money for construction and 7,890 pounds for "victualing" the 53-man crew for each armed boat. Franklin estimated that the Pennsylvania Navy cost more than eight times as much as the erection of fortifications and chevaux-de-frise. Not only was local naval construction expensive, but the product was inferior. Humphreys admitted that the first effort was, as the name of one galley implied, an experiment "which gave an opportunity for improving on the two first that was built." Philadelphia privateer captain John Macpherson was more harsh. "Superannuated and tiresome" but "reputed to be well skilled in naval Affairs," Macpherson told the Defense Committee that these new boats were unfit to carry guns. He wanted the rudder sweeps lengthened and a second lateen sail added to improve handling in the strong Delaware River tidal currents.[27]

Franklin ignored such criticism of his expensive little navy. After all, John Adams and other naval-minded members of the Continental Congress praised "the forwardness of Dr. Franklyn's Row Gallies." Franklin also stopped meeting with the local Defense Committee as his attention turned more to Continental affairs, particularly those of the Committee of Secret Correspondence. He left defense preparations to Vice President Robert Morris, who developed an institutional framework for the first American naval shore establishment, with shipyard, recruiting office, paymaster's office, supply and procurement office, naval planning board, and even a "Cannon Committee" (a forerunner of the U.S. Navy Bureau of Ordnance). The latter committee contracted with Morris's close friend George Taylor, manager of the Durham Iron Works, to provide each galley with fifty iron round shot for the main gun and one hundred smaller shot for swivel guns.[28]

Recruiting officers and men for the Pennsylvania Navy became Morris's primary concern. He convinced local ship captains James Josiah, Charles Alexander, Nicholas Biddle, and Thomas Read to serve on the tiny warships. Soon all would receive Continental Navy commands. Morris established a standard system of recruiting crews based on the militia associations, applying to militia leaders for men who "may engage to go as Minute Men on Board the Boats when required." The Defense Committee drew up articles of war for officers and crew and established a pay scale. On the advice of fellow Defense Committee members and business partners

Nixon and Nesbitt, Morris allowed "each good and able seaman fifteen Shillings as a Bounty for their entering into the Service on board the armed Boats."[29]

The Defense Committee provisioned the tiny fleet. Owen Biddle gave his brother Clement the contract to furnish "provisions, Rum or Beer." There was no competitive bidding, and Clement Biddle set his own price at sixpence value of "roots and vegetable," and 10½ pence per ration of bread, beef, mutton, or pork. As prices for military and naval supplies soared, the Defense Committee rationalized its supply system by creating a "commissary of the Magazine and Military Stores" and advertising competitive bids. Nathaniel Falconer, James Bringhurst, and Francis Wade submitted lower prices than Clement Biddle, who lost the naval provisioning contract. But Owen, who managed the saltpeter works for the Defense Committee, hired him to superintend the powder mills.[30]

As Pennsylvania organized a local naval establishment during the late summer of 1775, the Continental Congress debated the creation of a national navy. Early discussion anticipated the long, bitter controversy over the nature of an American navy, between navalist proponents of a large deepwater fleet and antinavalist advocates of a small coastal defense force. Rhode Island delegates Samuel Ward and Stephen Hopkins, who had recently organized their own state navy, introduced a plan for a Continental Navy "of sufficient force for the protection of these Colonies." Their big-navy plan drew immediate opposition. "It is the maddest idea in the World to think of building an American Fleet," Maryland delegate Samuel Chase argued. "We should mortgage the whole Continent." Christopher Gadsden of South Carolina expressed the more moderate antinavalist position. "I am against the Extensiveness of the Rhode Island Plan, but it is absolutely necessary that some Plan of Defense by Sea should be adopted."[31]

The Continental Committee on Trade entered the debate in October 1775, warning that naval protection of American maritime commerce was vital to the economic survival of the colonies. Consequently,

Congress appointed Gadsden, Deane, and New Hampshire merchant shipowner John Langdon as a committee to procure and outfit two armed seagoing vessels. Deane insisted that Congress first take "our Connecticut and Rhode Island Vessels into Continental pay" and "That N. London [Connecticut] harbor is well situated for the rendezvous of an American Navy." (Deane asked Connecticut naval supplier Thomas Mumford, "is it not worthwhile for N. London to labor to Obtain the advantages of such a Collection of Navigation spending their Money there?"[32])

In October 1775, as British naval activity increased along the American coast, Congress convened a second naval committee, adding John Adams, North Carolina merchant shipowner Robert Hewes, Richard Henry Lee of Virginia, and merchant ship captain Stephen Hopkins of Rhode Island, but no Pennsylvania delegate. The New England majority members immediately dispatched Deane to the north to procure four warships for a Continental Navy. He failed. "I did not leave New York until Saturday Morning and then with the Mortification of effecting Nothing," he wrote to his wife Elizabeth. After returning to Philadelphia, Deane at last sought help from Pennsylvania Defense Committee President Benjamin Franklin.[33]

Consulting with the Philadelphia naval establishment marked a turning point in American naval policy and led directly to the creation of a national navy. Deane initially found the Pennsylvania defense mobilizers reluctant to cooperate because of the "unhap-py Dispute" between Connecticut and Pennsylvania over ownership of the Wyoming Valley frontier region in Western Pennsylvania. Deane agreed not to push Connecticut's claims to the territory and stopped his campaign to make New London a rival to Philadelphia as the center of an American naval establishment. In return, the Pennsylvania Defense Committee assisted Deane and the Continental Congress in raising a navy.[34]

Local Defense Committee members owned four merchant ships then unloading cargos on the Delaware River front just a few paces from where the Congress met, and agreed to sell them to form a Continental Navy. In late 1775 Congress first acquired the ship *Black Prince*, built by Wharton and Humphreys for Defense Committee members Morris, Nixon, and the Whartons. Philadelphia Ship Master John Barry commanded the 400-ton merchant brig. Renamed *Alfred*, the ship dropped down to Wharton and Humphreys shipyard where master shipwrights Coates and Humphreys directed carpenters and sawyers to strengthen gun decks and cut additional gun ports in the hull to mount twenty 9-pounders and ten 6-pounders. The conversion took forty-two days, used 1,056 feet of two-inch plank, and cost the government 360 pounds. Barry superintended rigging so that *Alfred* could carry more sail, and Southwark chandlers John and Nathaniel Falconer completed outfitting.[35]

Congress next bought the ship *Sally*, which had been transferred to the Defense Committee by its owner and committee treasurer John Maxwell Nesbitt. Renamed *Columbus*, Nesbitt's merchant ship went to Humphreys's Southwark facility below Christian Street, where Coates and the same ship carpenters and laborers who had built *Experiment* and *Washington* and converted *Alfred* now modified *Columbus* to carry twenty-eight guns. The brig *Defiance* (renamed *Andrea Doria*) and another *Sally* (renamed *Cabot*) were sold to Congress in November 1775 to form an American navy. Humphreys's carpenters, sawyers, and laborers began a thirty-four-day conversion on the 112-ton *Andrea Doria* to carry sixteen small cannon and performed some work on *Cabot*, although this conversion probably occurred at Willing and Morris's wharf upriver. A few weeks after the acquisition of the four larger vessels, Rhode Island merchant John Brown's armed sloop *Katy*, rated to carry ten guns, arrived in the Delaware River, was purchased by Congress, and renamed *Providence*. The Wharton and Humphreys shipyard performed a six- day overhaul on the tiny Rhode Island vessel.[36]

The Continental fleet rendezvoused at the wharves of Thomas Cuthbert and Willing and Morris to raise crews and receive additional provisions from ship chandler James Wharton. Clement Biddle delivered gunpowder from the Provincial works and James Craig finished rigging and sails. At the last minute, the Continental Congress ordered guards placed on board the naval vessels to prevent pilfering of supplies or sabotage by local Tories sympathetic to Britain. Meanwhile, fearing that ice was about to close the Delaware River, Congress ordered the "Fleet of the United Colonies" under the command of Rhode Islander Esek Hopkins, navalist Stephen Hopkins's brother, to leave Philadelphia in December 1775 and cruise south against British shipping. "Naval preparations are Now entering upon with Spirit and Yesterday the Congress chose a Standing Committee to superintend this Department, of which I had the honor to be Unanimously chosen one," Deane told his wife excitedly. They had given birth to a fleet, a Navy Department, and in the process, quietly, to the Philadelphia Navy Yard.[37]

2 ★ CONTINENTAL SHIPYARD

IN DECEMBER 1775 the Continental Congress authorized construction of thirteen fast heavily-armed cruisers. This first American naval construction act overshadowed earlier preparations by Philadelphia merchants and shipwrights to organize a river defense with armed row galleys or convert merchant vessels for service in a makeshift Continental Navy. These new warships would carry between twenty-four and thirty-two carriage guns and feature innovative hull design and sail and rigging configuration for maximum speed at sea. The growing rebellion and increased British naval threats to American ports and commerce had forced Congress to adopt a deep water fleet appropriate for an independent maritime nation at war.[1]

Congress distributed frigate contracts to seven different colonies. Rhode Island received two, and delegate Samuel Ward boasted that it meant "A Grant from Congress of 120,000 Dollars in Advance on our Acts." Samuel Adams of Massachusetts thought that the frigate business assured large contracts for Berkshire and Hampshire County craftsmen to manufacture duck cloth and canvas, while New Hampshire representative Josiah Bartlett promised increased naval business for his colony. "It is proposed that one or two persons well Skilled in Ship building, or approved Integrity, be forthwith appointed to provide the materials, Employ workmen, oversee the Business, to keep Exact & Regular accounts of the whole, to Draw on the Marine Committee of Congress for money to Carry on the Business, and to be accountable to said Committee for all which they will be handsomely Rewarded."[2]

The New Englanders' celebration was premature. Congress gave Philadelphia shipbuilders far more frigate business than the colonies to the north. "This Port has Double the Number of any other Colony, which was Claimed as due to our Ship Carpenters, who are more Numerous and we hope will prove on this Occasion, to have greater Abilities in their profession than their Neighbors," Continental officials informed Southwark shipwright Joshua Humphreys, whom they selected to direct the frigate program. Humphreys knew how to build armed vessels and convert merchantmen into warships. Moreover, longtime contractor with Humphreys shipyard Robert Morris, now the leading maritime official in the Continental Congress, promised to send frigate business his way.[3]

Morris held a "Masterly Understanding" of the business of mobilizing a navy for war and had an "honest Heart" in advancing the Continental cause, John Adams observed, but harbored "vast designs in the mercantile Way, and no doubt pursues mercantile Ends, which are always gain." Morris admitted as much. He expected rewards for managing naval and maritime affairs. "If the Congress mean to succeed in this Contest, they must pay good Executive Men to do their business as it ought to be & not lavish millions away by their own mismanagement." Morris had developed such executive skills already as president of the Pennsylvania Committee of Safety and Defense, where he organized a rather complex set of military and naval institutions.[4]

Morris rose rapidly in the bureaucracy. Elected to Congress from Pennsylvania in November 1775, he

Leading Philadelphia merchant ship owner Robert Morris financed the Wharton and Humphreys shipyard and organized the first American naval shore establishment in Southwark between 1775 and 1777. Section of a painting by Charles Willson Peale, Print and Picture Collection, Free Library of Philadelphia.

became chairman of the powerful Secret Committee of Commerce (and Trade) that Congress authorized to procure arms, ammunition, and war supplies overseas. After passage of the Naval Act of 13 December 1775, Congress appointed Morris to the Marine Committee, a quasi navy department in charge of building the first national naval force. Also, he joined Pennsylvania Defense Committee mentor Benjamin Franklin on the Continental Committee of Secret Correspondence that served in many ways as a foreign office and central intelligence organization. Morris received secret arms shipments and confidential information from his network of commercial and naval agents overseas. He kept such information from Congress. "It is unnecessary to inform Congress of this Intelligence at present," the Committee of Secret Correspondence confided, "because Mr Morris belongs to all the Committees that can properly be employed in receiving & importing the expected Supplys from Martinico, St. Eustatius or Cape Francois."[5]

Morris built a naval establishment primarily to protect the trade that he knew the Continental government needed to survive in the struggle with Great Britain. As president of the Pennsylvania Defense Committee succeeding Franklin, he had exerted influence over naval affairs by delivering four merchantmen to the Congress, including three of his own. Morris made certain that contracts to convert these merchant vessels for Continental service went to business partner John Wharton's shipbuilding enterprise. As Commissioner of Frigates, Morris established a Continental account in January 1776 with Wharton and Humphreys to procure timber and materials, before news of the frigate construction program reached other colonial shipyards. The Marine Committee delivered the first contracts for the 32-gun frigate *Randolph* to the Wharton and Humphreys shipyard and for the 24-gun ship *Delaware* to Warwick Coates and Joseph Marsh, affiliated with Wharton and Humphreys. Coates and Marsh shared timber sheds, molding loft, steam ovens to bend plank and board, and shipbuilding facilities with the other Southwark shipwrights.[6]

Most likely, Morris wanted to give all four lucrative Philadelphia frigate contracts (each worth $66,000) to the emerging Southwark naval establishment. However, as vice president of the Defense Committee, he had watched President Franklin balance the politically sensitive Southwark and Kensington shipbuilding and maritime communities by granting them equal shares of the contracts for armed boats. Consequently, Morris gave two frigates to Kensington shipbuilders Jehu, Benjamin, and Emanuel Eyre and John Rice. The Eyre brothers received a contract to build the 32-gun frigate *Washington*, and Rice the smaller 28-gun frigate *Effingham*. Veteran shipwright Rice, who constructed the Pennsylvania Navy armed boat *Dickinson*, owned Delaware riverfront properties between those of Emanuel and Benjamin Eyre in Kensington and undoubtedly shared shipbuilding facilities—much as Coates and Marsh worked with Wharton and Humphreys in Southwark.[7]

To coordinate construction of four warships at two different locations and provide guidance to frigate builders in six other colonies, Morris created an American naval shore establishment "divided into different Departments." These included an executive board, a supply agency, a timber commission, and an accounting and inspection office. Thus Philadelphia's merchant shipbuilding community, affiliated closely with Willing, Morris and Company, again dominated Continental Navy organization. Morris directed operations from a Board of Naval Commissioners comprised of close business associates John Nixon, John Maxwell Nesbitt, and John Wharton, who jointly owned shares in ships, shipbuilding property, warehouses, and landings along the Southwark waterfront.[8]

Delaware riverfront businessmen Nathaniel Falconer, James Craig, William Davis, and John Wharton's older brother James became commissioners "for providing Rigging, ship Chandlery and other stores." The Timber Commission held the most direct Southwark connections to the emerging naval shore establishment, including merchant shipbuilder Samuel Penrose, Wicaco Lane boatyard and wharf owner David Thompson, and John Wharton's former shipyard partner Francis Grice. Waterfront merchants Benjamin Fuller, Isaac Hazelhurst, Clement Biddle, and Thomas Fitzsimmons served as Commissioners of Inspection and Correction (Accounts), completing the first offices of this quasi-naval shore establishment. These commissioners existed for "the Service of the United Colonies" and to maintain "good Understanding with all these boards of Commissioners, with the Superintendents, Clerks and with the other ship Carpenters, and all other Trades men." The commissioners also fixed dimensions for frigates and ordered ship plans made to guide builders at the various ports.[9]

With Wharton and these other Southwark maritime businessmen on the Frigate Commission, it was no surprise that the 24-gun ships authorized under the first naval act followed the plan for the *Hero* built in 1762 by James and Thomas Penrose and handed down through the firm of Penrose and Wharton to its successor Wharton and Humphreys. The origin of the design for a 32-gun ship that became the basic cruiser for the American fleet seems less certain. Frigate builders, particularly in New England, modified dimensions because plans from Philadelphia often were received too late to cut molds before keel laying or framing of the warships. None of the frigates duplicated the original Philadelphia design for the *Randolph*. Moreover, ship captains at six shipyards altered rigging and increased the number of guns to thirty-six on some vessels.[10]

However, Southwark shipwright Joshua Humphreys, Jr., most certainly designed America's first frigates. John Adams considered Humphreys the leading ship designer in Philadelphia. "I then went to the House of one Humphreys an ingenious shipwright and found him making a Model of a seventy-four Gun Ship," Adams reported. "He has nearly completed it. You see every Part of the Ship, in its just proportion in Miniature."[11]

Building a ship model (which according to family tradition Humphreys carved with his pen knife) differed from designing and constructing an actual man-of-war. Yet Humphreys demonstrated by his rapid completion of *Randolph* during the early summer of 1776 a thorough familiarity with the design of this sophisticated vessel. Nearly all the pieces for

the warship lay in "the hands of Joshua Humphreys Junr," suggesting further that the Continental commissioners had adopted his design. Meanwhile, Kensington shipbuilders lagged far behind in completion of frigates *Washington* and *Effingham*. The Kensington yards recruited 150 volunteers from the residents and militia associations of the Northern Liberties to carry lower deck beams and upper deck timbers on board the frigates. Neither would ever be completed. Morris complained bitterly that *Effingham* remained useless. "I don't know how it happens, but there is no Guns made nor making for that ship."[12]

Morris knew full well how it happened. He favored his long-time business associates' shipyard, now referred to as the Continental Yard, in providing timber, gunpowder, naval stores, cannon, and money to pay workers. He urged the Continental Yard clerk John Ashmead to stay and take care of material for *Randolph* rather than join the Quaker Light Infantry "on the proposed expedition to the Jerseys." When Wharton's brother James, the ship chandler, complained that Durham Forge was late in delivering 300 tons of pig iron to *Randolph* for ballast, Morris's Committee told ironmonger George Taylor to rush iron to the Southwark shipyard without regard to the original contracted price. "If you cannot afford it under 8.10—perhaps they would give it, but don't let them know that I mentioned it," confided Morris's agent Clement Biddle. With such deals, *Randolph* and *Delaware* were made ready for launching by early July 1776, barely seven months after keel laying. Apparently Wharton and Humphreys planned to launch the frigates simultaneously. On 13 July *Delaware* slid smoothly into the water, but *Randolph* stuck on its shipways and its launching was delayed until the following day, "to the great disappointment of the Builder, John Wharton, and a numerous set of Spectators."[13]

Naval business expanded along the Southwark waterfront in early 1776 as sentiment in Congress for a complete break with Britain grew more vocal and stimulated increased defense preparation. Master carpenter Arthur Donaldson directed the sinking of chevaux-de-frise in the main ship channel, while

Simon Sherlock's carpenters worked on the Provincial ship *Montgomery*. It was Continental Navy mobilization, though, that dominated the riverfront on the eve of the Declaration of Independence.

Captain John Barry brought the swift Bermuda-built *Wild Duck*, purchased recently from Baltimore merchants, to the Wharton and Humphreys Continental Shipyard for conversion to carry sixteen carriage guns as the Continental brigantine *Lexington*. Close in *Lexington*'s wake arrived the armed schooners *Wasp*, *Hornet*, and *Sachem* and the schooner-rigged tender *Fly*, rated to carry six guns. Humphreys supervised caulking, scraping, painting,

and strengthening of bulwarks and decks with fresh pine board for each vessel. *Hornet* and *Fly* had collided during operations against British shipping off Bermuda and required major repairs. So Humphreys spent nineteen days superintending work in order to get these tiny warships back to sea.[14]

Once overhauled, Continental warships rendezvoused upriver at Thomas Cuthbert's wharf, where recruiters raised crews sufficient to man the squadron. Finding men proved difficult, so the Marine Committee raided the Pennsylvania Navy's armed galleys for officers and crewmen or impressed British sailors imprisoned in city jails.

The Continental frigate *Randolph* (left), shown in battle with the British 52-gun *Yarmouth*. *Randolph*, designed and built by the Wharton and Humphreys shipyard in Southwark, was the only one of thirteen frigates laid down under the Naval Act of December 1775 to see action in the Revolutionary War. From McClay, *History of the Navy*, 1: 85, drawn by J. O. Davidson, 1891. Naval Historical Center Photo NH 1102.

Randolph, for one, relied on British prisoners to fill out the ship's complement. The Marine Committee also requisitioned guns and gun powder from the Pennsylvania Defense Committee to arm the first Continental naval squadron. At the same time Clement Humphreys and Anthony Morris supplied porter to the ship crews and Jonathan and Joel Evans added cabin stores. James Craig's ropewalk delivered cordage, John ap Owen sails, and William Rigden yellow paint for gun ports along the black hulls of the Continental fleet.[15]

Though Philadelphia's warships received the most attention, Morris facilitated Continental naval enterprise in other ports, after the Declaration of Independence in July 1776 propelled him to the center of naval planning for the entire war effort. He dispatched Southwark ship captain Nathaniel Falconer, "a Gentleman well acquainted with Maritime Matters," to assist New Hampshire in the construction of the 32-gun frigate *Raleigh* at James K. Hackett's Portsmouth shipyard, and to investigate why Sylvester Bowers's shipyard in Rhode Island charged Congress twice the contracted price for construction of frigates *Warren* and *Providence*. Morris acquired gunpowder, timber, pine tar, and naval stores for New England and Chesapeake shipbuilders. He arranged most naval contracting for the newly independent states with arms merchants John Brown of Rhode Island, William Hewes of North Carolina, Francis Lewis of New York, and Thomas Mumford of Groton, Connecticut.[16]

Congress decided in late 1776 to greatly expand the fleet, authorizing construction in November of three 74-gun ships, six frigates, and a packet boat. "We are bent on building some Line of Battle Ships immediately," Morris wrote to Silas Deane in France, claiming that the additional work nearly blinded him and destroyed his health. He thought it fair that he should take a 2½ percent commission as compensation on all business connected with expanding the Continental Navy. "I have much in my power or under my influence both Public & private, and my desire and design is to serve justly & faithfully every interest I am connected with & at the same time as I take the trouble my Friends can't begrudge paying so small a consideration."[17]

Congress provided Morris with a Board of Assistants (later the Middle Department Navy Board) for the "systematical Management of naval affairs." The new agency was designed to monitor Morris's conduct of Continental naval business, but the merchant entrepreneur easily bypassed Congressional checks by naming close business associates John Wharton and John Nixon to the board. They gave contracts at once to Wharton's Southwark firm to build the packet boat *Mercury*, 16-gun ship *Saratoga*, and probably one of the three 74-gun ships-of-the-line for which Wharton's partner Humphreys had made a model and cut molds for framing. So Congress added shrewd Bordentown, New Jersey, maritime lawyer and composer of patriotic songs Francis Hopkinson to the vital Board of Assistants to monitor the other two and, not incidentally, to provide legal expertise on maritime law as it applied to the prize courts that condemned and sold cargos captured by privateers.[18]

At that moment, Congress viewed privateers as a more effective way to wage maritime war against the powerful British fleet than by deploying frigates, and so far Morris and his assistants had failed to get to sea. Surprisingly, Morris had remained silent about the employment of privateers. His partner Thomas Willing's opposition in Congress to licensing privately outfitted and armed ships to plunder British trade probably discouraged Morris. Eight months after Congress approved privateers, Morris insisted that "I have not hitherto had any Concerns in privateering & even at this day my Partner Mr. Willing objects positively to any Concern." Everyone else seemed involved. "Thousands of schemes for Privateering are afloat in American imaginations," John Adams explained. Franklin participated eagerly in this irregular naval warfare soon after arriving in France on a mission for Congress by outfitting his own fleet of privateers. He wrote to Morris that privateers were the only way to "make us respectable on the Ocean." Inevitably, Morris sponsored privateers, admitting that "My scruples about Privateering, are all done away."[19]

While Morris expanded his naval functions in late 1776, British activity to the north threatened the new government in Philadelphia. British Army commander General Sir William Howe and his brother Vice Admiral Lord Richard Howe drove General George Washington's army out of New York and through New Jersey to the Delaware River. British-Hessian garrisons occupied Trenton and Princeton in December, and rumors spread that the British army would follow Washington across the Delaware, march south, and descend on Philadelphia. Though some delegates claimed that these rumors were nothing but lies spread by the Pennsylvania Council of Safety (which in 1776 had replaced the Committee of Safety and Defense), the threat seemed real enough to force Congress to pack up and leave Philadelphia for sanctuary in Baltimore. Samuel Adams of Massachusetts insisted that, if they stayed in Philadelphia, "the People of Pennsylvania, influenced by Fear, Folly or Treachery, would have surrendered their Capital to appease the Anger of the two [Howe] Brothers, and atone for their Crime in suffering it to remain so long the Seat of Rebellion."[20]

Before leaving Philadelphia, Congress suggested that Morris burn the frigates and other Continental vessels on the Delaware River to prevent capture by enemy forces when they occupied the city. This threatened destruction of Philadelphia's frigates prompted Morris to remain in Philadelphia, to save the ships and manage "a Constant Magazine of Necessarys for the Army, Navy & All the other States." He asked Congress President John Hancock for permission to stay behind, and the Congress granted Morris authority to direct the entire Continental defense supply organization. To this end, Morris established an executive committee office on Front Street above Cedar, where every day between 10 and 3 o'clock he carried out public business. He drew large amounts of money from the Loan Office to pay for war material, and decided where it should be sent. He coordinated the Continental logistics organization through the ever-reliable Nixon, Nesbitt,

Wharton, Wharton's nephew Carpenter Wharton, and James Mease, a local merchant associated with Willing, Morris and Company.[21]

Morris's primary mission was to save all the Continental armed vessels on the Delaware River. "I think we ought to hazard every thing to get the Ships out," he told Hancock, and offered a scheme to save the Continental Navy. Morris planned to provision *Randolph* and ready *Delaware* for the sea. He had Humphreys repair the armed sloops *Independence*, *Hornet*, *Fly*, and *Mosquito* so that they could join the bigger frigates to escort all merchant ships then in port downriver through the British blockade and into the open sea. *Andrea Doria*, just arrived with a load of arms and clothing from St. Eustatius, would be overhauled at the Continental Shipyard in order to add twenty guns to the planned convoy. Also, Morris plotted to find some way to get the unfinished frigates *Washington* and *Effingham* to safety upriver near Bordentown.[22]

General Washington's surprise attacks on British forces at Trenton and Princeton in late 1776 and early 1777 lifted the immediate military threat to Philadelphia and gave Morris time to get the frigates out. But things did not go as planned. The Council of Safety failed to support his recall of tradesmen to work on *Delaware*. Frigate captains Thomas Read and John Barry volunteered for duty with Washington's artillery on land. Other local officers in the Continental service simply vacillated. "I have scolded the Officers like a Buster-Whore for their dilatoriness," Morris explained. More damaging, cannon for the warships cast in local forges failed to pass test proofs; inferior iron and poor workmanship caused the guns to split when fired under the British navy's standard powder charge. There were other setbacks as well. "You will doubtless be surprised that our Navy is not farther advanced," Morris wrote Deane, "because you are unacquainted with many of the difficulties which have retarded its Progress, particularly the Want of Sea Coal for our Anchor Smiths, the Disappointments in our first attempts to Cast Cannon, and above all, the frequent Calling out of our Militia in a manner which would permit an

Exemption of the necessary Workmen."[23]

Friends in Congress assured Morris that no one blamed him for failure to get a Continental fleet to sea. There was unhappiness with the Philadelphia merchant shipowner's assumption of so much power, however. Marine Committee leader Richard Henry Lee questioned the "perfidy" of Morris's business connections and hinted that one of Morris's closest associates was a "vile Scotchman" who hated America, Congress, and the idea of independence. Lee built an anti-Morris and anti-Philadelphia faction that included Richard's brothers Francis, a Virginia delegate, and Arthur, Congress's agent in Paris. This so-called Virginia Family Compact allied with Marine Committee members William Whipple of New Hampshire and Samuel Adams of Massachusetts to speak against Philadelphia's apparent control over development of an American navy.[24]

Congress gradually diluted Morris's influence by establishing new military supply depots in Virginia, Massachusetts, and Carlisle, Pennsylvania, the latter to replace arsenals at Lancaster and York managed by Carpenter Wharton, an intimate Morris associate suspected of enriching himself at public expense. Congress further sought to decentralize authority over the naval establishment by creating another navy board in Boston to "superintend the Building, fitting out and manning the Ships of War ordered to be Built there." This Eastern Department Navy Board deprived Philadelphia of control over naval supply and ship construction policy.[25]

Erosion of Morris's authority continued. The Marine Committee ordered him to strip anchors and cables from Philadelphia's frigates and send them by cart to Baltimore so that 28-gun frigate *Virginia* could get to sea first. Committee president Hancock appeased Morris, assuring him that "you can with much more ease replace them there than we can procure them here." In fact, naval supplies, anchors, iron cannon, arms, and ammunition were running perilously short in Philadelphia as the local Council of Safety drained stockpiles of war supplies accumulated by the defunct Defense Committee to build defenses for the city and river against an expected British invasion.[26]

In the end, these defense measures failed to prevent British occupation of Philadelphia in late 1777. Some elements of the local naval shore establishment seemed ready to serve the British when they entered the city. Captain Andrew Snape Hamond told fellow British officers of warships damaged by the chevaux-de-frise in the Delaware River that "assistance of Shipwrights, or other Artificers will be given by Mr. Thompson, shipbuilder at Philadelphia." There was little interruption in naval business during the British occupation. Hamond overhauled *Roebuck* at Cuthbert's Wharf, while *Vigilant* and *Camilla* were outfitted at "Penroses Wharf." The number of local shipwrights, chandlers, and captains accused of treason by the Supreme Executive Council of Pennsylvania after the British withdrew in 1778 suggested the extent of collaboration. Pennsylvania's wartime government convicted Continental Timber Commissioner David Thompson, Kensington shipwright William Williams (builder of the Pennsylvania Galley *Hancock*), ropewalk owner James Craig, ship outfitter Joseph Stansbury, and Deputy Commissary General Carpenter Wharton. The post-occupation Pennsylvania government also accused Southwark ship captain Robert White, an architect of Philadelphia's river defense system, and Fort Mifflin stonemason John Palmer of treason.[27]

Nevertheless, the British abused some Delaware riverfront naval facilities, destroying Joseph Marsh's Southwark boatyard and Queen Street shipwright Francis Grice's waterfront property. Kensington shipwrights Emanuel, Benjamin, and Jehu Eyre, active in the resistance to British occupation, lost timber supplies and ships on the stocks above Hanover Street. Meanwhile, British raiders ventured upriver and destroyed the Kensington frigates *Washington* and *Effingham* on the Jersey shore two miles below Bordentown, where they had been brought and scuttled in late September 1777 when the British entered the city.[28]

Destruction of the Kensington frigates coincided with a series of disasters that befell the original thirteen Continental frigates, all captured, destroyed, or incapacitated by the British blockade. *Randolph*, which had escaped to sea before occupation, blew up off Barbados in a battle with the British 64-gun ship *Yarmouth*, killing Captain Nicholas Biddle and most of his men. *Delaware* fell into British hands when it ran aground while attacking enemy earthworks on Philadelphia's waterfront. New York frigates *Congress* and *Montgomery* were burned in the Hudson River to prevent capture. The 28-gun ship *Virginia* ran aground as she tried to slip past British blockaders in the Chesapeake Bay and was captured. Congress stopped construction on new warships that had been authorized by the Naval Act of 20 November 1776. This included a 74-gun ship under construction in New Hampshire and two frigates on the ways at Gosport, Virginia. "If we were disposed to build a Navy and were to proceed at this slow rate and the enemy were to continue to capture and force us to destroy our frigates as they had done for a twelve month past," worried Marine Committeeman William Ellery of Rhode Island, "it seems to me we should never have so many frigates afloat as there are States in the Union."[29]

At this nadir of naval development during the Revolutionary War, Morris remained powerless to revive the American naval shipbuilding program. He resided on his Mannheim farm about a dozen miles from the Continental government-in-exile in York, Pennsylvania. No one consulted him on naval affairs, while the Virginia Family Compact (Arthur, Francis, and Richard Henry Lee) spread rumors that Morris "pocketed the Public Money" and "paid for all his private purchases out of the public Treasury."

Samuel Adams, William Whipple, and the Virginia clique ran naval affairs during the summer of 1778 after Congress returned to Philadelphia. What funds they could raise from prize money obtained by auctioning off cargos captured by American privateers, they channeled through the Marine Committee to the Eastern Department Navy Board in Boston to support New England shipbuilding or through the Family Compact to Chesapeake Bay maritime interests. Richard Henry Lee guaranteed $20,000 to Gosport shipbuilders to resume construction on one of two frigates authorized in 1776. At the same time, the Marine Committee rejected a petition by "Sundry People in Phila." to repair the partially burned-out hull of the *Washington*, which had been refloated near Bordentown and brought downriver to the city. Despite ordering a great quantity of live oak from Georgia to rebuild the frigate, Congress disposed of the hulk at auction.[30]

Naval and maritime business suffered greatly in post-occupation Philadelphia. Runaway inflation, profiteering, and social and political unrest plagued the city. Shortages of material and funds deprived local shipbuilders of new construction contracts, and "the extravagant price of Spirituous Liquors, & the extreme difficulty, if not impossibility of procuring a supply for our Navy" meant that Philadelphia merchants and chandlers lost this enterprise as well. Pressed for money, the Supreme Executive Council of the Commonwealth of Pennsylvania, the radical government that replaced the Council of Safety, decommissioned the State Navy, dismissed officers and crews, and in December 1778 auctioned off the entire fleet at the Coffee House. The Supreme Executive Council advertised a "for Cash only" sale to the public of ten armed galleys, brig *Convention*, schooner *Lydia*, and armed sloops *Speedwell*, *Sally*, *Industry*, and *Black Duck*. On 18 December the Council ordered Nixon, Nesbitt, and Benjamin Fuller to settle accounts and close the books on the State Navy.[31]

Maritime trade suffered. The Congress and the Pennsylvania Supreme Executive Council instituted various regulations and an embargo on Philadelphia merchant shipowners that prevented ships from leaving port. Philadelphia merchants begged permission for their ships to sail to Bermuda or the West Indies. Finally, Nesbitt obtained certification to clear sloop *Mars* laden with provisions for the French fleet now that France had joined the American fight against Britain. For the most part, Philadelphia ships stayed in port. Unemployed sailors grew restless, roaming the waterfront, vandalizing ships, and rioting in the streets. There was no work at the Wharton and Humphreys shipyard either, and

Humphreys applied for the "office of Wharfinger & Weighmaster of Hay at the Draw bridge in this city." Upriver, Kensington shipbuilder Emanuel Eyre settled for a job with the new Pennsylvania government, investigating local merchants and war contractors for "forestalling and enhancing the Price of flour."[32]

Philadelphia seemed on the brink of general anarchy. The Continental Congress managed affairs badly, particularly ignoring naval defense of the Continental capital. "Very few of the Members understand even the State of our naval affairs, or have Time or Inclination to attend to them," New York delegate John Jay worried. The Virginia Family Compact dropped Morris from the Marine Committee. John Wharton resigned in disgust from the Eastern Department Navy Board, refusing to turn over the Board's records to Congress. Trying to correct matters, Congress replaced the Marine Committee with a Board of Admiralty and resumed naval construction at the Wharton and Humphreys Continental Shipyard, contracting to build the 16-gun ship *Saratoga* and tiny packet boat *Mercury* authorized by Congress before British occupation. Humphreys laid keels in December 1779 and launched the vessels in April 1780. By then the Board of Admiralty had nearly ceased to function, and its business was taken over in September 1781 by Robert Morris, who had in the meantime been appointed Continental Superintendent of Finance and now became the Agent of Marine.[33]

Morris, with few ships to manage and almost no naval budget, spent considerable time settling naval accounts and trying to pay debts to naval suppliers. With the surrender of British forces at Yorktown in October, the immediate need for an American naval shore establishment quickly faded. The Philadelphia naval establishment survived during the last days of the Revolution by servicing privateers. Humphreys's carpenters overhauled New England privateers and repaired the 13-gun "Pennsy. ship" *Rising Sun*, commanded by Stephen Decatur (senior) and owned by Southwark Port Warden Francis Gurney and other Delaware riverfront investors. Humphreys refit the privately owned 14-gun ship *Commerce* for Thomas

Randall and Company of Philadelphia, a firm that included privateers Stephen Girard and Thomas Truxtun. After the Treaty of Paris of 1783 recognized American independence, Truxtun joined former French Consul-General John Holker, Jr., and members of the disbanded Continental naval establishment in Philadelphia to give more business to Humphreys's former Continental Shipyard.[34]

A flurry of maritime activity followed the Treaty of Paris. The Articles of Confederation government, the loosely coordinated set of committees that served as the national government from 1781 to 1789, encouraged local economic development. At first, Philadelphia merchant shipowners and shipbuilders welcomed this decentralized business environment, pursuing highly speculative commercial ventures. They ordered new ship construction and overhaul of older merchant vessels for risky but immensely profitable trading cruises to the West Indies, Mediterranean, and Far East.[35]

Humphreys invested heavily in this postwar boom. He helped John, Nathaniel, and Benjamin Hutton to develop shipbuilding facilities along the Southwark waterfront. He leased part of the former Wharton and Humphreys establishment to ship carpenters John Delavue and Henry Rhile, retaining a one-quarter share in every shipbuilding contract received. And there were many. Rhile and Delavue built *Congress* for Philadelphia merchant Thomas Irwin and refit *Commerce* for James Collins, Thomas Truxtun, and John Holker. They raised merchant vessels for Donaldson and Cox, Mease and Caldwell, Isaac Hazelhurst, and "sundry merchants." Humphreys constructed a ferryboat for Holker and *London Packet* for Robert Morris and twenty-three other Philadelphia shareholders; he purchased huge amounts of timber to construct these vessels from the Penrose family's riverfront business, opening a mast yard on Trotter's Wharf at the foot of Catharine Street.[36]

Suddenly the Philadelphia shipbuilding bubble burst. There was "great stagnation in mercantile affairs [and] of course it was felt by the shipwrights," Humphreys recalled. "At this time I had two vessels on the stocks for sale, and several oth-

Continental Naval Officer Thomas Truxtun worked with Joshua Humphreys, Jr., in Southwark to outfit privateers and build merchantmen for the China trade, and in Baltimore to launch the frigate *Constellation*. After a painting by Archibald Robertson. Print and Picture Collection, Free Library of Philadelphia.

indebted to the Penroses that their business failed and Humphreys dissolved the partnership. Humphreys, Truxtun, and other shareholders sold ship *Commerce* at a substantial loss.[37]

The unstable financial situation and depressed market under the Articles of Confederation led Southwark shipbuilders and shipowners to favor creation of a stronger central government with powers to organize national finances, collect revenue, and provide naval protection for maritime trade. A constitutional convention to reorganize the central government met in Philadelphia in 1787, where New York representative Alexander Hamilton urged other delegates to create a "powerful Marine." Under Hamilton's scheme, the South would supply naval stores, New England raise crews, and the Middle Atlantic region build "dockyards and arsenals." As center of a moribund Continental naval shore establishment, Philadelphia expected to become the main dockyard and arsenal for Hamilton's "powerful Marine."[38]

ers of my profession had vessels on at the same time at this end of the Town." The Articles of Confederation provided no central mechanism for the collection of revenue to support a naval establishment or protect American maritime traders. The British interfered with Atlantic and West Indian trade, while the Barbary States of North Africa continually threatened American shipping off the Straits of Gibraltar and in the Mediterranean Sea. The cost of marine insurance soared, and overextended investors lost heavily in the West Indian and European business. Only the risky Sumatran and China trade showed a profit. Shipbuilders, who had borrowed money to build and repair ships for these same Philadelphia investors, failed to pay bills to suppliers. Delavue and Rhile became so deeply

3 ★ UNITED STATES SHIPYARD

PHILADELPHIA'S maritime and ship building community welcomed the new U.S. Constitution. Former Pennsylvania State and Continental Navy administrators paraded in July 1788 to celebrate local ratification. John Nixon rode on horseback at the head of the mile-long procession that wound its way from the Southwark border above Cedar Street to Market Street in the center of the city. Close behind rode former Frigate Commissioner Thomas Fitzsimmons and Marine Committeeman George Clymer. On a horse-drawn carriage decorated as the "Grand Federal Edifice" sat John Wharton, John Maxwell Nesbitt, and other shipowning merchants involved in founding the Continental Navy during the late War of the Revolution. Shipwrights Francis Grice and Joshua Humphreys marched behind Kensington's leading shipbuilder, Emanuel Eyre, who carried a craft guild flag that bore the image of a ship on the stocks.[1]

Confidence in the federal government renewed ship construction activity along the Southwark waterfront. Humphreys contracted with Levi and Jehu Hollingsworth to build brig *Hetty*, with John Donaldson for brig *Patty*, and with Andrew Clow and Company for the "good and substantial" ship *Ceres*, agreeing to pay "£6 special lawful money of Pennsylvania per ton." Conyngham and Nesbitt ordered ship *Ann*, while Southwark ship captain and wharf owner Rickloff Albertson had his neighbor Humphreys construct a large schooner for the coastal trade. Truxtun, still devastated by the loss of *Commerce*, returned to Southwark and supervised the construction of a large China trader at his

friend Humphreys's shipyard. Robert Morris reappeared with money to ready ships *Asia* and *Canton* for the growing China market and financed a complete overhaul for the badly rotting former Continental frigate *Alliance*, recently purchased for the China trade.[2]

Humphreys expected that ratification of the U.S. Constitution and location of the federal capital in Philadelphia would stimulate naval business. He leased and purchased waterfront properties, including Windmill Island in the Delaware River opposite Willing, Morris and Company's wharves and store house at the foot of Cedar Street and riverfront lots at the foot of Catharine Street. By 1790, Humphreys was operating an upper and a lower shipyard, just in time to participate in a federal shipbuilding program that began in August when Secretary of the Treasury Alexander Hamilton secured $10,000 from Congress to construct and equip "so many boats or cutters not exceeding ten" for a revenue service. Shipbuilders along the Atlantic coast scrambled to obtain the first U.S. government ship construction contract. As usual, New Englanders were most assertive. "Our harbor is as well calculated for Navigation & our River for Shipbuilding as any perhaps in the United States," New Hampshire customs agent Joseph Whipple told Hamilton. Boston port collector Benjamin Lincoln claimed that Massachusetts ship carpenters were superior to all others and were working already on building Canadian revenue cutters.[3]

Hamilton assigned the first contract for revenue cutter *Scammel* to its Portsmouth, New Hampshire,

Joshua Humphreys (center) oversees construction of the subscription frigate *Philadelphia* on the stocks in Southwark below Old Swedes Church, on part of the property that became the original U.S. Navy Yard in 1801. William Birch print, 1800.

yard. Other contracts went to Baltimore, Charleston, Savannah, and Wilmington, North Carolina shipbuilders. Hamilton planned to build "one in Philadelphia for the bay of Delaware and the coast adjacent to the Capes May & Henlopen," and Southwark ship carpenters launched the revenue cutter *General Greene* in 1792. No new naval business followed, however, as the federal government, located temporarily in the city, struggled to reorganize its finances and establish offices. Management of naval affairs remained under Secretary of War Henry

Knox, a former Chief of the Continental Army Artillery. Meanwhile, a Republican faction in Congress comprised of anti-Federalist friends of Thomas Jefferson resisted creation of strong central government agencies under the Constitution and opposed construction of warships larger than a revenue cutter for coastal patrol.[4]

The Anglo-French maritime conflict that accompanied the Wars of the French Revolution spilled into American waters and forced the George Washington administration to consider building a navy to protect

U.S. neutrality and freedom of the seas. It was attacks on American ships by Algerian corsairs rather than conduct by France or Britain, however, that first caused the United States to raise a naval force and brought Humphreys to the center of a movement to create a national navy. The Barbary State of Algiers seized eleven American merchantmen between October and November 1793, including the merchant ship *President* (not to be confused with frigate *President*) out of Philadelphia. Southwark ship carpenters had built *President*, and the Penrose family owned and commanded it. Therefore the capture brought home to the Southwark maritime community the urgency of raising a navy to protect American trade and travel on the seas.[5]

A few weeks after *President*'s capture, Humphreys wrote Robert Morris, now U.S. Senator from Pennsylvania, that the country needed a navy. Morris asked his wartime naval constructor to provide details for such a naval force, and the Southwark shipwright introduced a plan for a new class of 44-gun ships that would be longer, wider abeam with "scantlings equal" to a 74-gun ship, and more heavily armed than the standard British 36-gun frigate. Humphreys advised that this uniquely American type of warship had to be constructed of the best live oak, Carolina pine, and red cedar wood, and that "all timber should be framed and bolted together before they are raised."[6]

The Washington administration's call for "a Naval Armament" to stop "the depredations committed by the Algerine corsairs on the commerce of the United States" led to heated discussion in Congress. Antinavalist Jeffersonian Republicans opposed development of a deep water fleet, while federalist navalists wanted a large naval establishment. Former Frigate Commissioner Thomas Fitzsimmons, now a U.S. Representative, presented the Federalist position, recommending construction of larger frigates. In the end, the Federalists used the Algerine crisis to secure passage of the Naval Act of 27 March 1794 that authorized President Washington to purchase "or otherwise equip and employ" six frigates.[7]

With no separate Navy Department, responsibility for development of the frigate program of 1794 fell to Secretary of War Knox. The former army officer immediately sought advice from Southwark shipbuilders Humphreys, John Wharton, and William Penrose, Continental Navy officer John Barry, and British-trained naval architect Josiah Fox, who had arrived in the city in 1793 to teach drafting to Penrose's sons. Members of this informal naval board of construction presented Knox with at least two different designs for an American frigate. Penrose and Fox wanted a small, less expensive warship modeled closely after the British 36-gun frigate, while the others liked the idea of a larger, more powerful 44-gun ship. "It would give us a superiority over any of the European Frigates & would render all their Frigates of little or no effect in a contest with us," Humphreys argued. "It would give us a lead in naval affairs that no smaller dimensions would afford us."[8]

Knox accepted Humphreys's design, and in so doing started a controversy that upset the first U.S. naval construction program. "I admit that Mr. J H produced some drafts & models of his own to the secretary of War," Fox agreed, "but on submitting them to the inspection of the principal Master Builders of Philadelphia from Swedes Church to the upper part of Kensington they were rejected by their unanimous voice." According to the former British naval architect, the U.S. government adopted his draft. Humphreys recalled events differently. The master shipwright claimed that Knox called him to the War Office on Market Street in February 1794 for the first of many discussions about the design and construction of an American navy. When Humphreys arrived, he found that Penrose had already introduced Fox's plan for a smaller frigate, but he was able to convert Knox to his 44-gun ship. "In a long conversation with Secretary Knox, I stated to him the advantages of building the frigates of extended dimensions, and making them the most powerful ship of that class."[9]

The secretary of war asked Humphreys to draft specifications and make plans for frigates, and instructed that "Mr. Fox who is under your direction will also apply himself closely to this business." Humphreys employed Fox, "a first rate draftsman and being late from one of the Kings Yards in England," to prepare draft drawings of the 44-gun ship, following Humphreys's exact specifications. "Instead of conforming to the instruction I gave him," Humphreys complained, "he drew the draft according to his own opinion, which was so foreign from my ideas that I set it aside and drew another myself, by which the 44-gun frigates *UStates*, the *President* and the *Constitution* were built." Humphreys ordered Fox and apprentice draftsman William Doughty, son of Humphreys Southwark shipbuilding neighbor James Doughty, to lay down the "44" in the mold loft, "making moulds for cutting timber by and other sets for the master builders in the different yards."[10]

John Barry (seen here with sword) and Joshua Humphreys converted and outfitted the first four merchant ships purchased in 1775 to form an American navy, built the frigate *United States*, and collaborated on ship design and naval developments. International Portrait Gallery, Print and Picture Collection, Free Library of Philadelphia.

Knox selected Humphreys to develop the first United States naval construction program because he showed great loyalty to Washington's Federalist administration. Moreover, many Continental Navy veterans, including John Barry, Thomas Truxtun, and Stephen Decatur (senior) considered Humphreys the most experienced and competent naval constructor in the country, as his election in 1794 to the Society of Naval Architecture testified. Knox had a harder time finding people to organize other aspects of the frigate program. The president wanted contracts distributed to six regional port towns at Portsmouth (New Hampshire), New York, Boston, Philadelphia, Baltimore, and Norfolk. To accomplish this task, Knox turned to Treasury Secretary Hamilton, who assigned Tench Coxe, "Commissioner of the Revenue, Relating to the procurement of Military, Naval, and Indian Supplies," to arrange contracts, purchases, and finances. Frigate construction, Coxe explained, was "a new, very expensive, and highly important operation of the Government," and required the "combined judgement" of proper officers from Treasury and War Departments. It meant formation of a shipbuilding bureaucracy of naval agents, constructors and officers to serve as shipyard superintendents and a "clerk of the yard" for each construction site.[11]

Acquisition of timber to build large wooden warships stood at the center of the federal government's first ship manufacturing program. Hamilton appointed Boston shipwright John T. Morgan to procure live oak in coastal Georgia and the Carolinas. "His business was to search for the timber, to superintend the cutting and forming it by the molds for the frigates, and to procure it to be shipped for the six, several ports at which the frigates were to be built," Coxe informed Hamilton. Unfortunately, Morgan hated the rainy "fever" season in the Southern coastal region, becoming seriously ill. The fever-ridden Morgan discovered that New England axe men sent to cut timber had not arrived and that the teams of oxen to pull timber wheels and carry cut wood to a schooner wharf for shipment north had all died. He found no one to remove timber at the government contract price. Local white laborers appeared too sick to

The frigate *Constitution* under repair at the Southwark Navy Yard in 1873. Naval Historical Center Photo NH 55583, courtesy of Capt. W. P. Robert.

work at the government wage, and Morgan rejected black laborers as "useless." "If I am to stay here till all the timber is cut," he wrote Humphreys back in Philadelphia, "I shall be dead."[12]

Desperately in need of timber to begin construction of a 44-gun frigate assigned to his yard, Humphreys dispatched shipyard superintendent John Barry to help Morgan procure timber. Gradually Humphreys took charge of the entire timber acquisition and naval construction process. Appointed on 28 June 1794 simply as "constructor or master builder of a 44-gun frigate to be built in the port of Philadelphia," he assumed the position of U.S. naval

constructor when the federal government rented his shipyard in August 1794, for $400.00 a month the first year and $500.00 each subsequent year until 1800. Humphreys laid the keel in December for a 1,500-ton frigate. He expanded shipbuilding facilities, purchased riverfront property from ship carpenter Jacob Cox on the east side of Front Street bordering Warwick Coates's and ship carpenter Arthur Donaldson's waterfront properties, and rented James Doughty's mold loft, where Fox and Doughty's son William outlined nearly 500 life-sized wooden molds for cutting timber to exact size for the contracted ship. Humphreys borrowed neighbor

Coates's great grindstones to sharpen sawyers' and carpenters' tools, and built a mast shed, a "simple air furnace" to boil salt water and pickle planking, and a large iron-plate oven to heat, season, and bend plank for the hull. Finally, the Southwark shipwright constructed a new wharf on open riverfront ground "near the Swedes Church."[13]

Humphreys assumed coordinative functions informally and spontaneously in order to standardize the American system of naval ship construction. He instructed naval agents in New York and Boston on frigate specifications and types of materials to be employed in construction, and other naval constructors on how to use the molds sent to them from his Southwark mold loft. "I consider it my duty to convey to my brother builders every information in my power." New England and New York naval constructors found Humphreys's guidance reassuring and addressed him as "the Naval Constructor of the United States." At the same time, the government named the Philadelphia frigate, laid down on the stocks in Humphreys's shipyard, *United States*, prompting many to refer to the local shipbuilding facility as the "United States Ship-Yard" and Humphreys's dock as the "United States Warf." The titles made it appear as though Humphreys acted as chief U.S. naval constructor and that his shipyard was the official U.S. government navy yard.[14]

Humphreys did little to dissuade such sentiment, using the position to advance his frigate and obtain contracts for shipbuilding material for in-laws and friends. He consorted regularly with top administration officials. President and Mrs. Washington observed progress on *United States*. Timothy Pickering, a former Continental Army quartermaster general who replaced Knox as secretary of war in 1795, ordered that "when letters arrive from other builders, they will be handed to Mr. Humphreys," and sought Humphreys's advice on naval policy, particularly about the number of frigates needed to defend American interests against the Algerian corsairs. Humphreys advised that "if two ships would be forwarded with all possible dispatch, they would be able in great degree to protect our trade in that quarter." Secretary of the Treasury Hamilton intervened on Humphreys's behalf when Pennsylvania Governor Thomas Mifflin threatened to send master carpenter Samuel Owner with the state militia artillery to suppress the Whiskey Rebellion in western Pennsylvania. "Joshua Humphreys represents that Owner has been employed in preparing models," Hamilton told the governor, "and that the absence of Mr. Owner at this time would be a material injury to the public service as he is one of his principal workmen."[15]

Intimacy with national politics had its drawbacks, however, and threatened to injure Humphreys's reputation. Purveyor of Public Supplies Tench Francis announced that Humphreys overspent and wasted government funds on *United States*. The loyal Quaker businessman Humphreys resented imputations of impropriety in government contracting and treated Francis's comments with "my utmost contempt." Humphreys insisted that he received no "compensation from the public or otherwise without rendering adequate compensation." Nevertheless, the Treasury Department further investigated the "United States Ship-Yard," suspecting that some "criminality" might be involved in expending over $7,000 in one month on the 44-gun ship and in employing the rented government shipyard for personal business. Humphreys fought back, testifying that he never used public property to store timber for private interests and always tried to save the United States money. "When we came to calculate on expense, I have in my yard exploded everything that was unnecessary."[16]

The attack on Humphreys reflected growing disapproval of Federalist foreign and naval policies, particularly the highly unpopular Jay Treaty that sought to ease Anglo-American tensions on the seas. Censure of President Washington's administration came from anti-Federalist, pro-French Republican opponents of an expensive naval establishment who expressed frustration over soaring construction costs on the frigates. Rising expenditures and long delays in naval shipbuilding troubled Washington as well. Secretary of War Pickering warned Humphreys that the president might cancel the program. Ratification of a "Treaty of Peace and Amity with

Algiers" on 7 March 1796 made such cancellation almost certain because, under terms of the Naval Armament Act of March 1794, work on the frigates must stop if the United States concluded peace with the Barbary corsairs.[17]

The Federalist government was reluctant to disassemble an institution once created, and Washington asked Congress to fund at least some of the warships or risk "derangement of the whole system." After a month of acerbic debate between Federalist proponents of continued work on the frigates and Republican opponents of a naval establishment, Congress voted to complete frigates *United States* in Philadelphia, *Constitution* in Boston, and *Constellation* in Baltimore. Though the government had saved *United States*, Humphreys lamented abandonment of the other frigates. "Had they suffered the whole six to be completed, we should have had a pretty little Navy in a short time." Humphreys needed to worry more about his own frigate, however, as the administration could not raise money to continue the work authorized by Congress. Domestic banks charged exorbitant interest for loans, while contractors refused to accept Treasury Department drafts. New Secretary of the Treasury Oliver Wolcott, Jr., who replaced Hamilton in 1795, warned that if they could not raise adequate capital soon—"the Frigates must stop!"[18]

The resourceful Hamilton supposed that Wolcott might "improve and extend" the revenue enough to build three warships by using clandestine and extralegal methods. The former treasury secretary told Wolcott to "circulate in whispers" requests for private loans from the leading shipowning merchants in New York and Philadelphia. "I know you want money, but could not the Merchants by secret movement be put in motion to make you a loan?" Washington was more forthright, finding that the solution to funding construction of the frigates lay in Congress establishing a permanent navy with an annual budget to increase the fleet. In his last message as president, Washington asked Congress to establish a U.S. Navy, and instructed Secretary of War James McHenry, a former Continental Army officer, to develop "a systematic plan for the creation of a moderate navy." At the same time, Hamilton advised McHenry privately to contemplate "commencement of a Navy."[19]

Pressed by Federalist leaders, McHenry informed Chairman of the House Committee on Naval Equipments Josiah Parker of Virginia that the administration would seek legislation for a naval establishment. Frustrated by the loss of a frigate contract for Chesapeake Bay, Parker seemed more interested in learning why, after nearly three years of funding, the Pennsylvania-dominated frigate construction program had not launched a single vessel. The committee chairman demanded estimates on the price of live oak and red cedar, funds necessary to create a live oak reserve, and "the probable Cost of a Site for a Navy Yard and the Buildings necessary thereto." McHenry turned to his naval expert Humphreys to supply answers, and Humphreys used this opportunity to promote development of a naval supply system, standard rules of construction, and a central navy yard. Without such an establishment, the Philadelphia naval constructor argued, shipbuilders "to the South" (Virginia) would submit "preposterous" ship dimensions and inflated cost estimates for construction of warships. Humphreys doubted that the antinavalist Congress would appropriate $55,950 to finish *United States*, never mind the $400,000 necessary to support a "Respectable Navy," but that if "we cannot immediately have a sufficient Navy, will it not be wise to prepare for one by providing materials and a proper Navy Yard where all ships should be built for the United States?"[20]

Influenced by such views, Secretary of War McHenry requested that Congress "secure early, a lasting fund of live oak for the future" and "purchase a site for a navy yard." Expecting that his "United States Ship-Yard" would become the first U.S. navy yard, Humphreys lobbied with Virginia Representative Parker and navalist legislators William Bingham and John Swanwick of Pennsylvania, the latter pair once partners in Willing, Morris and Company. Certainly, Humphreys influenced the Committee on Naval Equipments, which recommended on 25 January 1797 that Congress authorize purchase and development of a government shipyard.

Swanwick and Bingham introduced such legislation, but antinavalist Republicans in Congress rejected the idea, viewing it as another effort by Federalists to build a permanent military establishment and further centralize power in the federal government at the expense of the states.[21]

At least Congress finally voted in March 1797 to sustain *United States*, *Constitution*, and *Constellation*. Humphreys and Captain Barry rushed work on the Philadelphia frigate, while local wood carver William Rush finished the ship's figurehead, a female figure symbolizing the "Young Republic." Humphreys's ship carpenters erected launching platforms, bilgeways, cross pieces, and fore-and-aft wedges. At the last instant, the veteran shipwright realized that he had erected the shipways at a dangerously steep incline and that *United States* might enter the Delaware River too rapidly, strike bottom, and suffer hull and keel damage. Workers sunk anchors in the ground in front of the 1,500-ton warship and attached heavy lines through the hawse holes to slow the descent of the huge craft after removal of restraining wedges.[22]

Over 25,000 spectators lined the Southwark waterfront on 10 May 1797 to watch Humphreys launch *United States*. Hundreds more filled dwellings along Swanson Street and choked the Delaware River with every imaginable type of small craft so that they could catch a glimpse of the most momentous event so far in Philadelphia's long history of shipbuilding. Cabinet Secretaries Pickering, Wolcott, and McHenry boarded armed brig *Sophia* and joined the flotilla on the river to observe the launch, which occurred at 1:15 P.M. *United States* entered the Delaware "without straining a hogging more than one-and-a-quarter inches," Humphreys reported, far less than similar-sized British warships. Despite a too rapid entry that caused the frigate to hit bottom, the launch proved a great success.[23]

Introduction of *United States* came at a most critical moment in American foreign affairs. Newly-elected President John Adams told Congress that worsening relations with France had reached a crisis. French privateers violated American neutral rights and freedom of the seas, while the French government ignored American pleas to settle disputes. Adams asked Congress to "augment [the] Existing Military Establishment" by expanding the army, fortifying ports, purchasing armed ships, and completing frigates. At first, Jeffersonian Republicans (led in the House by financial expert Albert Gallatin of Pennsylvania) opposed such expensive war preparations and blocked a measure to outfit *United States*. Continued French violations of American maritime rights during the summer of 1797, however, helped the Federalists overcome Republican opposition to "An Act providing a Naval Armament" for the completion, manning, and equipping of all three frigates.[24]

Funding the frigates should have been a period of triumph for Humphreys. Instead, it turned into another of the bureaucratic feuds that characterized the growing pains of the nascent U.S. naval establishment. Shortly after Congress authorized completion of the frigates, Josiah Fox, now a clerk in the War Department in charge of naval business, ordered his former boss Humphreys to go to Baltimore at once to see to the "safest manner" for launching the 36-gun frigate *Constellation*. The normally methodical, workmanlike Quaker shipbuilder lost his temper, telling Fox that in the future he would take such orders only from the secretary of war himself. "I cannot receive hereafter or attend to my directions from you, while you style yourself Naval Constructor, you must know that my station in the service of the United States require no direction from a Naval Constructor, you also know that I am at the head of that Department," and "whenever the Secretary deems my services no longer necessary you may to other persons assume such title as your vanity may suggest."[25]

Humphreys went to Baltimore to arrange the launch of *Constellation* only after Secretary of War McHenry asked him to accompany his friend Truxtun, who was to assume command of the frigate, and naval constructor David Stodder to the launch site to "assess ascent of keel [and], depth of water." Humphreys instructed Stodder on how to arrange the launching ways, sharing his experience with

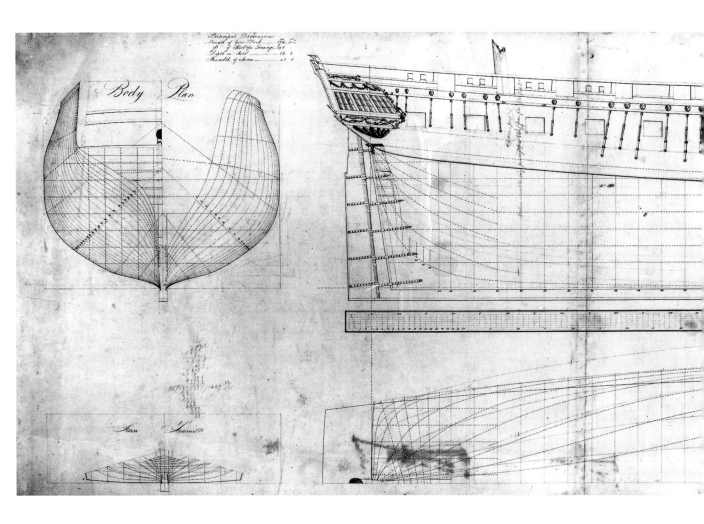

United States. The Philadelphia shipwright admitted quite frankly that his method was flawed, and corrections on *Constellation* helped launch the frigate on 7 September 1797 without the "smallest Accident." Truxtun wrote Humphreys, who had returned to Philadelphia, that "a Better Launch I never saw." Thus buoyed, Humphreys offered to travel to Boston to assist naval constructor George Claghorn, who was having serious problems getting the 44-gun frigate *Constitution* to slide down the shipways into the water. "I cannot help feeling for the situation of the Frigate, as well as for Col. Claghorn whose situation must be mortifying." Most disconcerting, President Adams had attended launching day ceremonies and came away badly dis-

appointed when *Constitution* refused to move down the ways. "If you should consider that I could offer any services to the Builder," Humphreys wrote McHenry, "I shall cheerfully obey your order."[26]

Claghorn succeeded in getting *Constitution* into the water in late October without assistance from Philadelphia. But the entire sorry affair of launching three frigates, one with too high an incline, the one with too little, and third with perfect entry, demonstrated the inefficiency that occurred by scattering naval construction around to distant yards with different naval constructors who altered ship plans and dimensions as they went along. There was no navy department or naval shore establishment to coordinate U.S. construction policy. Humphreys's own

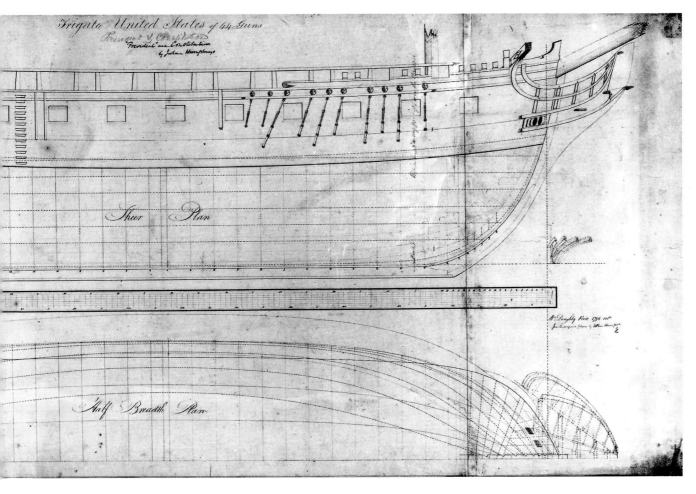

Sheer and hull plans for the 44-gun frigate *United States*, which was built between 1794 and 1797 at the Southwark shipyard and established the site for the first Philadelphia Navy Yard. ISM, Lenthall Collection, loan from Franklin Institute; photograph by Todd Bauders.

Southwark yard, rented by the federal government for $500 a year, suffered from the informal nature of the naval establishment. Guards were so drunk that they slept through their watch or left their muskets in sentry boxes as they went into town to carouse. Curiosity seekers climbed over the hull of *United States* and pilfered lumber and other items from the unfenced yard. Humphreys worried particularly that antinavalist Republicans in Philadelphia might burn the unguarded frigate to prevent completion.[27]

Without the purchasing and procurement functions of a proper navy yard, the outfitting of *United States* came to a standstill. Gun carriage parts delivered by John Haines of Burlington County, New Jersey, were "all wrong." Masts were cut to improper dimensions. Too much of one size of timber arrived by sloop in the river and had to be stored in a makeshift timber yard. Humphreys built a wharf and a temporary lumber storage yard on Mud Island, a few miles below his Southwark yard. Here, in 1795, the War Department assumed authority from the Commonwealth of Pennsylvania over a partly finished earth and stone fortification called Fort Mifflin. No sooner had Humphreys "stowed" five cargos of timber on the island, than military engineers assigned to finish the defensive works purloined the live oak supply so that they could build gun platforms and gun carriages at Fort Mifflin.[28]

Constructing tribute vessels for Algiers in late 1796–97 revealed further weaknesses in the American naval shipbuilding process. As part of the Treaty of Peace and Amity with Algiers, the United States agreed to build four ships for Hassan Bashaw Dey of Algiers, in return for the release of American hostages held in the North African state. The construction of treaty ships fell under the direction of the State rather than the War or Treasury Department. Secretary of State Pickering asked for cost estimates for the construction of warships for Algiers, but admitted that the Barbary State had yet to provide a list of specifications for such estimates. Once the State Department received ship dimensions from Bashaw, Pickering sent out the contract for the 36-gun frigate *Crescent* (designed by Fox) to Portsmouth, New Hampshire, shipbuilder James Hackett, and gave Humphreys authority to receive "sealed Terms" from Philadelphia shipbuilders for the remaining three tribute vessels granted to Algiers by the 1796 treaty.[29]

Humphreys distributed contracts to build Algerian tribute vessels among family and friends. Apprentice shipwright Samuel Humphreys, Joshua's son and later chief naval constructor of the United States, helped Nathaniel and Benjamin Hutton design and build the "corsair" schooner *Skjoldebrand*, a 20-gun ship rated as one of the best produced by the Humphreys naval shipbuilding firm. Twenty-year-old Samuel Humphreys also helped his father prepare plans for 300-ton tribute brig *Hassan Bashaw*. Kensington shipbuilder Samuel Bowers, who had assisted Humphreys on *United States*, received the contract for 18-gun tribute schooner *Lelah Eisha*. As keels were being laid for these treaty ships, a yellow fever "Contagion" that seemed to grip Philadelphia nearly every summer during the late eighteenth century struck the shipyard, killing Humphreys's chief carpenter, yard clerk, and many shipyard laborers. Humphreys left the city for the relative sanctuary of his farm near "the Nine Mile Stone on the Haverford Road," and Pickering, McHenry, and federal officials fled north to Trenton, New Jersey. Only Nathaniel Hutton and a handful of carpenters and joiners stayed to raise the frames for *Skjoldebrand*.[30]

Autumn frost brought the U.S. government and shipbuilders back to Philadelphia anxious to resume work on the treaty-mandated tribute vessels. Instead, they found that in their absence France and the United States had arrived at the brink of an undeclared naval war. During the summer of 1797, President Adams had dispatched three envoys to Paris to negotiate a settlement of outstanding maritime grievances. But France, locked in a titanic worldwide struggle against Britain, refused to accept the American peace mission, instead sending agents (referred to in correspondence as X, Y, and Z) to demand a bribe as a way to open negotiations. In March 1798 Adams released the XYZ correspondence to the public, revealing details of the affair. At the same time, the French government decreed that it would seize American ships carrying supplies to Britain without warning. Reluctantly, Adams asked Congress to pass measures to protect "our seafaring and commercial citizens." Federalist Senator Samuel Sewall of Massachusetts introduced a resolution on 8 March to create a separate Navy Department, while Secretary of War McHenry considered "the propriety of making a provision for a permanent navy yard and gradual or prompt purchase of timber etc. proper for building and equipping Ships of different rates &c."[31]

Jeffersonian Republicans claimed that Adams (and behind the scenes Alexander Hamilton) fabricated the crisis with France in order to obtain a naval establishment. Republicans in Congress "vehemently" opposed the construction of sloops-of-war and were "still more opposed" to construction of 74-gun ships-of-the-line as proposed by Hamilton. Also, anti-Federalist representatives objected to creation of a separate navy department or development of a government navy yard. Bitter debate between Federalists and Jeffersonian Republicans over the direction of American naval and foreign policies spilled into the streets of Philadelphia and down into the shipbuilding section of Southwark. In one instance, Joshua Humphreys's rugged eldest son Clement beat a Philadelphia *Aurora* newspaper editor because of his strident Republican antinavalist, anti-Adams sentiments. Fortunately, the elder Humphreys used more peaceful means to protect his naval shipbuilding business and defend the Adams administration's foreign and naval policies. The Quaker shipwright led local pro-Adams rallies, circulated a petition in support of a navy, and served as toastmaster at a grand Federalist dinner in honor of the president's strong stand against France.[32]

Philadelphia shipbuilders and merchant shipowners contributed money and influenced passage of legislation in 1798 to increase naval armament and the pathbreaking Naval Act of 30 April that created a separate executive department and cabinet post for a United States Navy. Former Willing, Morris and Company merchants William Bingham, now a Federalist U.S. senator, and William Swanwick, a pro-navy Republican U.S. representative, led the fight for a Navy Department. Passage of this act created a naval establishment on paper only. Secretary of War McHenry remained in charge, complaining that "Neither the President has mentioned the subject yet to me, nor any other gentleman." Adams asked old friend George Cabot to become the first navy secretary, but the retired Massachusetts merchant refused. The president turned next to Maryland merchant Benjamin Stoddert, who agreed to serve, but wanted a month to settle private business before coming to Philadelphia to take office.[33]

McHenry realized that someone else had to organize the new naval establishment and get the naval force to sea to protect American coast and commerce against the French—"there being no chance that we shall have a secretary of the Navy in time." Raising the Provisional Army and planning fortification of the coast overburdened McHenry. That left the aged Treasury Secretary Wolcott. "The purchase, building & providing of the Ships falls upon me," Wolcott told Hamilton, "& you know that my duties are enough to employ a mind more active & vigorous than I possess." He "despaired of obtaining a tolerably fit character" to organize the naval

establishment and fight an undeclared naval war. All along, the intensely loyal Federalist shipwright Humphreys readied the United States Navy for war with France, becoming in June 1798 quasi-secretary of the navy. He inspected, purchased, and converted private ships, "calculated to carry from 16 to 22 guns." He reestablished familiar connections. Twenty-three years earlier, his first conversion job for the Continental government had been *Alfred*, the Willing, Morris and Company ship *Black Prince*. Now, on 3 May 1798, Humphreys purchased the 504-ton, square-sterned West India merchantman *Ganges* from Willing, Morris and Company's successor firm of Willing, Willing, and Francis.[34]

In June the Adams administration sent Humphreys to Baltimore to examine merchant vessels for

Federalist Secretary of the Navy Benjamin Stoddert instructed Joshua Humphreys to buy the Southwark property and establish a Philadelphia Navy Yard in 1800 before the Jeffersonian Republicans took office and canceled the purchase. Print and Picture Collection, Free Library of Philadelphia.

purchase and conversion, where he bought 400-ton ship *Adriana*, bringing it back to Philadelphia for conversion into 20-gun cruiser *Baltimore*. The Southwark shipbuilder's most important acquisition in 1798 occurred at home on the local waterfront with sleek 320-ton *Hamburgh Packet*, built in his own yard in 1794. Humphreys converted the Atlantic packet into the fast 20-gun cruiser *Delaware*, which under Captain Stephen Decatur cleared the Delaware Bay and Jersey coast of French privateers. *Delaware* captured the French raider *La Croyable* off Egg Harbor, New Jersey, and brought the prize back to the United States Ship-Yard for conversion into the U.S. warship *Retaliation*.[35]

Secretary of the Treasury Wolcott gave Humphreys responsibility for designing and constructing a new class of larger revenue cutters to replace the tiny fleet assembled by former Secretary of the Treasury Hamilton. "Having decided to place the whole of the business in relation to the building and equipping of the said Cutters under your direction— I will thank you to take immediate measures for procuring the Sails, Anchors, Cables, Rigging, Cannon with their Carriages, and all other kind of Stores, whether warlike or otherwise necessary for fitting them for Sea," Wolcott instructed. Humphreys asked his son and former apprentice Samuel to design and build cutters *Diligence* for South Carolina and *Eagle* for Georgia. At the same time, Joshua Humphreys reviewed proposals for cutters and ships from other shipbuilders. He held authority to accept or reject designs and make contracts. Encouraged by growing influence to shape the U.S. naval establishment, Humphreys once again pressed the federal government to acquire a proper navy yard where all designs, drafts and contracts could be stored "in order that the plan of the most approved and useful vessel may be resorted to in future."[36]

At the peak of Humphreys's influence, Stoddert arrived in Philadelphia to take office in early July 1798 as the first U.S. secretary of the navy. His presence threatened Humphreys's position. Unlike earlier government officials, Stoddert held no special attachment to what he called the "small clique of Philadelphia shipwrights." A prominent member of the rival Chesapeake maritime community, Stoddert favored the Chesapeake region and New England over the Delaware River for naval shipbuilding and as sites for U.S. navy yards. He consulted more with Baltimore naval agent Jeremiah Yellot and Annapolis naval agent William Marbury than with their Philadelphia colleagues. He also sought advice and guidance on naval business from New England naval agent Stephen Higginson, and dispatched Humphreys's long-time nemesis and rival Josiah Fox to Virginia to build frigate *Chesapeake* and develop a Gosport (Norfolk) naval shipyard as the major base of operations in the undeclared war against France.[37]

But Stoddert had connections with the Philadelphia maritime community as well. He had served under Robert Morris and Clement Biddle during the Revolution as a deputy forage master in the city. Indeed, Stoddert employed Biddle as his legal advisor in 1798 and hired Morris's former bookkeeper Garret Cottinger (unemployed when his boss went to debtors' prison) as chief clerk of the Navy Department. Stoddert also made Humphreys his "principal naval constructor" and manager of the United States Ship-Yard which held all specifications, cost estimates, and contracts for the three frigates and lists of merchantmen to be purchased by the government. More important, Stoddert wanted Humphreys to mobilize Philadelphia's Federalist "Junto" to subscribe $64,000 for construction of the 44-gun frigate *City of Philadelphia*.[38]

During the summer of 1798 one of the worst yellow fever epidemics of the decade made Humphreys indispensable. Stoddert fled his new Walnut Street naval office, leaving Humphreys behind to get "the little fleet in the city" ready for sea. "I fear the French Cruisers will be on our coast before we are

Letter from Joshua Humphreys to John Barry, 26 June 1798, showing the close relationship between the shipbuilder and naval officer, which began in 1775 when the two converted *Black Prince* into the Continental warship *Alfred* and culminated in the development of the frigate *United States*. ISM.

Philadelphia June 26. 179.

Sir

Under the Authority of the Secretary of the Navy I have to
request you will be so obliging as to note and transmitt to me a
true state of your Ship after being at Sea

First   The height of your Guns above water

For what time can your ship carry Provisions & water for her crew

What difference of draft of water is the Ships best Sailing

Which is the best Sailing of the Ship, whether by or large and
at what rate - - - - - - - - -

Is she capable of carrying a heavy press of sail without lay-
-ing over too much - - - - - - -

Does she pitch in a heavy Sea or is she lively

Is she easy on her rigging or otherwise

does she work well, & quick when in stays, is more distance
necessary than other Ships require

Note the exact distance and rake between the masts, the
distance, the Fore & Mizen masts are from the Perpendicular
also the quantity of Kentledge you have on board, if
you have a sufficient quantity or otherwise; In
fact be pleased to point out every good & bad
quality your ship possesses

As soon as you have had sufficient opportunity of
trying the Ship and clearly made up your mind, with
your reasons in support of your opinions I will thank
you to make me the Communications as they may be of the
utmost consequence in the Construction of other ships now
in contemplation

                    I am with Respect

                         Joshua Humphreys

John Barry Esquire

prepared for them," Stoddert wrote from the U.S. summer capital in Trenton, New Jersey. "I hope you can get as many carpenters as you can employ to advantage on the *Ganges* and the more you can employ without being in each others way, the sooner the ship will be prepared." Humphreys had no intention of staying in the plague-ridden city either, and rushed to Pont Reading in Haverford Township, Delaware County, about seven-and-one-half miles west of the Schuylkill Bridge, where he conducted Navy Department business a mile from his estate at Buck's Tavern on the "Turnpike Road" (Lancaster Pike). Humphreys moved his United States Ship-Yard downriver to Marcus Hook—halfway between Chester, Pennsylvania, and Wilmington, Delaware— where Philadelphia merchants had long tied up ships to unload cargoes when ice or storms prevented passage further upriver to the city.[39]

The Marcus Hook Naval Shipyard prepared the U.S. Navy for the Quasi-War with France. Humphreys established a mast yard, timber supply, and naval storehouse there. Nine-pounder cannon cast at nearby Foxall Foundry lay on the wharves at Marcus Hook awaiting Humphreys's carpenters to construct wooden gun carriages for mounting on U.S. Navy warships *Ganges*, *Sophia*, *Delaware*, *Eagle*, and *Diligence* and the Algerian tribute vessels *Hassan Bashaw* and *Lelah Eisha*. By September, seventy-five Southwark carpenters, joiners, caulkers, and laborers had come south to "the Hook" to work on these ships. Humphreys dispatched Thomas Wharton, nephew of his former partner, to keep accounts as "Clerk of the Navy Yard now at the Hook." A U.S. Marine guard set up barracks on the wharf next to *Ganges*.[40]

The yellow fever soon reached Marcus Hook. "So fatal has the disorder been in this little place," Humphreys reported, "that nine persons have died in thirty-six hours." The disease raged through the Marine Barracks on the waterfront, and Humphreys relocated the guard "to the back part of the hook." The pestilence devastated warship crews and struck the ship carpenters, most of whom returned home to the city to die. The only physician available at the Marcus Hook Navy Yard came from *Ganges*, and

Humphreys would "not trust a dog to his care." By November, the frosts that reportedly purged the city of yellow fever brought the navy yard back to the Southwark waterfront, and Humphreys ordered *Ganges* and *Sophia* upriver to complete conversions. Humphreys returned to the city, as well, "althou' my fears of the fever have not totally subsided." Here son Clement recruited naval crews, and son Samuel and shipbuilding partner Nathaniel Hutton raised "the Frigate [*Philadelphia*] building for merchants in this part," on the stocks just below Old Swedes Church.[41]

When Humphreys reentered Philadelphia, he found instructions from Secretary of the Navy Stoddert waiting for him to study the cost, draft a plan, and estimate facilities needed to build a 74-gun ship. Stoddert hoped to build twelve of these large warships. Such a fleet would make the United States a naval power. It was a perfect warship to cruise distant stations, control sea lanes, and defend coasts against enemy forces. Heavily armed with two main gun decks, and very seaworthy, the ship outclassed single-gun deck frigates and outsailed heavier 120-gun vessels. Construction of a dozen 74s would give the fledgling U.S. Navy the ability simultaneously to command coastal waters and protect American interests on the high seas.[42]

Assisted by son Samuel, apprentice William Doughty, and master shipwright Benjamin Hutton, Joshua Humphreys drafted plans for a 74-gun ship in time for Stoddert to submit them to Congress in early 1799. Humphreys's model called for a ship of nearly 1,800 tons, with 178-foot keel and carrying 28 guns on the main deck, 30 on its upper deck, 12 on its quarterdeck, and 4 on the forecastle. The Southwark shipbuilder estimated that each ship would cost $320,000 to build, and confided to Stoddert that an additional $15,000–$20,000 would be needed to develop each naval dockyard, like the Royal Navy Yard in Woolwich, England, to construct and maintain the mighty 74s. Humphreys suggested that the government purchase land for such dockyards in Georgia, Virginia, Rhode Island, Massachusetts, and Amboy, New Jersey, and for the establishment of three principal naval shipbuilding yards in New

York, Philadelphia, and the City of Washington.[43]

Humphreys campaigned quietly to make Philadelphia the premier site for a navy yard. He told Stoddert that no shipyard south of the Potomac River should ever be given 74-gun ship contracts because the region lacked skilled labor. Nor should the government consider shipyards located north of New York harbor, because the region lay too far from the supply of live oak timber. Timber for *Constitution*, for instance, had to be transported at very great expense from Georgia to Boston. The Delaware River shipyard owner outlined minimum requirements for a U.S. Navy Yard. "A city that would be desirable for a Navy Yard, taking it for granted none to be south of the Federal City, should be but a small distance from a large commercial city as contiguous to others as possible, in order to have occasional supplies of workmen, and where naval architecture is carried on in an extensive way," Humphreys explained. "It should be within reach of good white oak timbers," and located in a safe freshwater anchorage, free from storms and the damaging effect of saltwater worms on wooden hulls. Such conditions existed only in Philadelphia. Over the coming months, Humphreys repeated these basic requirements for a navy yard until he received assurance that the U.S. government would purchase land and establish a navy yard on the Southwark waterfront ninety miles up the Delaware River from the sea.[44]

4 ★ "Second and Next Best" Navy Yard

SECRETARY of the Navy Benjamin Stoddert left Philadelphia in June 1800 when the federal government was transferred to the new capital in the District of Columbia on the banks of the Potomac River. Delaware River shipbuilder Joshua Humphreys worried that the move meant the end of his role as principal United States naval constructor. He wondered whether Stoddert wanted him to finish sending the sets of molds to frame six 74-gun ships authorized by Congress in February 1799 to naval agents in New Hampshire, Massachusetts, New York, Virginia, and Washington, where Stoddert's friend William Marbury managed naval procurement. "Is it your pleasure that I attend to the transportation of the Live oak to the different Navy Yards where the 74 Gun ships are to be built?" Humphreys inquired. "What is the duty you will generally require of me?"[1]

Stoddert enjoined his chief naval constructor to continue contracting for timber and to ready the 44-gun frigate *United States* for sea. Stoddert directed Humphreys to examine vessels of war as they arrived in the Delaware River in increasing number following the Quasi-War with France. Humphreys's United States Ship-Yard assumed the institutional role that characterized the Philadelphia Navy Yard for the next 175 years by overhauling, laying-up in reserve, or disposing of warships at the end of hostilities. So many ships arrived in late 1800 that Humphreys rented additional dock space, leased Robert McMullen's wharf to complete subscription frigate *Philadelphia*, and borrowed Joseph Marsh's dock just below the United States Ship-Yard to

repair frigate *Constellation*. Unfamiliarity with the river bottom along this older wharf caused a major naval accident. "The *Constellation* grounded opposite Marsh's wharf, heeled, and filled with water, and lay sunk for near three weeks," Humphreys wrote to his son Samuel, then in Georgia inspecting live oak timber for construction of 74-gun ships. The 20-gun brig *Patapsco* sustained considerable damage trying to cant and raise *Constellation*.[2]

Such difficulties reinforced Humphreys's desire to develop a government navy yard on the banks of the Delaware. Fortunately, Stoddert supported the idea of establishing navy yards along the Atlantic coast and dispatched Humphreys to the Potomac River "to Establish a Navy Yard at that place where one of the 74s are to be built." The Southwark naval constructor surveyed the area with Stoddert's Maryland business associate and Georgetown naval agent William Marbury, and selected Buckman's Point as the best site. Marbury preferred the "Exchange Ground" Landing along the East Branch of the Potomac River, where he held substantial properties with Stoddert and Maryland merchant-land speculator Uriah Forrest. Not surprisingly, Stoddert chose Marbury's site, but appeased Humphreys by promising to hire Philadelphia carpenters and Southwark wharfbuilder W. Thomas Davis to develop a Washington Navy Yard. "I want the thing well done," he told Humphreys, "as it ought to be with a view to a permanent establishment, and can rely only on you."[3]

Humphreys went along with the Maryland clique's navy yard scheme, expecting Stoddert to support the establishment of a Delaware River navy yard. He did. A period of growing crisis for the Federalists helped Humphreys finally secure a local navy yard. The Quasi-War with France had ended, threatening to cut appropriations for further expansion of the naval establishment. Worse, the Federalist administration of John Adams disintegrated under the daily criticism of Alexander Hamilton, Secretary of State Timothy Pickering, and Secretary of War James McHenry. The internecine struggle forecast a Jeffersonian Republican victory in the November presidential election. Stoddert, who stayed loyal to Adams, worried that the antinavalist and frugal Republicans would abandon plans to build 74-gun ships and a system of naval dockyards. It became particularly urgent to buy land for navy yards before the Federalists lost power. As a first step, Stoddert ordered Humphreys to survey sites along the New England coast. "It appears desirable to some of the members of Congress to the Eastward, that some places in the district of Maine should be examined with a view to the construction of a dock, particularly Portland & Wiscassett [Casco Bay]." Stoddert advised Humphreys to reveal, "in secret," his first choice for the best navy yard location.[4]

Humphreys used this site-inspection tour to promote Philadelphia as the location for a U.S. navy yard. He appeared objective, but in his report to Stoddert he undermined other sites. For instance, while praising Charlestown, Massachusetts, Humphreys worried that fog, ice, and shoals restricted access to the harbor. He recalled that during the Revolutionary War two British 74s trying to enter the harbor had struck bottom, foundered, and sunk. Of course the Southwark shipbuilder said nothing about an identical grounding and loss of a large British warship (HMS *Augusta*) on the Delaware River during the late war. In the same vein, Humphreys reported that Newport, Rhode Island, was a wonderful natural spot for a navy yard, yet too vulnerable to direct attack from the sea. "There is no situation within the U.States more convenient to favour the depredations of any enemy than Rhode Island, nor from which our trade would be more liable to destruction from the excursions of their cruisers."[5]

Whether or not Humphreys intentionally slanted his report on navy yards to advance prospects of establishing one on the Delaware River, he received instructions to purchase local riverfront property a few days after submitting his views. The Southwark shipbuilder made a show of impartiality in locating a site. He sounded the river off the Lazaretto, a quarantine hospital a few miles downstream from the city, but dismissed the site "because of Shoals." He studied the Marcus Hook area, where he had operated a temporary naval station during the recent troubles with France, declaring it too distant from the labor supply in Philadelphia and unhealthy with "the fall fever." Finally, Humphreys sounded the river from Race Street north along the Kensington waterfront to Frankford Creek, announcing that conditions seemed perfect for establishment of a navy yard at the confluence of Frankford Creek and the Delaware River.[6]

Perhaps Humphreys recommended Frankford Creek as his first choice for a U.S. navy yard because he knew beforehand that property owners would charge an exorbitant prices for this prime location and that uncertain titles to the land would have to be searched in Britain. Either situation presented an unacceptable delay to Stoddert, who had to close a deal before the Federalists left office. As expected, the navy secretary rejected the Frankford Creek area in favor of what Humphreys called "the second and next best place, indeed the only one that was to be had at that time anywhere near the city" in the Southwark neighborhood adjacent to his United States Ship-Yard.[7]

No sooner had Humphreys received Stoddert's decision on the Frankford Creek site than he introduced the secretary to three Southwark riverfront properties, including a lot with 150 feet of river frontage on the southern side of Wicaco Lane (Prime Street), once the site of Franklin's Association Battery and now owned by John and William Allen. "This is the most desirable lot of the whole," he observed. The land lay above waterline at high tide and overlapped United States Ship-Yard property. Humphreys already leased the lot with Southwark shipbuilders Warwick Coates, Joseph Marsh, and the Huttons from one-time naval contractor William Allen, current title holder now residing in Hunterdon County, New Jersey, to store timber for 74-gun ships and Algerian tribute vessels on the "Battery Grounds."[8]

Immediately below the Battery Grounds lay Philadelphia merchant shipowner George Clymer's old lot, now held by Humphreys's business associate and attorney Anthony Morris. This lot presented over 240 feet of Delaware River frontage, mostly underwater at high tide. Humphreys admitted that this property had to be filled with dirt in order to serve as a government naval facility. A similar problem confronted Southwark merchant Luke Morris's lot, abutting that held by Anthony Morris. It offered fifty feet of riverfront wetland entirely submerged at mean high tide. Another difficulty stemmed from three partly graded streets—Church, Swanson, and Meadow—that ran nearly north to south along the river through these properties, and which the Southwark District Board of Commissioners wanted to keep open to the public. Humphreys would have to obtain a resolution asking the Pennsylvania State Legislature to close these streets before any work could begin on developing the property for the U.S. government.[9]

A further obstacle to founding a navy yard in the Southwark neighborhood appeared when Humphreys reported that the depth of water in front of these lots was barely sufficient to launch a 74-gun ship—lying only a foot deeper than at the riverfront spot where the *United States* had struck bottom and damaged its false keel upon entering the river. Nevertheless, the Southwark businessman negotiated with property owners for these riverfront lots. Stoddert at first wanted only the two lots with the higher ground, then changed his mind and asked Humphreys to acquire all three. The navy secretary seemed near panic. "So many difficulties present themselves respecting the fixing on a Site for a building Yard on the Delaware, that I am really at a loss how to act." Once again, Stoddert considered the "high ground extending from the River Delaware to Frankford and the Point Roads," and once more

rejected it because of unacceptable delays in securing a clear title.[10]

Time was growing short for the Federalists. Stoddert had to act and called Humphreys to Washington in early November 1800 to close the deal on the Southwark properties. "It being determined that a building yard and Dock for seasoning Timber for the Use of the Navy of the United States shall be established in or near the City of Philadelphia, I have decided grounds of Messrs. Anthony & Luke Morris, and Mr. Allen in the District of Southwark, continuous to, and adjoining Federal Street, containing about 540 foot front on the River Delaware and thence back to Front Street, as being the most eligible because to be obtained immediately which is not the case with the place above the City."[11]

Humphreys settled quickly with the Morris family in December 1800, paying $14,000 for Anthony Morris's lot and $11,000 for Luke Morris's wetland. Stoddert diverted money from the Congressional appropriation of 1799 to build six 74-gun ships in order to purchase the Southwark properties. Meanwhile, William Allen's lawyers tied up transfer of title to the government, holding out for more money. Humphreys concluded the deal at last in January 1801, securing the Allen family lot for $12,000. But Southwark District Commissioners and the Pennsylvania State Legislature refused to close the streets that ran through the property. Stoddert could wait no longer, as the U.S. House of Representatives declared Thomas Jefferson the victor in the disputed presidential election of 1800. Consequently, Humphreys forwarded titles to the three lots on 26 February 1801, a few days before the Federalist administration left Washington forever.[12]

Humphreys's crew started moving timber supplies and naval stores from his adjacent private lot onto the proposed government site before the legal transfer had been completed. Stoddert misappropriated several thousand dollars from an unspent ship construction fund to cart heavy timber pieces cut earlier for the 74-gun ships. All along, local residents pilfered public property from the proposed navy yard lots, and Humphreys begged for money to erect a stone and brick wall around the perimeter. If he

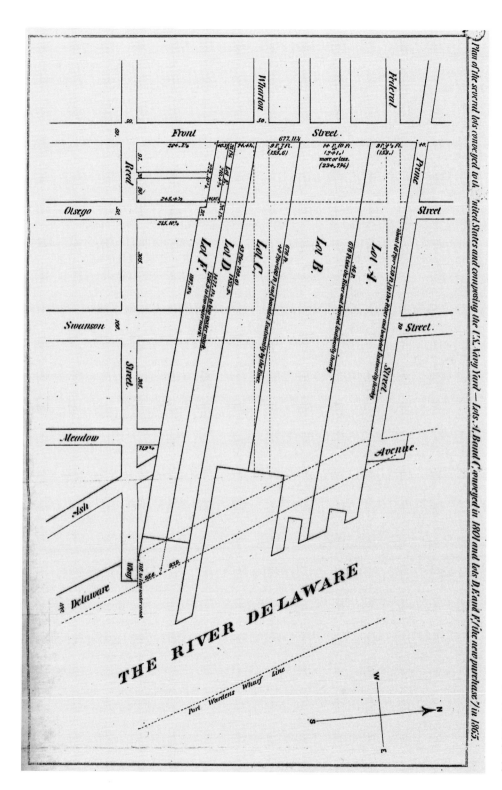

Title transfer map of 1875 showing the original three lots (A, B, and C) purchased by Joshua Humphreys in 1800. ISM.

could not do that, the naval constructor wanted to build a board fence and storage shed to prevent the "very great destruction of arms & stores &c." He wanted Marines to guard the new property. "There is now in the Navy Yard a small two-story Brick House and two wooden buildings which will want some repairs," Humphreys informed Stoddert. "I suppose they will contain twenty or thirty marines, who may be some protection to the light articles and cannon that belong to the Navy, which probably will be immediately deposited there."[13]

Humphreys began leveling and filling in December 1800, despite frozen ground and without specific instructions from Washington. He obtained "cellar dirt at 5 to 10 cents per horse cartload" from the Southwark Commissioners as they leveled streets in the district. His ship carpenters built a large scow to carry gravel from the Jersey Shore opposite to a dock that master wharf builder Davis had erected on the property before its official transfer to the government. Humphreys asked Stoddert to purchase Snowden and North's ropewalk, immediately south of the government site, for $14,000 "if 'tis intended to carry on the rope making business in the navy yard." All the while, Thomas and Benjamin Hutton were developing the new navy yard, asking the U.S. government to reimburse them for expenses in moving the "Counting House and Stores to Navy Yard and fitting them up."[14]

The cost of establishing a navy yard in Philadelphia overwhelmed Stoddert. He could "give no directions at present respecting expenses to be incurred in improvements," but on the last day of his administration he found "additional monies" to "proceed to have erected a Sufficient wharf for building and Launching Ships of 74 Guns, a Dock for docking timber, which ought to be large enough to contain the timber for two 74 Gun Ships & more, if to be made without very great expense." Humphreys used the money to repair and outfit warships before the incoming Jefferson administration could cancel expensive overhauls. He spent $3.50 per day to lay up *Philadelphia* at Robert McMullen's dock and $4.00 per day to repair *Constellation* at Marsh's wharf. It cost $1,000 to raise *Constellation* from the river bottom, $2,000 to repair damage to the frigate, and $150 to restore *Patapsco*, also injured in canting *Constellation*. Humphreys paid $3,000 for timber loaded on *George Washington* for shipment as tribute to Algiers and $1,500 in back wages to his son Samuel and three assistants when they returned from Georgia with a fresh load of live oak timber cut for the 74s.[15]

The Federalist naval constructor used up the additional monies, and petitioned the incoming Jeffersonian Republican administration for back wages and for postage for letters sent on government business. Humphreys sensed trouble when Navy Department Accountant Thomas Turner questioned these expenses. Nevertheless, he was encouraged by Samuel Smith, a Baltimore merchant who once commanded nearby Fort Mifflin on the Delaware River and now served as the acting chief administrator of the Navy Department while Thomas Jefferson tried to find someone to replace Stoddert. Smith sent Humphreys money to cover conversion costs on the purchased merchantman *George Washington* and hinted that the new president might revive the 74-gun ship program. "Much very much is to be done towards putting the Navy in a situation that may conduce to its Respectability While I Act," Smith promised.[16]

At last, Jefferson selected Navy Department ad-interim administrator Samuel Smith's brother Robert, a Maryland attorney, as secretary of the navy. The easygoing Smith supported Humphreys and agreed to pay for a clerk and master workman to facilitate repairs on the *Constellation*. The new secretary promised to send the decaying *Constitution* to Philadelphia for a complete overhaul and hire a new timber inspector for the live oak reserve in Georgia. Greatly encouraged, Humphreys advised Smith on how to improve the naval shore establishment by erecting a great shipbuilding wharf at the new navy yard for the proposed 74-gun ships. "It will take a considerable time before a wharf can be built & sufficiently settled to raise one of those heavy ships on."[17]

The navy secretary took no action on Humphreys's advice because Jefferson canceled the 74-

gun ship program and ordered frigates laid up under sheds and ships purchased during the Quasi-War sold. Work stopped on the shipbuilding facility in Southwark. Ominously, the president ordered an investigation of Stoddert's use of 74-gun ship funds to acquire the local navy yard. A seven-member investigative committee, which included William Penrose, a Jeffersonian Republican and Humphreys's enemy, arrived at the Southwark naval facility. Humphreys took the "gentlemen on the subject of the Navy Yard" on a tour of the waterfront and seemed certain that they planned to abandon the Southwark site. "Tomorrow or next day we shall close the report respecting the Navy Yard which I believe will be in favour of the ground at Kensington."[18]

In the end, the Jefferson administration retained the Southwark property for a U.S. Navy Yard, but discharged Humphreys as naval constructor in October 1801 as part of a general housecleaning of all Federalist office holders. "As it is not intended that either of the 74's shall be commenced until all the timber is duly prepared & properly seasoned, the station which you hold as Navy Constructor has become unnecessary," wrote Navy Secretary Smith, "and I am under the necessity though very reluctantly, of informing you that your services will be dispensed with after the 1st of November next up to which period you will be pleased to make out your account & transmit it to the Accountant for Settlement." Smith ordered the Federalist shipbuilder to turn over all public property under his authority to local naval agent George Harrison.[19]

Humphreys left the Jefferson administration with a partly developed navy yard. An old log house and stone kitchen, once occupied as a block house magazine by early Swedish settlers of Wicaco, stood on an ancient wharf. Generations of private Southwark shipbuilders had employed it as the central counting house for their businesses. A two-story frame house, a workshop, and a brick building fitted up as a Marine barracks—all used earlier by local shipbuilders—dominated the property. A joiner's shop and storehouse, owned by the federal government and moved south from Humphreys's rented shipyard

in early 1801, stood on the northern boundary of the navy yard property. The only new structure was a tiny armory constructed for the Marine guard by Captain Franklin Wharton. "These are all the buildings in the Navy Yard," observed Humphreys.[20]

The navy yard along the Southwark waterfront stored thousands of board feet of timber gathered in Georgia by Samuel Humphreys. The wood remained on the southeastern corner of the yard, the lowest point, where it "lay in a very bad situation and will soon spoil, part in and part out of the water at high tide." Piles of gun carriage "stuff" rotted. An old wharf was too fragile to serve as a heaving-down pier for ship repair or new construction. Consequently, Samuel Humphreys and Nathaniel and Benjamin Hutton constructed brig *Syren* and *Gunboat* No. 3 for the Jefferson government in their private shipyard nearby, most likely at the former United States Ship-Yard. The government navy yard built no large ship until the War of 1812, although it may have assembled several tiny gunboats.[21]

The sorry condition of the navy yard during the first years of Thomas Jefferson's presidential administration reflected the small-navy policy of the Republicans, particularly the antinavalist ideas of economy minded Treasury Secretary Albert Gallatin. Spending for the U.S. Navy dropped from $2 million in 1801 to just $946,213 a year later, the lowest appropriation since before the Quasi-War with France. Naval appropriations were barely sufficient to develop the Washington Navy Yard along the Potomac River, where Jefferson ordered seven frigates laid up or to maintain a small squadron in the Mediterranean to protect American commerce against Barbary corsairs. Officers and crews were dismissed, naval constructors discharged, and facilities placed in caretaker status. Gallatin wanted more cuts. He insisted that the previous Federalist administration had wasted vast sums in gathering naval stores and misappropriated funds to purchase expensive waterfront properties for six navy yards.[22]

Jefferson stopped development of navy yards on Gallatin's advice, and Congress asked for details about the Federalist acquisition of these facilities. Investigators discovered that the John Adams government had spent $137,666 for six waterfront properties without "express provision made by Congress for establishing navy yards for building the first six frigates directed by law." Timber for warships had been purchased and delivered to six different building sites and lay in the open, exposed to weather and water damage. If Congress decided to sell the federal property, it needed another $158,000 to move the wood and place it under covered timber sheds.[23]

Secretary of the Navy Smith decided that Federalist officials had caused a great "evil" by placing piles of expensive timber out in the open at navy yards, thus forcing an "unsuspecting public" to confront the additional expense of moving the timber or keeping all the navy yards. "The evil, however, did not stop here," Smith continued. After yards were established and timber delivered, there was insufficient space to store the material and the federal government had to develop new buildings, make improvements, and move original timber stores all around the navy yard grounds. "The expense of this unnecessary kind of labor, arising solely from the want of sufficient room in the yard, amounted to several thousand dollars in building the Frigate *United States* at Philadelphia!" Smith singled out *United States* because he realized that the frigate's builder was responsible for the great Federalist timber "evil" and had maneuvered Stoddert into procuring the Southwark location rather than a better one in Kensington, a Jeffersonian Republican political stronghold. "There is reason to believe that the site of the navy yard at Philadelphia ought to be changed," Smith decided.[24]

Humphreys countered this attempt to close the navy yard by telling Federal officials and friends in Congress that Southwark was a better site than Kensington. "The Navy yard is situated in Wicacoa Cove supposed to be the best Harbour in this port" with ground "handsomely elevated." Wide streets connected the place to city business and commercial centers and to the "Laboratory" on the Schuylkill River where gunpowder, shot, and cannon could be manufactured and rushed instantly to warships tied up at the government naval facility in Southwark. Most important, the district contained hundreds of skilled ship carpenters, joiners, riggers, and common laborers ready at a moment's notice to construct or repair ships.[25]

The Jefferson administration kept a U.S. navy yard in Southwark, but provided no additional monies for improvements. The president expanded the Washington Navy Yard instead, directing brilliant Philadelphia architect and engineer Benjamin H. Latrobe to design a covered dry dock sufficient to protect twelve frigates from the weather. This idea prompted the former Philadelphia naval constructor to go on the attack again, claiming that if the government placed wooden warships out of the water they would decay. "No dock whatever ought to be established at Washington, from the circumstance of its laying so far from the sea coast & from the difficulty and length of time it takes large vessels to get there." Humphreys ignored the same flaw inherent in his navy yard, located ninety treacherous miles up the Delaware from the seacoast.[26]

Congress rejected Jefferson's dockyard scheme. Undiscouraged, the pro-French president decided to construct a flotilla of small gunboats like those used by France during the Napoleonic Wars rather than build frigates or ships-of-the-line inspired by the British and favored by Federalist shipbuilders. Jefferson secured appropriations for over 200 small warships. *Gunboat* No. 3 was built in Southwark by Nathaniel Hutton, and other local shipbuilders constructed *Gunboat* Nos. 116–35. Designed by Humphreys's former assistants Josiah Fox and William Doughty, these gunboats were fabricated by private contractors in Kensington and Southwark. Possibly, some of these tiny gunboats were assembled at the Navy Yard, where Philadelphia naval agent George Harrison served as superintendent. More likely, the government yard simply berthed the gunboats and served as a base for a flotilla of gunboats sent to protect the Delaware Bay and River against increasing British threats to American maritime interests.[27]

British seizure of sailors from the U.S. frigate *Chesapeake* and Jefferson's adoption in 1807 of an unpopular economic embargo led the United States toward war with Britain. The crisis forced Congress to approve appropriations for the improvement of the six Atlantic coast navy yards. Washington secured the largest share; Philadelphia stood fifth out of the six yards, well behind New York, Boston, and Norfolk. With severely limited government funding, the Navy Yard in Southwark added only a dock to store timber. Surprisingly, the Federalist shipbuilding firm of Joshua Humphreys & Son received large naval contracts from the Jeffersonian Republican administration to overhaul *Wasp*, *Essex*, and gunboats laid up at the Navy Yard. In fact, rumors circulated in 1810 that Congress planned to close the government facility and presumably give all new work to private firms.[28]

The outbreak of the War of 1812 saved the Navy Yard and made Joshua Humphreys's talented son Samuel the nation's premier naval constructor. The younger naval constructor's career received a boost in February 1813 when President James Madison named Philadelphia representative and former ship captain William Jones his secretary of the navy.

Jones consulted at once with Humphreys and other Delaware River shipbuilding friends and business associates. He assigned the 44-gun frigate *Guerriere*, authorized by the naval expansion act of January 1813, to Kensington shipbuilders Joseph and Francis Grice and asked Southwark shipbuilder Charles Penrose "to unite with Mr. Humphreys" in the construction of a 74-gun ship at the navy yard downriver.[29]

In early 1813, however, the government naval facility on the Delaware River was in no condition to build a large warship. The place had changed little since establishment. Marines occupied the original buildings. A wooden shed employed as a naval hospital was nothing more than "the black hole of the Yard." The joiner's shop moved from Humphreys's shipyard years before remained standing, but the old Swedish log house and stone kitchen had been pulled down, as had two timber sheds erected around 1806 to cover the planks and beams gathered for construction of an original 74-gun ship

Gunboats of the type used by President Thomas Jefferson's gunboat navy. A squadron of these boats operated from the Navy Yard in Southwark during the War of 1812. Naval Historical Center Photograph NH 91345.

authorized under the Naval Act of 1799 and timber purchased in Baltimore for repairs to the "rotten worthless hulks" of old frigates. Jefferson's gunboats, outfitted in 1813 by Master Commander James Biddle for the Delaware River Flotilla, crowded the low-lying riverfront as they awaited repairs to damage suffered in an engagement in Delaware Bay with a blockading British warship. Penrose's carpenters converted the purchased timber transports *Camel* and *Buffalo* into block sloops for river defense by fabricating a solid enclosed bulwark and iron grating on each vessel to keep off boarders. With every foot of the Navy Yard taken, six barges and the armed galley *Northern Liberties* arrived on the Southwark waterfront to load gunpowder from "Dupont's Manufactory" of Wilmington, Delaware.[30]

The War of 1812 revealed that the local naval shore establishment lacked proper construction facilities, a system of organization, or "precise rules to guide conduct." Naval agent Harrison served as navy yard superintendent. Commodore Alexander Murray, "an amiable old gentleman" nearly deaf from a cannon explosion accident, was senior naval officer in the port of Philadelphia, commanded the gunboat flotilla, but held little authority over the operation of the Navy Yard itself. Marine Officer of the Yard Anthony Gale functioned as de facto navy yard commandant, treating the post as his personal fiefdom. He refused to take orders from Murray, bullied local ship carpenters when they entered the gates to work, and "hath built a Tavern opposite the upper Gate by the Barracks, which I fear will become a nuisance." This grog shop failed to satisfy the Marine guard or dozens of seamen crowded into the tiny Navy Yard hospital, who left without permission for the more exciting entertainment houses in the neighborhood next door. Naval Surgeon Edward Cutbush reported that most came home intoxicated and with diseases contracted "not in line of duty."[31]

Navy Secretary Jones wanted to reform the Philadelphia naval establishment. He instructed storekeeper Robert Kennedy to keep a careful monthly record of public property and on 8 July 1813 made Commodore Murray the first commandant of the Philadelphia Navy Yard with full authori-

Alexander Murray, first commandant of the Philadelphia Navy Yard, 1813, and veteran of the Revolutionary, Quasi, and Barbary Wars. Engraving by Edwin, after a painting by J. Wood. Print and Picture Collection, Free Library of Philadelphia.

ty over the Marines. Murray prohibited all "improper intercourse & communication" with the Southwark neighborhood that pressed in on the government compound. The move was extremely unpopular locally. "Speculators in lots adjacent to the Navy Yard and other interests are memorializing the judges of the court of quarter session to open Federal Street through it," Secretary Jones warned. Attorney Alexander J. Dallas, a leading Philadelphia Jeffersonian Republican, believed that the national government could not prevent this incursion since the original purchase had been between private parties and the Pennsylvania legislature had never voted to transfer title to federal authorities. Dallas warned that legal complications might cause "utter destruction of the public establishment." He wanted the U.S. government to close the navy yard, sell the site, and acquire a new one in Kensington.[32]

Southwark District and Philadelphia County Commissioners assessed the Navy Yard property heavily during the war, including water pump, street, poor, health, and county taxes. The Navy Yard remained the only U.S. government facility to pay local taxes, and probably faced inevitable closure had not Secretary of the Navy Jones wanted Humphreys to build a 74- gun ship there. Jones kept the Navy Yard busy during the War of 1812, but provided no coherent plan for expansion as the facility grew in response to fears that the British might attack Philadelphia. Such concern became near panic during the late summer of 1814, when the British burned the capital in Washington and bombarded Fort McHenry in Baltimore harbor. Already blockaded tightly by enemy warships, Philadelphia lay open to invasion from the sea. Consequently, a local Committee for Defense of the Delaware used the Navy Yard as a signal station to warn residents in the event the British came upriver as they had in 1777. Navy Yard carpenters constructed gun carriages and platforms at Fort Mifflin on Mud Island and fortified W. Thomas Davis's Pier (Fort Gaines), located in the middle of the Delaware River between Fort Mifflin and Red Bank, New Jersey.[33]

Service by Navy Yard workers for the Committee of Defense delayed keel laying and framing of *Franklin*, Philadelphia's first 74-gun ship. The Navy Yard became a wartime center for other activity, however. Commodores Murray and John Rodgers and Captain David Porter met from time to time as a war board at the Navy Yard to discuss strategy. Murray wanted the 44-gun frigate *Guerriere*, launched on 20 June 1814, brought downriver from Kensington to the Southwark waterfront to protect the government shipyard. Others opposed the idea. Joshua Humphreys, for one, worried that if the enemy arrived within range of *Guerriere*'s guns they would place the Navy Yard under bombardment and destroy *Franklin*, which lay unfinished on the stocks.[34]

Cessation of fighting raised questions about whether the "74" or some other class of warship should serve as the backbone of an American fleet and brought the usual postwar Congressional inquiries into the future development of the U.S. naval establishment. The ubiquitous elder Humphreys reflected on lessons of the recent naval war, telling U.S. Representative Adam Seybert of Philadelphia, who considered the elderly Southwark shipbuilder to be "a perfect master" of naval affairs, that the Navy suffered from a lack of system. To provide such systematic organization, Humphreys introduced the concept of an advisory board, comprised of a naval officer, constructor, and civilian administrator "well cultivated and versed in History and Politics." The Southwark shipbuilder claimed that warship design and construction supervised solely by naval officers was a flawed process that reduced efficiency and raised construction costs, and urged Seybert to introduce the idea of a civilian-naval advisory board to Congress. He was too late. Congress created a three-member Board of Navy Commissioners on 7 February 1815 without a civilian administrator or naval constructor.[35]

This was a critical moment in the institutional growth of the American naval establishment. For nearly thirty years until 1842 when the Navy adopted the bureau system, this Board of Navy Commissioners advised every navy secretary on warship design, construction and repair policy, often making the government shipbuilding program conservative and reluctant to accept technological innovation. Commissioners John Rodgers, Isaac Hull, and David Porter immediately threatened to close the Navy Yard in Philadelphia. Board President Rodgers, the most influential officer in the postwar navy, favored instead opening new navy yards in Newport, Rhode Island, and on the York River in Virginia, and construction of a dry dock at the Charlestown Navy Yard in Boston harbor. Rodgers objected specifically to the "unexceptionable" site for the navy yard in Southwark. "You cannot carry to Philadelphia more than eighteen feet of water, over which no ship of the line or heavy frigate could be taken without the agency of camels [floating dry docks sunk under a

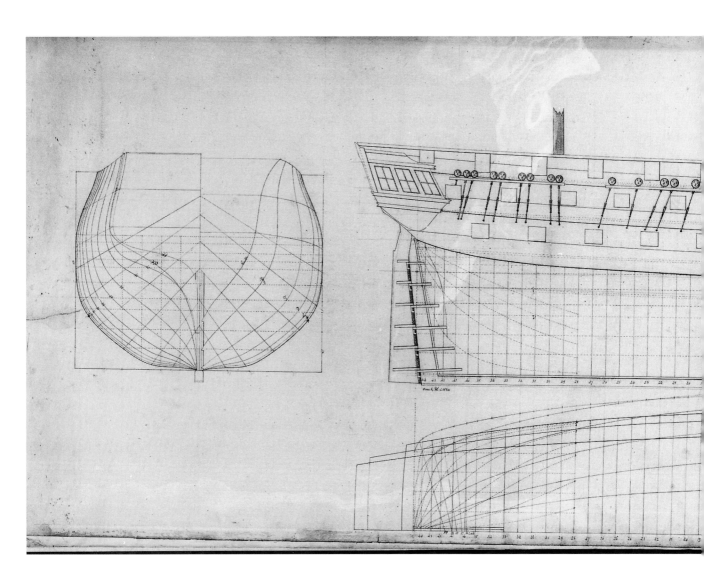

ship; water is then pumped out to raise it] or other buoyant preparations."[36]

This latest threat to the Navy Yard on the Delaware River mobilized Humphreys once more to defend his Southwark purchase. He asked President James Madison's financial expert Alexander Dallas of Philadelphia to protect their city's navy yard against outside attacks. The former naval constructor stressed the importance of the facility to the local economy, insisting that it was the best, most economical place in the United States to design, build, and alter warships. "I say it with truth that there is more scientific and practical knowledge in Naval Architecture in this port than in any of the U. States." The Delaware Valley abounded with the nation's most skilled ship carpenters, joiners, caulkers, mast makers, riggers, and mechanics. "Most all of the steady sober best workmen have acquired a

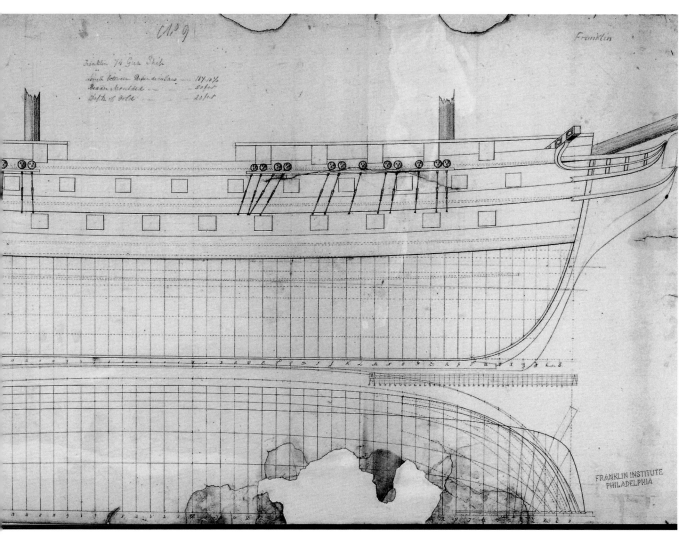

Sheer plan and cross section for 2,243-ton, 74-gun ship *Franklin*, launched on 21 August 1815. Joshua Humphreys, Jr., designed the ship and his son Samuel Humphreys constructed it at the Navy Yard in Southwark. ISM, Lenthall Collection, loan from Franklin Institute; photographed by Todd Bauders.

lot of ground and a house, which they do not like to leave to go any distance from home so that if the Navy Yard is removed to Marcus Hook or any other distant place you cannot procure but Workmen without an advance of wages."[37]

Jones's successor as secretary of the navy, Benjamin W. Crowninshield, a Boston merchant who took office in January 1815, decided not to disturb the Navy Yard in Southwark. Here the Navy Depart-

ment auctioned off gunboats and barges bought during the recent war and finished construction on *Franklin*. The Madison administration wanted the big ship-of-the-line to prevent future blockades of harbors and coasts similar to that imposed against the United States by Britain during the War of 1812.

As the Delaware River's first 74-gun ship approached launching day, however, an accident nearly led to the warship's destruction while still on

Construction of the *Franklin* saved the Navy Yard from closure after the War of 1812. ISM, H. A. K. Martin Watercolor 1109.

the stocks. A white pine staging used to work on the outside of the hull went up in flames from hot coals thrown by a kiln placed too close to the launching platform. "During the building of the Franklin 74, the Brow Stage was set on fire from the furnace of this stove," Naval Constructor Samuel Humphreys recalled. The Navy Yard's first double-decker warship survived the fire and was launched into the river on 21 August 1815, before a crowd estimated at 50,000 Delaware Valley residents. "Nothing indeed could be finer than the manner in which the *Franklin* glided down her ways, and entered the Delaware," Philadelphia diarist and legislator Samuel Breck recorded. "Graceful, slow and erect, she made one of the finest launches imaginable."[38]

5 ☆ Building a Wooden Sail and Steam Navy

PRESIDENT James Madison initiated an unprecedented peacetime naval expansion program with the Naval Act of 29 April 1816, which provided $1 million to begin construction on twelve 44-gun frigates and nine 74-gun ships. The keel of 74-gun Ship No. 1 (*Columbus*) was laid down in May at the Washington Navy Yard, where nearby the Board of Navy Commissioners oversaw construction.[1]

The Board determined the character and development of the Navy Yard in Southwark. It ordered Joshua Humphreys's son Samuel and former apprentice William Doughty, now the naval constructor at the Washington Navy Yard, to prepare plans for 74-gun ship no. 2 (*North Carolina*) to be built in Philadelphia, authorizing Humphreys to employ forty ship carpenters, mechanics, and laborers to frame the 1,800-ton ship-of-the-line. The *North Carolina* contract initiated the first major expansion at the Navy Yard in Southwark. Carpenters raised sheer legs to lift large pieces of timber from schooners. Humphreys purchased additional teams of oxen and timber wheels to move the heavy wood for a 74-gun ship from the wharf to a steam sawmill on the southern boundary of the Navy Yard. Plenty of timber arrived from the South, but much of it had been cut to the wrong size and had to be stored in a wet basin open to the weather, where the wood became too rotten to use in ship construction.[2]

Work on *North Carolina* stalled in early 1819. "We are partly at a stand for the want of Live oak—the 4th transom & all the Fashion pieces are yet wanting to complete the stern frame," Humphreys reported. "The deficiency in the 74s frame prevents the employment of the full complement of Carpentry authorized by the Commissioners of the Navy." The commissioners were unable to pay wages, in any event, as initial postwar enthusiasm for large warship construction faded. Faced with economic depression in 1819, Congress cut government spending, and the Board tried to save money by reducing wages for carpenters, top man sawyers, and bottom pit man sawyers. As winter approached, the government simply discharged workers. Master and Commander of the Navy Yard Captain Wolcott Chauncey reported in February 1820 that only five carpenters remained to build *North Carolina*. Even these found outdoor work on the ship too dangerous in the windy, icy Delaware Valley winter, as ship carpenter William Bills slipped from the upper inside staging of the main deck ports and plunged to his death in the deep hold below.[3]

With the spring thaw, some craftsmen and laborers returned to complete *North Carolina*'s hull and deck by early 1820. Copying the methods used to raise 74s at the British naval dockyard in Woolwich, Humphreys employed great white pine sheers and a series of heavy ropes, tackle guys, and blocks driven by a hand-operated "barrel crab" to raise the 10-ton stem frame onto the keel and "heave up" the 12-ton stern frame. The Navy Yard launched *North Carolina* without incident on 7 September 1820, although the worst yellow fever epidemic in Philadelphia since the 1790s kept the usually large crowd away from observing the popular Delaware River event. *North Carolina* was not completed in Philadelphia. A fully outfitted 74-gun ship drew too much water to cross the Fort Mifflin bar just downriver, so the big

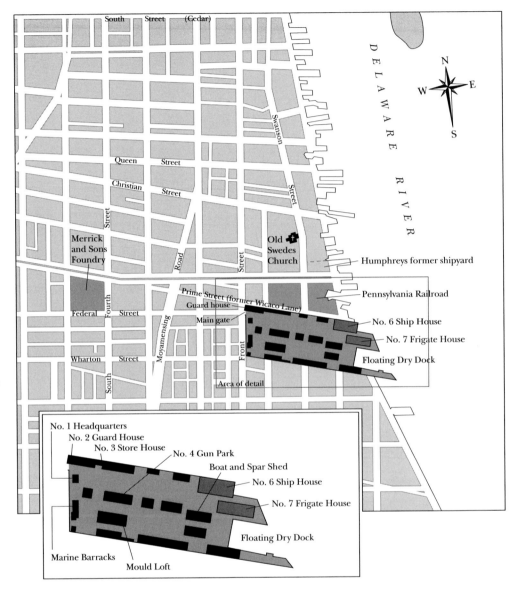

1801–75. Between 1801 and its closure in early 1876, the Southwark Navy Yard developed ship houses, a floating dry dock, and nearly fifty craft shops and support facilities to construct and repair wood and sail vessels. But growth of the surrounding residential and industrial area prevented the expansion necessary adequately to serve the new steam and steel navy.

warship was "only so far completed there as to enable her to proceed to Norfolk, at which place she was fitted for sea."[4]

As Humphreys's crew took up the wooden ship-ways and stacked keel blocks from *North Carolina*, the Navy Yard in Philadelphia faced an uncertain future. It was a period of transition. The only commandant that the facility had known, Commodore Alexander Murray, died and was replaced on 28 November 1821 by Commodore William Bainbridge.

Although a hero of the past war, Bainbridge was blamed by some for earlier losing the local frigate *Philadelphia* in Tripoli harbor during the Barbary Wars. The new commandant spent the next three years complaining that Philadelphia was a most disagreeable post and trying to arrange a transfer to the Charlestown Navy Yard in Boston. Moreover, national naval policy was undergoing changes that disturbed local navy yard development. *North Carolina*'s launching coincided with new President

James Monroe's antinavalist Jeffersonian decision to abandon expensive 74s as the backbone of the fleet in favor of smaller, faster warships to cruise against West Indian pirates, chase slavers off the African and Brazilian coasts, or show the American flag in Mediterranean and East Asian waters. Monroe canceled 74s and slowed work on 44s. Southwark started to build the 198-ton armed schooner *Dolphin* for pirate suppression duty in the West Indies. At the same time, construction stopped for the moment on the 44-gun frigate *Raritan* and 74-gun (later 120-gun) ship *Pennsylvania*, laid down already at the Navy Yard in 1820. The Monroe administration wanted covered ship houses built on the riverfront to protect the keels and frames of these two large wooden warships.[5]

Architect's pencil draft of proposed navy yard expansion in Philadelphia to accompany the Board of Navy Commisioners' instructions to Commandant William Bainbridge, 28 June 1827. HSP, Graphics, Plan of the Philadelphia Naval Yard (Bb615A732).

Emphasis on gradual construction of bigger warships under the cover of ship houses greatly influenced the landscape of the Navy Yard in Southwark. Government labor and private wharf builders dug foundations in 1821 for a frigate house over *Raritan*, and Philadelphia craftsmen directed by Wood Street house carpenter and Philadelphia County Commissioner Philip Justice constructed a 270-foot-long, 84-foot-wide and 103-foot-high ship house over the keel and frame of *Pennsylvania*. Carpenters erected an A-shaped mast sheer and masons built a stone wharf in front of the ship houses to berth a leased vessel that was housed over in 1822 to serve as the first naval receiving station and barracks.[6]

Gradually, a navy yard took shape on the grounds behind the ship houses. Timber sheds, workshops, a mast and spar shop, and an extension to the brick smith shop were constructed by private contractors. A Marine officers house, facing Front Street south of the Marine barracks, opened in 1821. A "block of houses" designed in 1827 occupied the Front Street part of the Navy Yard "10 feet back of the east line of the marine wall." These brick structures stood two stories high with dormer roofs and marble door and window sills. Navy yard laborers installed wooden gutters for drainage, stone-paved streets, and water pipes. By 1839, a high brick wall enclosed the U.S. government property, separating the facility from the surrounding Southwark neighborhood that was filling rapidly with German American, African American, and recent Irish Catholic immigrant working class folk. U.S. Marines and ten civilian night watchmen were supposed to secure the property by keeping out local residents, but the guard often opened the gate "under every pretext [so that] public property might be embezzled during the night."[7]

Cut off from the outside world, life at the Navy Yard during the decades following the War of 1812 reflected the peacetime routine of an early nineteenth-century American military installation. "Quiet inoffensive" craftsmen and laborers assembled daily to maintain buildings, grounds, and *Raritan* and *Pennsylvania* under the ship houses. The number of yard employees doubled to 200 during an election year "when members of Congress who ask situations for their needy constituents" increased the work force. Also, seasonal conditions affected the size of the work force as "the Master Painter, Sparmaker, Boat Builder, leader of Laborers were selected by the Commandants of Yards and [then] discharged [for the season] when the work was finished." So few men labored in the cold weather on *Raritan*'s "inboard works" that the frigate required "about 12,000 days' work to complete."[8]

Few were employed on *Pennsylvania* at any time as a parsimonious Congress provided annual funding for gradual construction by a small crew of ship carpenters. Commandant Bainbridge found that during his absence from Southwark in 1823–27 no deck planks had been laid on the big warship. So little progress had occurred that when the Board of Navy Commissioners considered making it into a 120-gun ship, the Navy Yard gladly sent its only copy of hull and deck plans to Washington. Moreover, Bainbridge discovered that the shipways under the warship had decayed and he recommended complete replacement. Instead, the Navy Board ordered the commandant to salt all the wood thoroughly to prevent further disintegration. Nearby, the frigate *Cyane*, captured from the British during the late war, rotted away. "The only means which have been taken towards the preservation of the *Cyane* have been to keep her hatchways closed, to prevent the rain,

Facing above Ship House (left) and Frigate House (center), constructed in 1821–22. Large wooden warships lay on the stocks under construction in the ship houses, some not finished for over a decade. NARAM; lithograph by William S. Pail from Henry M. Vallette, *History and Reminiscences of the Philadelphia Navy Yard*.

Gun Park and Timber House. NARAM; lithograph by William S. Pail from Henry M. Vallette, *History and Reminiscences of the Philadelphia Navy Yard*.

&c.,&c, getting into the ship, and occasionally giving the outside a coat of cheap paint."[9]

A small civilian staff managed shipyard business. The Navy Yard employed Naval Constructor Samuel Humphreys, Master Joiner James Keen, Master Blacksmith William Myers, Clerk Benjamin Weeks, and Storekeeper Robert Kennedy. Kennedy complained about the poor pay at the local yard, and the Board of Navy Commissioners in Washington was quick to respond. "The duties of the Store Keeper at the Philada. Yard are much less arduous than those of the Store Keeper at Norfolk, or New York, or Boston—hence the Board do not conceive that he ought to receive as much pay." The naval force at the Southwark facility was not much bigger. A commandant, master commander, two lieutenants, boatswain, purser, surgeon, and a dozen seamen and boys manned the place. Naval officers had to rent rooms in the city because of the "peculiar circumstances of this establishment there being no dwelling houses or quarters for any of the Sea officers attached to it, which places them on a different footing from officers attached to other Navy yards." Only local officers such as John Renshaw of Delaware or Philadelphians James Biddle, George Campbell Read, and Charles Stewart liked duty on the Southwark waterfront.[10]

The Navy Yard became increasingly isolated and alienated from the local community. "Unfortunately, the people at Southwark at present entertain unfriendly feelings toward both Navy Yard and state legislature," observed one local representative.

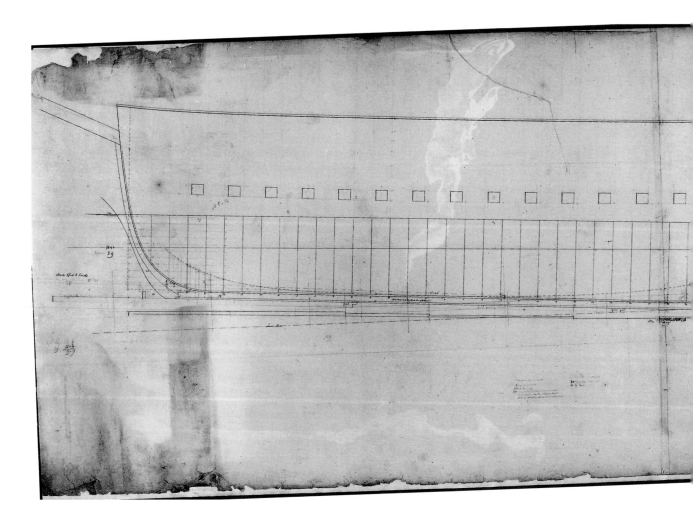

Strained relations between local community and Navy Yard arose from a longstanding dispute over property and taxation rights. Right after the War of 1812, Southwark District and Philadelphia County commissioners raised navy yard taxes to an unprecedented level. The Poor Tax increased from $130 to $216 annually, County Tax from $100 to $227, and Road Improvement Tax from $250 to $325. Commandant Murray offered to exchange an 8-foot strip of government land along Prime Street and provide free access to an old wharf at the yard, if the community agreed to give up further tax claims on the Navy Yard properties. The controversy continued. Naval agent George Harrison, who held the ambiguous position of navy yard superintendent, suggested that the Navy Department pay $700 in back taxes. The U.S. government refused, further straining local relations. Secretary of the Navy Samuel L. Southard, a tough former New Jersey prosecutor, visited Philadelphia in 1827 and warned local authorities that they had no legal right to tax a government naval facility. "We are not subject to the

Launching plan for the *Pennsylvania*, the largest wooden and sail warship ever built for the U.S. Navy. Designed as a 74-gun ship and modified by the Board of Navy Commissioners to carry 120 guns, *Pennsylvania* lay on the Ship House stocks in Southwark from 1821 to 1837. ISM, Lenthall Collection, loan from Franklin Institute; photographed by Todd Bauders.

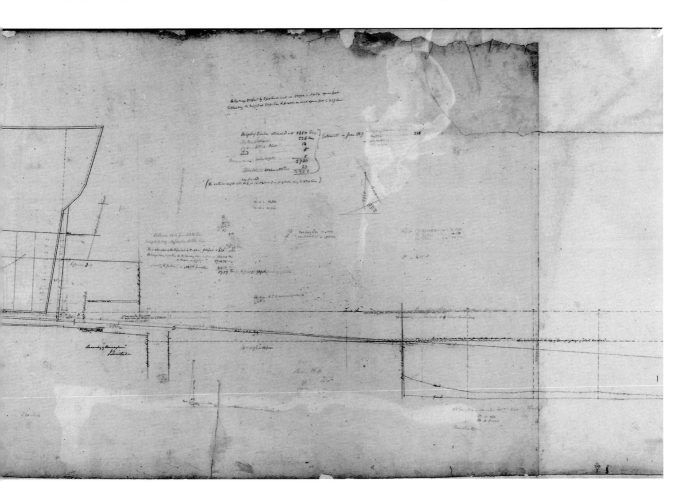

same inconvenience elsewhere, and it would be better to abandon the yard and station altogether than to submit to it much longer." To reinforce the threat of closure, President John Quincy Adams followed his navy secretary three days later to visit the Southwark waterfront.[11]

Philadelphia businessmen worried that the U.S. Navy might leave the city. Chamber of Commerce member John Vaughan, a local merchant shipbuilder, tried to obtain a tax exemption for the Navy Yard to keep government business in the Southwark District. The Navy Yard employed nearly 200 local mechanics and laborers, and area merchants and shopkeepers contracted to provide beef, bread, pork, tobacco, tea, sugar, soap, and other stores to outfit sloop-of-war *Vandalia*, launched on 26 August 1828. Nevertheless, Southwark commissioners and city officials continued to demand street access and taxation rights on Navy Yard property, forcing Commandant James Barron to ask Alexander Dallas, now vice president, to intervene on behalf of the federal government.[12]

Navy Yard tax disputes continued, but Barron had better results with another local property issue—the transfer of a title from the city to the federal government for land more than a mile west of the Navy Yard on which to develop a naval asylum along the Schuylkill River. Secretary of the Navy Southard instructed Surgeon Thomas Harris to purchase riverfront property in 1826 at Gray's Ferry Road and 24th Street. Noted Philadelphia architect William Strickland constructed a magnificent three-story marble and granite building on the location to serve as a seaman's retirement home, naval asylum, and burial ground. Barron took over the property for the government in 1833, dedicated the asylum grounds and wharf on the Schuylkill River, and instituted a regular stagecoach service for visitors traveling between navy yard and naval asylum. A decade later, Navy Yard Commandant Charles Stewart and Asylum Superintendent James Biddle established the first U.S. "Naval Academy" there.[13]

Local issues remained a constant bother, but national politics also buffeted the Navy Yard in Southwark, particularly during the tumultuous Jack-

sonian era, when President Andrew Jackson destroyed Philadelphian Nicholas Biddle's locally-based Bank of the United States. The Jacksonian Democrats dominated city politics in 1827 just as ailing Commodore Bainbridge returned reluctantly for his second tour as commandant of the Navy Yard. Bainbridge's connections to once prominent local Federalists and now anti-Jackson Whigs, including the Biddles, made the new commandant an enemy of Jacksonian Democrats from the beginning. Bainbridge contributed much to his own troubles with the Jackson administration. Wracked by painkilling drugs and close to retirement, he snapped at every order coming from Washington. He intended to run the Navy Yard his own way, ordering out ships when he pleased. "On exercising my judgement in sending vessels down the River before their final orders are received," he explained, "I have no small experience in fitting out and in commanding vessels of the Navy sailing from this Port."[14]

The Navy Department tolerated such conduct, but the U.S. Treasury Department was not so forgiving. Amos Kendall, fourth auditor of the Treasury, insisted that Bainbridge observe Jackson administration policy of the "principles of accountability in each yard" by carefully recording expenditures and curbing expenses. Kendall denied the commandant's request to pay Captain George C. Read's travel expenses to take a carriage a few blocks from his center city residence to the Navy Yard for duty as a member of a court martial board. Bainbridge became furious with the denial of travel money for his close friend and entered into an exchange of sarcastic notes with Navy Secretary John Branch, mostly critical of the Treasury Department auditor. "I perceive the finger of resentment from Amos Kendall, and I have yet to learn his claim on the nation to do the official injustice, for personally he shall not do it with impunity." Unfortunately for Bainbridge, Kendall was more than a minor government clerk, operating as one of President Jackson's most loyal cronies and contributors to the Democratic Party war chest. Within days of learning about Bainbridge's attitude toward Kendall, Jackson relieved Bainbridge of his command.[15]

In any case, Jackson preferred Captain James
Barron as commandant. Like the president, Barron
was a veteran of a duel, having killed Stephen
Decatur in one. The conservative Whig naval establishment hated Barron, as it did the Democratic
president. To show his particular favor for Barron,
President Jackson visited the Navy Yard on the
Delaware River. Administration support for the
local establishment stopped at this personal appearance, however. Southwark received the smallest
annual operating budget of any U.S. navy yard. It
secured only $11,750 for general improvements in
1835, while Boston obtained $199,575 and Norfolk
$167,000. Even the recently established naval station
at Pensacola, Florida, received a $64,000 annual
budget. Barron had to fight for money to pay for a
house carpenter to repair the dilapidated Ship
House and for a naval constructor to supervise the
launching on 14 September 1836 of the store ship
Relief. Nevertheless, the resourceful commandant
made the most of his limited budget between 1831
and 1837, erecting the Mast and Boat House, replacing rotten foundations for Ship Houses, tinning over
the Frigate House, building concrete slips, installing
a new floor in the mold loft, and constructing an
Engine and Hose House for fire fighting apparatus.[16]

Barron confronted the kind of graft, corruption,
and racial tensions at the Navy Yard that characterized local and national politics throughout the Jacksonian era. For instance, he had to remove a navy
yard clerk for running a theft ring at the government
facility. A Southerner, Barron wanted to remove
African-American recruits from the local naval shore
establishment as well. He was troubled by the influx
of blacks at the Naval Rendezvous that opened in
1832 to enlist "Prime Seamen and ordinary Seamen"
for the badly undermanned ships of the fleet (a
recurrent problem that plagued the U.S. Navy
throughout the first half of the nineteenth century)
He urged Secretary of the Navy Levi Woodbury to
examine "the practice to enlist Negroes for the
Navy" and decide "to what extent you are desirous
to encourage or to limit this practice, or what proportion of colored persons may be entered to any
specific number of white seamen."[17]

Unpopular Captain James Barron overcame a tight budget
and racial tensions as commandant of the Southwark Navy
Yard between May 1831 and July 1837 to expand the naval
shore establishment in Philadelphia. Contemporary engraving, Naval Historical Center Photo NH 56817.

Barron operated the Naval Rendezvous for several
years, dispatching new recruits from Philadelphia to
Baltimore and across the river to New Jersey, where
forty seamen and four officers traveled overland to
New York in early 1835 on the Camden and Amboy
Railroad. "The Camden and Amboy railroad company agreed to take them for two dollar per man with
an additional charge of 25 cents each for meals."
This steam and iron technology fascinated Barron,
unlike many of the more conservative naval officers
who had little faith in the still unreliable steam
engines and machinery. Barron could not ignore
such technological innovation during the 1830s. He
was stationed at the very center of a revolution in
steam navigation and locomotion. Steam-powered
locomotives on iron railroads passed on Prime and
Broad Streets within a few blocks of the Navy Yard,

carrying passengers, freight, or coal from the nearby Schuylkill region. In 1836, iron machinery manufacturer Samuel Vaughan Merrick and Pittsburgh steam engineer John Henry Towne opened the Southwark Iron Foundry within walking distance of the Navy Yard. Locomotive builder Matthias L. Baldwin stopped by the government facility to discuss new steam engine design.[18]

Steamboats of all types filled the river in front of the Navy Yard. Steam tow boats pulled wooden and sail ships against the tide or over the Fort Mifflin Bar. In 1837, Barron's last year in Philadelphia, a steam-powered ice boat began breaking the ice that paralyzed the Navy Yard every winter. A new steamboat wharf at the foot of Dock Street started a regular service between Philadelphia and Wilmington, Delaware, while by the middle of the decade shipyards in Kensington launched steamers regularly. The Navy Yard itself berthed one of only two steam vessels in the naval service, *Sea Gull*, a New York steamboat purchased by the Navy in 1823 for pirate suppression in the Caribbean and roofed-over in 1825 to serve as a receiving ship in Philadelphia. Thus Barron dabbled in such technology, sending models and ideas to the Navy Department on everything from innovative iron and zinc tackle blocks to a "half view and draft," complete with detailed cost estimates, for "armed Steamer vessels for harbour defence."[19]

Barron's relief Commodore Stewart, however,

View of Philadelphia from the Navy Yard. A new steamboat heads upriver against the current to Philadelphia, passing the covered ship houses and the obsolete *Pennsylvania*, the symbols of the end of the all wood and sail U.S. Navy. HSP, Naval Yards, Society Print Collection, Box 29, Folder 15.

This 1846 Currier and Ives lithograph of *Pennsylvania* under full sail with 140 guns in open ports glorified the last days of the all wood and sail navy. Rated to carry only 120 guns, *Pennsylvania* never sailed with full armament, making just one voyage from Southwark to Norfolk, where it was laid up from 1838 until burned in 1861 to prevent capture by the Confederate rebels. ISM.

wanted nothing to do with steam navigation. He shared the view of Secretary of the Navy James K. Paulding that war steamers were dirty, noisy sea monsters. Stewart asked the Navy Department to leave him alone on the subject "so as to free myself from any responsibility resulting from future constructions, armament and equipment of the National Steamers." If the department wanted information on steam vessels, Stewart advised, it needed to subscribe to the British *Nautical Magazine of the Naval Chronicle*.[20]

Stewart embodied traditionalist thinking that retarded the application of iron and steam power to warships; hence he should have drawn the displeasure of the Delaware Valley's many iron and steam engine and machinery entrepreneurs and steamboat builders. But Philadelphia loved Stewart, bestowing gold medals, swords, and other gifts on him during his unprecedented four tours as Navy Yard commandant (1837–60). He exemplified the Delaware Valley's long relationship with the American Navy that included the likes of John Barry, Thomas Truxtun, and Stephen Decatur. Born in the city in 1778, the son of an Irish immigrant shipmaster, Stewart attended the Protestant Episcopal Academy with young Decatur and other future U.S. Navy officers. His entire life centered on the Southwark waterfront

and ships built there. Stewart served first on the Southwark frigate *United States*, commanded the Southwark-built brig *Syren* and 74-gun ship *Franklin*, helped Decatur burn the Southwark frigate *Philadelphia* in Tripoli harbor during the Barbary Wars, and owned the Southwark-built schooner *Transfer*.[21]

Stewart used Navy Yard resources to protect and promote the ships of Thomas Cope and other Philadelphia merchants and "underwriters" against the pirate ships that supposedly lurked off Delaware Bay to intercept and plunder merchant vessels. Stewart endeared himself to the antislavery Quaker community in Philadelphia as well, when he presided over launching from the Navy Yard on 8 October 1839 of the swift 16-gun sloop-of-war *Dale* built specifically to suppress the African slave trade. The commodore's reputation was enhanced further by his supposed connection with the launch of the 120-gun ship *Pennsylvania* on 18 July 1837, although Stewart's role was limited to moving America's largest wooden and sail warship down river for its sole voyage to Norfolk, where it lay until 1861 when the Union Navy burned it to avoid capture by Confederate forces. Nevertheless, the estimated crowd of 150,000 people who attended the launch ceremony, and subsequent stories and paintings of the event, immortalized the launch and with it Stewart's place in Philadelphia Navy Yard history.[22]

Naval Constructor John Lenthall oversaw launching of *Pennsylvania* in a "slow, easy and graceful" fashion. Lenthall remained at the Navy Yard under Stewart's command, designing the *Dale* in 1839, while quietly promoting the type of steam engine development his commandant hated. Lenthall joined Southwark Foundry entrepreneurs Merrick and Towne in 1839 to secure a government contract to build the experimental, side-wheel armed steamer *Mississippi*. It was a difficult course to take in Stewart's yard. Though Merrick and Towne quickly manufactured two side-lever steam engines and machinery, work on the hull lagged far behind as the Board of Navy Commissioners, influenced by Stewart, pulled Lenthall off the project and sent him to Norfolk to survey the decaying sailing frigate *United States*. At the same time, the Board ordered Merrick and Towne to suspend all work on *Mississippi*'s innovative steam boilers.[23]

The Board of Navy Commissioners eventually returned to the "Sea Steamer building at the yard," where the delays on *Mississippi* characterized the trials of the U.S. Navy's period of transition from sail- to steam-powered warships. Chief Naval Constructor Humphreys and U.S. Navy Captain Robert F. Stockton of New Jersey, a wealthy steamship entrepreneur, encouraged the conservative Board to push ahead with construction of the side-wheel naval steamer on the Delaware River. Board President Charles Morris admitted that delays had

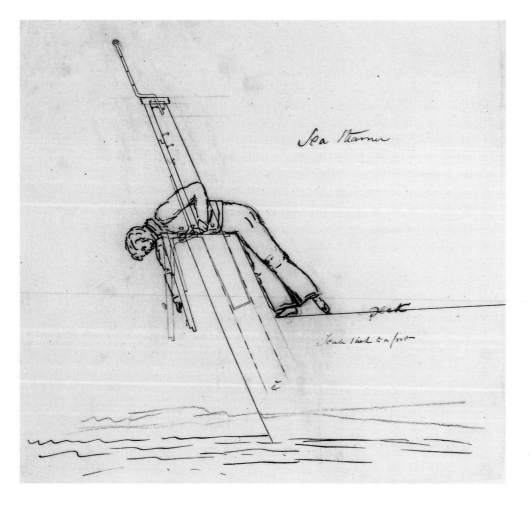

This sketch of a carpenter leaning over the side of Philadelphia Navy Yard's first steam-powered warship, *Mississippi*, suggests the draftsmen's increasing frustration with delays in authorizing construction of one of the Navy's first side-wheel steamers. ISM, Lenthall Collection, loan from Franklin Institute; photograph by Todd Bauders.

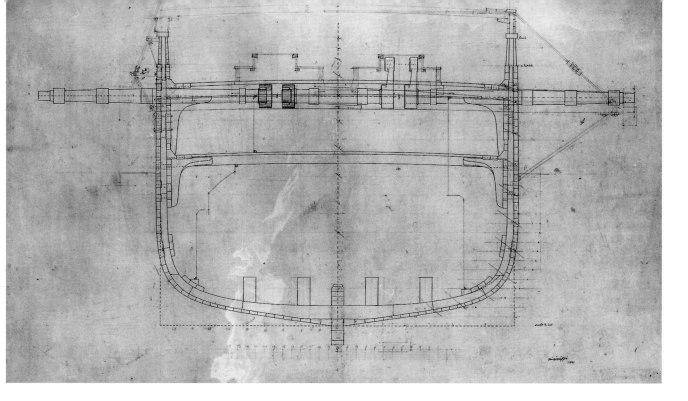

Cross section of shaft and machinery for the *Mississippi*, launched 6 May 1841. ISM, Lenthall Collection, loan from Franklin Institute; photographed by Todd Bauders.

occurred because the "character of this vessel is novel," including both steam and sail. Morris recommended resumption of work when he learned that Philadelphia constructors and engineers had designed watertight iron bulkheads and a system of valves and air funnels to regulate the escape of steam and smoke, and had relocated the mizzen mast so that it allowed space for machinery below deck. Local naval architects also moved the main mast, so that the sails would not catch fire from contact with the ship's smoke "chimney." At the last minute, the Board compromised the original design by ordering the Navy Yard to eliminate bulwarks for mounting a swivel carriage gun.[24]

The Navy Yard launched *Mississippi* on 6 May 1841, before an immense gathering along the river banks. "With every hour the crowd in and about the yard increased, and about by twelve o'clock every eligible situation within a quarter mile on either side of the Navy Yard, and on the Jersey shore opposite, was crowded with people, with eyes intently bent upon the front of the great ship house." After

launching and outfitting, *Mississippi* held steam trials on the river from New Castle, Delaware, to Fort Mifflin, setting a speed record against the tide upstream of one hour and fifty minutes and consuming three tons of coal. The steam Navy had arrived, born in part at the Navy Yard in Southwark. Within weeks of *Mississippi*'s launching, the Navy Department gave the local shipyard another steamer contract—this one designed by Stockton and named *Princeton* after his home town in New Jersey. The *Princeton* was propelled by a novel 11,900-pound "helicoidal six-armed propeller," invented by brilliant Swedish naval engineer and architect John Ericsson. Other Ericsson innovations included a telescopic smokestack that could be retracted while under sail power and an engine that burned clean Schuylkill Valley hard anthracite coal. "Steam is to be the main propelling power upon Erickson's plan," the Navy Board explained to Commandant Stewart. The Board ordered Lenthall to design the wooden hull for the revolutionary screw steamer, and Merrick and Towne received the contract to forge and cast machinery and engines at the Southwark Iron Foundry.[25]

Princeton promised to transform the U.S. Navy into the world's leader in warship design and propulsion technology, but a conservative and frugal Board of Navy Commissioners appeared reluctant to start construction. In 1842, the Board ordered Captain Read, who had replaced Stewart as Navy Yard commandant in 1841, to stop work on the propeller steamer. "You will go no further at present in the construction of the Steamer *Princeton* than to lay keel & making preparation to build her." The Board wanted Read to focus instead on construction of the wooden sailing frigate *Raritan*.[26]

Raritan's launching on 13 June 1843 illustrated how launch-day defined the relationship of the U.S. Navy Yard to the Delaware Valley community. Days before, newspapers advertised steamboat excursions to the Navy Yard to "take a favorable position opposite the Ship House to give passengers a full view of the Launch." Nearby ropewalk, pier, and wharf owners erected stands and charged admission for people to observe the launch, while Acting Commandant Frederick Engle raised temporary platforms between the two ship houses for families of shipyard workers. Prominent federal, state, and city officials and dozens of local society ladies entered the Ship House and boarded the vessel before launching. Military bands played patriotic tunes for the expectant audience. As launch hour neared, the entire riverfront community awaited the firing of cannon from the Navy Yard Battery to signal ships in the river to steer clear. "By two o'clock, every point from which the launch could be viewed—all the wharves from far up in the city down to the long wharf of the Southwark Canal, and from thence down along the banks of the Delaware the lower point of Smith's Island, Kaighn's Point and the Jersey shore, and every prominence, the windows, of the houses in the Navy Yard, and the masts and rigging of every vessel within reach—were literally alive."[27]

Most launches from the Navy Yard in Southwark went smoothly enough. Occasionally a launch resulted in personal tragedy as spectators fell into the dangerous tidal currents of the Delaware and drowned, but for the most part the event promoted a sense of community between U.S. Navy and Delaware Valley. However, *Raritan*'s launch raised disturbing questions about the future of the Navy Yard in the Southwark neighborhood. The 44-gun frigate slid down the shipway much too rapidly, slipped its chain cable restraint, and raced into the middle of the Delaware River, narrowly missing dozens of small boats and striking a sandbar. The frigate twisted grotesquely against the river obstruction and drifted uncontrolled downriver for more than an hour until the steamboat *John Fitch* managed to attach a line and tow *Raritan* back to the Navy Yard, where sheers could begin lifting the great wooden masts into place.[28]

The *Raritan* launching cleared the way for a return to work on the *Princeton*, already delayed in 1842 by a major reorganization of the Navy Department. The first reorganization since establishment of the Board of Navy Commissioners in 1815 created a system of technical bureaus for yards and docks, construction, repair and equipment, and ordnance. However, this reform reflected rather than contributed to the revolutionary changes in naval technology exemplified by the *Princeton*. Commodore Stockton inspired *Princeton*, used his own money on construction, and directed the launching ceremonies at the Navy Yard on 7 September 1843. "She can go in and out of port at pleasure, without regard to the force or direction of the wind or tide, or the thickness of the ice," Stockton enthused. Nevertheless, troubles plagued *Princeton* from the beginning. Machinery broke down regularly, experiments in cracking the Delaware River ice failed, and a wrought iron gun exploded on board during test firing demonstrations before President John Tyler and assembled guests on the Potomac River, killing Secretary of the Navy Thomas W. Gilmer, Secretary of State Abel P. Upshur, and other officials. *Princeton* returned to Philadelphia, where the remaining heavy shell gun was unloaded and stored on the Southwark waterfront.[29]

Princeton's launching in 1843 left the Navy Yard with one construction project—20-gun sloop-of-war

Germantown, the last exclusively sail-powered warship built at the U.S. government shipyard in Southwark. The 939-ton vessel rose at a leisurely pace in the Frigate House, and was not launched until 22 August 1846. There was little new work after this launching. The Navy Department laid off laborers and cut the wages of skilled craftsmen, many of whom left for higher paying jobs in private shipyards. Routine maintenance suffered and the yard fell into disrepair. Foundations for the Saluting Battery collapsed, the reservoirs for containing water on top of Ship House broke open, loose pieces of timber and empty oil casks littered the grounds. A brooding, sickly Commodore Jesse Duncan Elliott, who served less than a year as commandant in 1845 before dying at his house in Philadelphia, felt that the Navy treated Southwark badly. "I would invite your attention to the fact that this Yard is not and never has been placed upon the same basis as to importance, with the other Yards in the U. States, either as to its operations or the employment of per-

manent leading mechanics, and I would suggest that as labor can be performed here considerably less than at the other yards, we may at least be placed upon an equal footing with them."[30]

Lack of work, employee layoffs, and general decay at the Navy Yard reflected and probably contributed to the terrible labor, ethnic, and religious unrest and violence that convulsed Philadelphia from the working class neighborhoods of Kensington to the shipbuilding and industrial community surrounding the government compound in Southwark. Anti-Catholic rioting during spring 1844 reached Queen Street within three blocks of the Navy Yard's main gate on Front Street. Reportedly, Commandant Read dispatched men from *Princeton* to suppress rioting, but Acting Commandant Engle insisted that "The marines & men from the Yard will not be sent out as a protective force, nor will they be sent at all unless the volunteers are overpowered." The Navy Yard doubled the watch and extended the Guard House to the "footway of the main entrance" to protect the

Ethnic and religious rioting within three blocks of the Southwark Navy Yard in 1844 forced the strengthening of the Guard House and Main Gate on Front and Federal Streets. HSP, Navy Yard, Philadelphia, Society Photograph Collection, Box 65, Folder 3.

federal property against possible attack.[31]

Brigadier General George Cadwalader of Philadelphia, who led the Pennsylvania militia forces that crushed the Southwark Riot of 1844, thought that more jobs at the local navy yard might prevent future disturbances. He organized a committee with Southwark Foundry owner Merrick, iron-engine builder Isaac W. Norris, Philadelphia shipbuilder John Vaughan, and Delaware munitions baron Alfred E. Du Pont to revive naval business along the river by building a "National Floating Dock" at the Navy Yard. The idea of establishing a naval dry dock on the Delaware River had intrigued Philadelphia businessmen from at least 1821, when the Navy Department decided to build dry docks in Massachusetts and Virginia. The Franklin Institute of Science and Technology sponsored a dry dock study in 1827, and the citizens "of the city and county of Philadelphia in favor of a dry dock" held a large rally in 1837. The local firm of Samuel D. Dakin and Rutherford Moody offered its patented "Platform Basin and Marine Railway" to the Navy Yard in 1843, and the following year John S. Gilbert presented a competing proposal for his "balance dry-dock for a U.S. Navy."[32]

Political controversy soon engulfed the entire Philadelphia naval dry dock affair. U.S. Representatives Lewis Charles Levin, leader of the Native American Party in Philadelphia (and backed by Whig factions in the city), introduced an amendment to a naval appropriations bill in 1846 to fund construction of "Deakin & Moody's floating dry dock, basin and railway for the Philadelphia Navy Yard." Levin's opponents in the Democratic Party backed Gilbert's scheme, accusing Levin of corruption. At the same time, Philadelphia U.S. Representatives Charles Jared Ingersoll undermined both proposals for a Southwark dry dock. "He will display the cloven foot by advocating another form of Dock, to be located at Kensington," Philadelphia dry dock lobbyist Woodburne Potter reported from Washington, D.C., "not that he thinks it can be had, but because the agitation of such a scheme will give him a better chance of success in that district for the 30th Congress." In the end, Congress narrowly defeated Levin's proposal in 1846 for an $150,000 appropriation to begin construction of a sectional floating dry dock in Southwark.[33]

The newly elected presidential administration of James K. Polk brought naval business back to the Delaware River and provided the conditions for construction of a dry dock at the Navy Yard. Polk's belligerent and expansionist foreign policies generated crises with Britain over the Oregon territory and with Mexico over control of the Rio Grande and California regions. The latter led to U.S. declaration of war against Mexico on 11 May 1846. The Navy Yard in Southwark became a center of war preparations against Mexico, outfitting the armed steamer *Princeton*, brig *Delaware*, and frigate *Brandywine* for duty in the Gulf of Mexico. The Southwark plant refit the experimental shallow draft river boat *Water Witch* with new steam engines and Philadelphia inventor Richard F. Loper's twin screw propellers that had been tested by the Franklin Institute. President Polk visited the Delaware River navy yard to inspect the flag-bedecked *Princeton* and revenue cutter *Forward*, just back from the war zone. Not coincidentally, after the president's tour, work started on a floating dry dock and a contract arrived to build the powerful side-wheel steamer *Susquehanna*.[34]

Susquehanna, the largest warship laid down so far at the Navy Yard in Southwark, displaced nearly 3,600 tons with a 250-foot keel. By comparison, the world's largest wooden sailing warship, *Pennsylvania*, weighed 3,105 tons and extended 210 feet along keel line. *Susquehanna*'s additional weight came from the increased use of iron braces for the 7-inch-thick white oak gun deck that had to support larger and heavier rifled pivot guns. The *Susquehanna*'s increased size required dredging of the river to more than thirty feet in front of the Ship House, where the side-wheel war steamer was launched on Saturday morning, 6 April 1850. Though successful, the launching caused some embarrassment and revealed once again what Joshua Humphreys had known in 1800, that the Navy Yard stood in a very

The U.S. Sectional Floating Dry Dock, launched in 1851, was not tested successfully at the Southwark Navy Yard until 1853 with the steamer *Fulton*. HSP, Navy Yard, Philadelphia, Society Print Collection, Box 29, Folder 14.

poor location. When *Susquehanna*'s large hull entered the Delaware River, it created huge waves that crashed against nearby piers, knocked down restraining fences, and tossed spectators into the treacherous tidal current. Waves disrupted boats on the river, crushing one under a wharf on the Southwark waterfront. Worse, *Susquehanna* pulled its underpowered steam towboat into the Washington Avenue (Prime Street) dock, forcing the merchant steamship *Osprey* to vacate its berth quickly in order to avoid a violent collision.[35]

Completion of the *Susquehanna* in 1850 marked an institutional watershed of sorts for the Navy Yard, as over the next decade the facility adjusted to new technologies and larger, more complex steam- and sail-powered warships. Water pipes, gas lights, and a telegraph system were added between 1850 and 1851, while the work force climbed to over 300

to rebuild wharves, timber shed, and craft shops. The large Ship House was moved 200 feet closer to the water, and the shipway strengthened for heavier vessels. The Navy Department installed new machine shops, while private contractor A. B. Cooley of Philadelphia used his patented steam dredging machine to cut a 38-foot channel in front of the ship houses and a basin for the new sectional dry dock, "launched" in July 1851.[36]

Introduction of the floating dry dock made the Navy Yard a fully functional industrial ship manufacturing and repair facility. Final acquisition was not easy, however. Charles Stewart, who returned in 1846 for the second of four tours as commandant, negotiated unsuccessfully with seven proprietors, "amongst whom are some elderly females," to obtain a large lot on the south side of the Navy Yard for a dry dock. Eventually, Dakin and Moody (contracted to build the floating dry dock) rented the property on their own and charged the government more than if it had negotiated directly with the landowners. Meanwhile, the Navy Yard removed a brand new pier and sea wall so that the private company could dredge the basin. Problems continued after completion of the dry dock, when one section failed to lower with the rest to allow a test docking of the big naval steamer *Saranac*. The contractor forced the section down and broke the pumping apparatus. *Saranac* was called away to protect American interests in Cuba during a crisis in 1851 before a second attempt at docking could be made. Dredging of the dry dock basin slowed, and the sectional floating dry dock was not tested successfully until October 1853 with the docking of the steam warship *Fulton*.[37]

Despite these early problems, the dry dock assured Southwark a major role in naval shipbuilding and overhaul in the years before the U.S. Civil War. The Navy Yard received contracts during the decade for *Wabash*, the first of five powerful *Merrimack*-class steam frigates authorized by Congress in 1854, and *Lancaster*, one of five *Hartford*-class screw steam sloops of war funded in March 1857. The Navy Yard also constructed *Martin's Industry* and other light house tenders and steam propeller *Arctic* to rescue Naval Surgeon Dr. Elisha Kent Kane's Arctic expedition (and later to lay telegraphic cable between the United States and Europe). At the end of the decade, the yard completed the screw sloops of war *Wyoming* and *Pawnee*, the latter the first fully operational twin-screw vessel in the U.S. fleet. With the floating dry dock, basin, and marine railway the Navy Yard overhauled Philadelphian Alexander Dallas Bache's Coast Survey ship *Bibb*, sloop of war *Vandalia*, steamer *Saranac,* and old frigate *Congress*, among others.[38]

Navy Yard expansion and prosperity during the

The U.S. Japanese Squadron included *Vandalia* (1828), *Mississippi* (1841), *Princeton* (1843), and *Susquehanna* (1850), built at the Southwark Navy Yard. ISM; from *Gleason's Pictorial Drawing-Room Companion*.

1850s arose, in part, from the influence of powerful Pennsylvania Democratic Senator and later President James Buchanan, from nearby Lancaster County, and from Philadelphia Representative Thomas Birch Florence. Son of a Southwark boat builder, Florence used his position on the House Naval Affairs Committee and intimacy with the Buchanan administration to obtain lucrative government shipbuilding and repair contracts for his Southwark election district and to increase the civilian work force at the Navy Yard from 322 in 1852 to over 1,400 by 1858. Two warships authorized by Congress on 12 June 1858 were laid down at the Southwark plant before final funding, rushed to completion, and launched when the Democratic administration confronted a difficult reelection battle. The first of these warships was launched so quickly into the ice-filled Delaware River in an unprecedented peacetime winter ceremony on 19 January 1859 that the Navy

Department had no time to name the vessel, simply designating the ship as "Sloop of War No. 1" (eventually *Wyoming*). Sloop of War No. 2 (later christened *Pawnee*) entered the water nine months after the first.[39]

Navy Yard business stimulated the entire Southwark shipbuilding community, so that it prospered during the 1850s despite economic dislocation and depression throughout the rest of the Delaware Valley. The rapidly industrializing waterfront district was incorporated formally in 1854 as part of the city of Philadelphia, and the Southwark yard was designated the "Philadelphia Navy Yard." Street railways extended to nearly every section, including one that stopped at the Front Street main gate. The Merrick Foundry employed 350 skilled iron workers, machinists, and laborers to repair engines and fabricate boilers and machinery for *Susquehanna*, *Wabash*, *Lancaster*, and steamers under construction or

Screw Sloop of War No. 1 (*Wyoming*), launched 19 January 1859, was one of the many contracts given to Southwark by President James Buchanan of nearby Lancaster County, Pennsylvania, to reward his neighbors and local political supporters. F. D. Roosevelt Collection; *USNIP* 88 (September 1962): 161.

repair. Merrick & Son built the iron side-wheel steamboat *Shubrick* and other iron-hulled craft at its riverfront facility at Reed and Front Streets below the Navy Yard.[40]

Meantime, a group of local investors chartered a company to build a coal depot with basins, docks, piers, and store houses just below Southwark on League Island at the confluence of the Schuylkill and Delaware Rivers. A Pennsylvania corporation (probably the Pennsylvania Railroad) planned to run a double-track railroad directly through Southwark to connect the island to the city. The League Island coal-depot scheme clearly presaged future institutional development of the Philadelphia Navy Yard. In the months ahead, continual expansion of the Southwark community and a civil war that greatly strained the Navy Yard's productive capacity revealed the inutility of the Southwark site and forced the U.S. government to relocate the Philadelphia Navy Yard to League Island.[41]

6 ☆ CIVIL WAR AND TWO NAVY YARDS

AFTER FOUR remarkably pleasant tours as commandant, Commodore Charles Stewart retired from the Philadelphia Navy Yard on the last day of 1860. Advanced age and recent turmoil in the city forced Stewart to give up his comfortable home town command. Quasi-military clubs known as Wide Awakes, who supported Republican presidential candidate Abraham Lincoln and a protective tariff to aid local labor and industry, battled violent Democratic city gangs called Minute Men, who sympathized with the Southern agrarian slaveholding states and regarded Lincoln and the Republicans as threats to states' rights. By late 1860, Stewart could no longer face the situation in Philadelphia. "I presume he thought himself too old to be ordering out Marines to put down bread riots or to keep Wide-Awakes and Minute Men apart," observed Stewart's relief, Captain Samuel Francis Du Pont.[1]

Du Pont welcomed the Navy Yard command. Forty-three years earlier he had served there as midshipman on board *Franklin* under Stewart, the officer he now replaced. Moreover, the new command lay upriver from the Du Pont family estate at Louviers near Wilmington, Delaware, where he expected to be with wife, family, and friends for the first time in over four decades of continuous naval service. But the Civil War cut short Du Pont's stay in Philadelphia. He served barely nine months at the Navy Yard and spent much of that time in Washington with close friend Alexander Dallas Bache, superintendent of the U.S. Coast Survey, as members of a board to organize the naval blockade of the Confederacy. Nevertheless, Du Pont had to mobilize his Navy Yard command back in Southwark as the first line of defense for the Union after Rebel forces seized federal facilities farther South. "The capture of Pensacola Navy Yard, and the terrible destruction, or rather want of destruction of the Norfolk one," he explained, "brings that of Philadelphia nearest the seat of war [but] few of those appliances and facilities which pertain to a first-class naval station."[2]

Du Pont had few vessels to patrol the Delaware River, escort troop transports to the Naval Academy at Annapolis, or protect the railroad ferry on the Susquehanna River and tracks and bridges between Philadelphia and Baltimore. Only the 400-ton side-wheel Coast Survey steamer *Water Witch* and a borrowed steam tugboat patrolled the vital waterways between the Delaware and the Chesapeake. Du Pont commandeered the merchant steamer *Phineas Sprague* and U.S. Treasury Department revenue cutter *Forward* and accepted the city's offer of an ice breaker. "I am now fitting out with all possible dispatch the Philadelphia City *Ice Boat*, strong and suitable, offered by the patriotic city which owns her to the Government," Du Pont informed Secretary of the Navy Gideon Welles.[3]

The sluggish *Ice Boat* proved a poor acquisition because it burned too much coal for extended cruises and housed thirty city employees who quickly consumed all the government supplies on board. Welles suspected that they held Rebel sympathies and preferred that Du Pont lease "five staunch steamers" capable of mounting 9-inch pivot guns.

The commandant located several side-wheel steamers at the Harlan and Hollingsworth Company in Wilmington, Delaware, and in Philadelphia rented the 1,114-ton Havana steam packet *Union* and Ocean Steam Navigation Company's 1,364-ton side-wheel steamer *Keystone State*, the "fastest one out of this port." Du Pont brought the swift steamer to the Navy Yard for mounting three 8-inch Dahlgren smooth bores, one fifty-pounder rifle and several twelve-pounder howitzers.[4]

Leasing private steamers was expensive. The steamer *Union* alone cost the U.S. government $7,000 rental per month. So the Navy Department instructed Du Pont to purchase ships in Wilmington

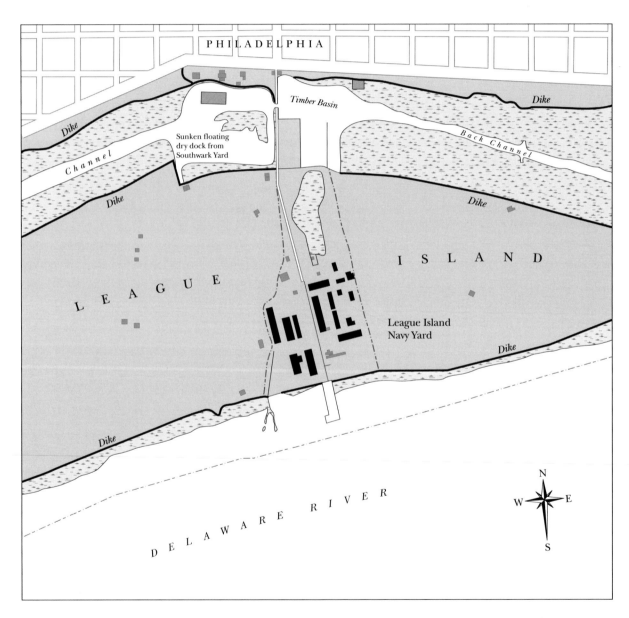

1868–75. The U.S. Navy operated two navy yards in Philadelphia simultaneously between 1868 and 1875, the old Southwark yard and the one on League Island, seen here with the buildings developed before the closure of the Southwark yard.

The small, overcrowded Southwark Navy Yard riverfront was unable to meet the demands of the Union Navy during the Civil War, contributing to the decision to move the yard south to League Island in 1876. The monitor in the background may be *Tonawanda*, the only ironclad completed at the Navy Yard. ISM.

The fast side-wheel steamer *Keystone State* (here sketched at Port Royal, South Carolina by Xanthus Smith, 1862) was purchased to chase Confederate blockade runners by Samuel Francis Du Pont, Philadelphia Navy Yard commandant from December 1860 to September 1861. Naval Historical Center Photo NH 60640.

and Philadelphia. He bought for $90,000 the 938-ton screw steamer *Phineas Sprague* and converted and armed it as the cruiser *Flag*. He paid $125,000 for *Keystone State*, which promptly cruised to sea and seized a Confederate blockade runner. He acquired nine additional private vessels after returning from meetings in Washington where the Lincoln administration organized the South Atlantic and Gulf Coast blockade squadrons.[5]

The sudden demand for repairs to steam engines and machinery during the first months of the Civil War greatly overtaxed the government work force at the Southwark yard. Before the crisis, private contractors had performed major steam engineering business and the Navy Yard had simply maintained engines and machinery. These tasks required few machine shops and fewer machinists. "The mechanics are so scattered that the visitor would suppose very little was being done there," observed one local newspaper writer. Not until the creation of the Bureau of Steam Engineering in July 1862 as part of a major reorganization of the twenty-year-old bureau system would the Navy Department address the shortage of machinists. The 1,500 employees hired in 1861 by the Navy Yard during the war emergency were first employed by the old Bureau of Construction, Equipment, and Repair on construction of the 1,457-ton wooden screw sloop-of-war *Tuscarora*, which was completed in fifty-nine days, from keel laying on 26 June to launching on 24 August.[6]

Tuscarora revived the spirits of local communities badly shaken by the outbreak of the war. Secessionist rioting against Philadelphia volunteers as they passed through neighboring Baltimore and the July 1861 failure of Union troops to destroy the Rebel army outside Washington at the first Battle of Bull Run demoralized the region. So the Delaware Valley celebrated appearance of the big warship. "For half a mile around, the river was dotted with small craft, and the war vessels laying off the Navy Yard were gaily decked with bunting," announced the strongly pro-Union *Philadelphia Public Ledger*. "On board the U.S. frigate *Susquehanna*, a full band performed one of the national airs."[7]

No sooner had *Tuscarora* entered the Delaware River than the Bureau of Construction, Equipment, and Repair laid the keel in the Ship House for the screw sloop *Monongahela* and completed the sloop-of-war *Juniata* in the smaller Frigate House next door. The Navy Yard laid keels for twelve and launched nine warships during the Civil War, averaging a launching every four months between August 1861 and March 1864. The 730-ton, side-wheel, double-ender gunboat *Miami* slid down the shipway into the water in November 1861, followed by the screw sloop *Juniata* in March and *Monongahela* in July 1862. The latter demonstrated Philadelphia's growing Unionist sympathies. "Tickets of admission into the Navy Yard had been liberally distributed, and every spot from which a favorable view of the launch could be obtained was taken possession of." Four months later, the Navy Yard launched *Monongahela*'s sister *Shenandoah*. The side-wheel gunboat *Tacony* appeared in May 1863, and steam sloop *Kansas* four months later. The last wooden-hulled steamer built at the Navy Yard during the Civil War, screw sloop *Yantic*, entered the Delaware in March 1864. Keels were laid for steam sloop *Swatara* and cruiser *Neshaminy* in 1864, but neither was completed until after Confederate surrender at Appomattox.[8]

Ten of the twelve new ship construction projects at the Navy Yard during the Civil War had all-wood hulls designed by John Lenthall, former naval constructor at the yard and now chief of the reorganized Bureau of Construction and Repair in Washington. Lenthall's closest friend, Captain Benjamin Isherwood, chief of the new Bureau of Steam Engineering, designed engines and machinery for these Philadelphia-built warships. Lenthall and Isherwood favored maintaining the Delaware River naval shipyard as the main wooden-hulled steamship construction facility, leaving development of iron warships to the navy yard in New York. Commodore Garrett Jesse Pendergrast, who succeeded Du Pont on 23 September 1861, also opposed using the overcrowded Navy Yard in Southwark for construction of ironclads. Du Pont concurred, drawing an angry response from Assistant Secretary of

Tuscarora (here photographed in Honolulu by H. L. Chase, 1874) was built in a record fifty days. The launching on 24 August 1861 revived the lagging spirits of the Delaware Valley after the early Union setbacks in the Civil War. R. Adm R. R. Belknap Collection, Naval Historical Center Photo NH 92341.

the Navy Gustavus Vasa Fox. "Du Pont neither understands nor appreciates [ironclads], he is of a wooden age, eminent in that, but in any engineering one, behind the times."[9]

Fox and Welles wanted all U.S. shipyards to build iron vessels, including the *Monitor*-type, a low-free-board ironclad ship with a heavily armored revolving gun turret that had saved the wooden blockading fleet from attack by the Rebel ironclad *Virginia* in March 1862 off Hampton Roads, Virginia. Already private shipyards in Wilmington, Chester, and Philadelphia were building iron ships. Iron foundries were plentiful, including that of Samuel Vaughan Merrick and Sons, located a few blocks from the Navy Yard. So it was reasonable for the Navy Department to expect Southwark to produce iron-

clads. This fact was brought home in June 1862 when the 4,000-ton armored steam frigate *New Ironsides* arrived at the Southwark naval facility. Cramp & Sons of Kensington had constructed the huge iron vessel upriver, but it was the Navy Yard in Southwark that "screwed" on the iron plates and made other alterations. Soon thereafter, the Navy Department contracted with the Navy Yard to build the 3,400-ton, two-turret "ironclad frigate" *Tonawanda*. Lenthall designed a wooden hull to support the iron plating, Isherwood provided engine plans, and Merrick & Sons Foundry fabricated boilers and iron machinery. To facilitate ironship construction, the Navy Yard "put up" new machinery for bending and planing armor plates for gun turrets and installed hydraulic rams and jacks on a marine

Twin-turret monitor *Tonawanda* (seen here by the U.S. Naval Academy, Annapolis, Md., c. 1870) was the only ironclad completed at the Southwark Navy Yard during the Civil War. Collection of the Pennsylvania Military Order of the Loyal Legion of the United States, Naval Historical Center Photo NH 90533.

railway to move the heavy ironclad hull.[10]

Problems plagued construction of the Navy Yard's first ironclad monitor from the beginning. Turret armor had to be cast and rolled in New York, shipped to Philadelphia, and bent and planed again, greatly delaying the shipbuilding process. An explosion at the Merrick & Sons plant on Fifth and Federal Streets killed eight and injured thirteen craftsmen, further interrupting construction. Advancing *Tonawanda*'s heavy ironclad hull from the marine railway below the ship houses to the sectional dry dock in the basin strained the yard's machinery. "After the vessel had been moved about three inches," an engineer reported in March 1864, "some portion of the hydraulic ram broke and the work had to be suspended." It took a week to repair the damaged yoke and key of the hydraulic ram. The new machinery moved the massive ironclad another fifty feet and pushed the structure onto the floating dry dock for work on the twin propellers and copper bottom. In the end, the Navy Yard "floated" the completed hull into the Delaware on 8 May 1864 rather than hold a formal launching ceremony.[11]

Unfazed by the troubled experience with *Tonawanda*, the Navy Department assigned construction of another ironclad to the Southwark facility. This was the "Three-Turreted Monitor" *Shackamaxon*, an armored monster, 100 feet longer and 1,000 tons heavier than *Tonawanda*. Work on the new ironclad went no more smoothly than it had

on the smaller armored vessel. White oak timber for framing arrived months after keel laying. Turrets, engines, boilers, and machinery built by Pusey and Jones of Wilmington, Delaware, lay in the open air around the yard. The Navy Yard finally placed *Shackamaxon*'s iron gun turrets and machinery under cover in sheds, but construction never resumed. *Shackamaxon*'s wooden keel lay on the stocks at the Southwark plant until in 1875, when facing closure the Navy Yard broke up the keel.[12]

Southwark's flaws as an iron shipbuilding plant stemmed partly from poor location. Surrounded by growing residential, business, and industrial neighborhoods, the Navy Yard had no place to expand for the new machine shops and forges needed to develop a modern iron ship manufacturing plant. Even traditional wooden shipbuilding functions overextended the Navy Yard. Fifty-two sheds, shops, and brick buildings were squeezed against the Delaware River to the east, Front Street to the west, Prime Street's railroad station to the northwest, and Reed Street to the southeast. Roads within the Navy Yard were too narrow for large equipment or material. To provide additional docking space during the war, the Navy Department leased wharves at the ends of

Facing Ship houses and lumber yard at the U.S. Navy Yard, Southwark. Naval Historical Center Photograph NH 001587.

Riveting tower, Southwark Navy Yard. These buildings crowded the tiny 17-acre yard, taking up space needed for machine shops and other facilities to build and repair ironclad warships. National Archives Mid Atlantic Region, Philadelphia Naval Shipyard, Office of Public Affairs, History, Box 2; from Henry M. Vallette, *History and Reminiscences of the Philadelphia Navy Yard*.

Reed and Prime Streets to convert and outfit prize steamers seized by the South Atlantic Blockade for service in the Union Navy. Too late, the government negotiated with private owners for an additional 4 5/8 acres adjoining the yard.[13]

Crowded wartime conditions increased the likelihood of industrial accidents. A general alarm sounded on 13 September 1863, warning that the Navy Yard was on fire. Flames engulfed storehouses along Prime Street and showered cinders against the great wooden and tin-sided ship houses on the waterfront. The fire roared through carpenters' shops, storerooms, paint and turpentine lockers, canvas and cotton waste rooms, and a gunnery loft filled with new Sharp's breech-loading rifles, manufactured nearby at the Sharp and Rankin factory on the Schuylkill River. The Navy Yard's three hand-operated fire pumpers manned by the Marine guard failed to contain the flames, which sent sparks onto row houses in the nearby community. Navy Yard Commandant Cornelius K. Stribling locked the gates and refused to allow city firefighters into the government compound, but the Weccacoe Engine and Hose Company of Southwark broke into the yard, doused the fire, and probably saved the community from a conflagration.[14]

The Navy Yard fire of September 1863 revealed as never before what Joshua Humphreys had known from the beginning, that Southwark was the "second best site" for a navy yard. The local neighborhood pressed in around the government shipyard during the Civil War. Military hospitals, refreshment saloons, canteens, military camps, railroad depots, and warehouses crowded along Ostego, Prime, Wharton, Front, Swanson, and Reed Streets. Most ominous, the constantly expanding Pennsylvania Railroad system laid new tracks from the Gray's Ferry iron bridge crossing down Washington Avenue (Prime Street) to the Delaware riverfront. The railroad company opened a station on Washington Avenue, erected a grain elevator, and negotiated for the property next to the Navy Yard to build a depot for a Philadelphia to Liverpool steamship line that would block the approaches to government piers and ship houses on the waterfront.[15]

Pennsylvania Railroad interest in the Navy Yard property in Southwark arose early in the war, when Commandant Du Pont held "daily intercourse" with leading Delaware Valley railroad executives to plan the defense of the region. Du Pont became familiar with Assistant Secretary of War Thomas A. Scott, vice president of the Pennsylvania Railroad, whom he considered a particularly "smart fellow." Most likely, Du Pont revealed at this time his views about the "one-horse status" of the Southwark naval facility, the need for a "first-class navy yard" in Philadelphia, and the possibility of using League Island, two miles downriver from the Southwark plant, for development of this larger government facility. Alexander Dallas Bache of the Coast Survey, who had earlier charted the river and surveyed the island, discussed the benefits of League Island with Du Pont when they met in Washington on the blockade planning board. Coincidentally, after Du Pont returned to Philadelphia from meetings with Bache, the City Select and Common Councils appointed a "Special Committee on Extension of the Navy Yard" headed by councilmen who held stock in the Pennsylvania Railroad.[16]

The Special Committee toured the Southwark facility in September 1861. After the visit, committee members introduced Du Pont's idea about a "first-class navy yard" in Council debates over whether the city should buy League Island and offer it to the U.S. government. City Council first considered ways to enlarge the Southwark site, but Special Committee Chairman Alexander G. Cattell, a local Republican Party leader and Pennsylvania Railroad director, contacted the owners of League Island to ascertain the acreage available and price of the property. The Pennsylvania Company for Insurances on Lives and Granting Annuities owned 409 acres of "fast land" on the island, 124 acres of marshland east of Broad Street and 67 acres of wetland west of Broad Street. Joseph C. Harris & Company also owned an acre lot and wharf at the foot of Broad Street used as a ferryboat landing for the island. The insurance company offered the entire property to the city for $300,000, and Harris wanted $20,000 for his lot.[17]

While city government studied an extension of Navy Yard properties, Secretary of the Navy Welles and Assistant Secretary Fox campaigned for "a suitable navy yard and establishment for the construction of iron vessels and iron armor." Welles's concern for development of a government-owned shipyard to construct ironclads began in March 1862, after John Ericsson's *Monitor* stopped the Rebel ironclad *Virginia* from destroying the wooden Union blockading fleet. The secretary asked Congress twice in 1862 for funds to purchase a site for establishment of a new naval shipyard. At this moment, Philadelphia Mayor Alexander Henry, a Republican Unionist, warm friend of Pennsylvania railroad interests, and facing a difficult reelection campaign against city Democrats, arrived in Washington to offer League Island to the Lincoln administration. Encouraged by Welles, Henry returned to Philadelphia and called a special emergency City Council meeting on 17 June 1862, where Pennsylvania Railroad and iron industry magnate John Price Wetherill pushed a resolution through Select Council, "tendering [League Island] to the Government of the United States for a site for a Naval Station and Depot." At the same time, Cattell convinced Common Council to endorse unanimously "An ordinance providing for the donation of League Island to the Federal Government."[18]

The deal went smoothly enough, and Welles kept his end of the bargain by asking Congress to accept League Island. But political and personal interests soon interfered with the city's offer of free land for a navy yard, tying up the transaction for six years. Partisan politics that plagued Welles's entire Civil War administration of the U.S. naval shore establishment blocked acceptance of League Island. "The interference of Members of Congress in the organization of the navy yards and the employment of workmen is annoying beyond conception," Welles complained. New England congressmen, anxious to receive the lucrative iron shipbuilding plant for their own section, opposed the Philadelphia site. Connecticut Senator Lafayette S. Foster hoped to influence fellow Connecticut Republican Welles to consider New London as the location for the big naval facility. Not to be outdone, Rhode Island Senator Henry B. Anthony urged Welles to look at "the water of Narragansett Bay" for an ironclad ship depot.[19]

Congress voted on 15 July 1862 to accept Philadelphia's offer, but only if the secretary convened a "board of naval and scientific gentlemen" to inspect League Island, New London, and Narragansett Bay and recommend the best site for an iron shipbuilding facility. Welles appointed Commodore Silas H. Stringham to head the board, learning too late that the Rhode Island naval officer "had a rival feeling as regards Philadelphia." The commodore thoroughly disliked Du Pont, an advocate of the Philadelphia site, whom Stringham blamed for his earlier removal from command of the South Atlantic Blockading Squadron. Not unexpectedly, the Stringham Board recommended by a four-to-two vote that the U.S. government establish its iron shipyard in New London, Connecticut. They selected New London over Philadelphia and an "indifferent" Narragansett Bay location because "League Island is situated at the distance of about one hundred miles from the sea, on a river, the navigation of which is often tedious and difficult for vessels of large class, especially in winter when obstructed by ice." The Stringham Board called League Island an unhealthy swamp that would cost the federal government an estimated $1 million to transport gravel and dirt from New Jersey to raise the normally dry land sufficiently above the high water mark to make the place habitable.[20]

A minority report made by Coast Survey Chief Bache maintained that the island was healthy, the water was fresh, and good fill was available for a low price from iron slag produced in Philadelphia. He insisted that the location was secure from enemy attack, near a skilled labor force, and close to supplies of coal, iron, and timber. Bache's arguments convinced Welles, who made up his mind to accept the gift of League Island from the city of Philadelphia, "unless Congress shall otherwise direct."[21]

Throughout the war Philadelphia attempted unsuccessfully to give away League Island. The

Republican-dominated Pennsylvania state administration secured General Assembly passage in early 1863 of "An Act Ceding to the United States of America the Right of Exclusive Legislation over League Island in the Delaware River, in the County of Philadelphia," thus clearing the roadblock that had confronted Joshua Humphreys's 1800 acquisition of the Southwark property. City Council advertised "proposals for the purchase of the stock held by the City in the Pennsylvania Railroad Company" to raise the $340,000 needed "to promote said acceptance, and to complete the conveyance of the title to said Island to the United States." Still no action from Congress. Facing a Confederate invasion near Gettysburg, Pennsylvania, during the summer of 1863, Philadelphia renewed its campaign to give the island to the U.S. government, because "recent events admonish us that it is right and proper that the City should be at all times placed in a position of strength and defense against any foe, and League Island affords a sure and impregnable protection against any attack by water."[22]

Gettysburg failed to stir Congress. As a last resort, City Council appealed directly to local residents in every ward and congressional district to hold public meetings and petition their representatives for a "National Navy Yard" on League Island. William D. Kelley of Philadelphia entered the fight, introducing in the House of Representatives a report for establishment of a "proper site for a Navy Yard for the construction of Iron-Clads." Known as "Pig-Iron" Kelley for his promotion of Pennsylvania iron and coal industries, the Philadelphia representative actually retarded development of an iron shipyard on League Island by circulating Stringham's report that had rejected League Island by a 4–2 vote. Worse, Kelley attached to the first Navy Department report of 1862 a damaging survey by a Pennsylvania Railroad civil engineer declaring that League Island lay too far below high tide on the river for the private company to use as a coal depot.[23]

Welles suspected that Kelley held up the League Island deal to force the government to release some of his political cronies who had been employed at the Southwark Navy Yard during the war and recently arrested for theft of government property. "Efforts are being made to aid a set of bad men who have been cheating and stealing from the government in Philadelphia," Welles confided. The arrests of Navy Yard employees between November 1864 and January 1865 marked the culmination of months of investigation by the puritanical War Department Special Commissioner Colonel Henry Steel Olcutt. Olcutt uncovered wide-scale corruption at the Philadelphia establishment, and as a result federal officials arrested master caulkers, joiners, plumbers, painters, storekeeper clerks, timber inspectors, and Naval Constructor Henry Hoover. "The discoveries and disclosures in the Philadelphia Navy Yard are astounding," Welles concluded.[24]

Philadelphia Federal District Judge John Cadwalader refused to prosecute, and few of the Navy Yard crime ring went to prison. However, the Philadelphia Board of Trade found the scandal useful to promote Navy Yard expansion and relocation to League Island. The Board of Trade and City Council gave dinners for Olcutt and his detectives, local naval officers, and Congressional investigators when they visited the Navy Yard. At the same time, Pennsylvania Railroad Company representatives promoted League Island, which they had secretly rejected as unsuitable for their own company use, warning federal officials that private business planned to develop the place. Welles passed the warning on to Congress, explaining that, unless it accepted the island at once, private interests would take the property "forever."[25]

When the Civil War ended in April 1865, the government had yet to reach an agreement about moving the Southwark naval facility to League Island. Nevertheless, the Navy Department laid up warships there. The U.S. Navy held no legal title to the property and paid no rent for its use, but berthed ten side-wheel and screw steamers of war there anyway. League Island also provided anchorage for monitors *Yazoo*, *Nahant*, *Passaic*, *Nantucket*, *Dacotah*, and *Tonawanda*, ironclad frigate *New Ironsides*, and sloop-of-war *St. Louis*, housed over for use as a receiving ship. The Southwark Navy Yard upriver disposed of the other elements of the Civil War fleet, auctioning forty-two ships between May and November 1865 alone, including the captured block-ade runners *Bermuda* and *Alabama* (renamed *New Hampshire* during the war).[26]

After the war, the Southwark naval shipyard maintained a work force at the wartime level of over 2,000 civilian employees, despite dramatic postwar cutbacks at other naval and military facilities, including the Frankford Arsenal in Philadelphia. The large force completed *Swatara* and *Nesha-miny*, laid down during the war and continued when Welles secured a postwar appropriation to finish these ships as cruisers to show the American flag on distant cruising stations or to confront war-related maritime problems with Britain and France. The local Republican Party machine guaranteed such high employment at the postwar Philadelphia naval

establishment. In 1866 "Pig Iron" Kelley urged Welles to hire 3,000 during the forthcoming local elections to ensure reelection of three Republican representatives from the area. Philadelphia Republicans taxed shipyard workers for "party pur-poses" as a prerequisite for keeping their govern-ment jobs. This situation disturbed Welles. "The Navy Yards must not be prostituted to any such pur-pose, nor will Committee men be permitted to resort tither, to make collections for any political party whatever."[27]

Welles wanted to remove the Delaware River navy yard from the clutches of Philadelphia politics by isolating it on League Island. "This is the most important matter affecting the Navy which is now pending," he confided. But the Senate "strangely delayed" legislation for acceptance of the island as a

Monitor *Nahant* laid up after the Civil War in the weeds and mud of the Back Channel that separated League Island from Philadelphia. Naval Historical Center Photograph NH 45634.

naval facility. Welles thought that the chaotic post-war political situation accounted in part for the delay. Lincoln's assassination left the Republicans leaderless, and the resurgent Democratic Party grabbed for office and control of policies and contracts for reconstructing the Union. These "high party times" permeated the naval committees in Congress with intrigue, corruption, and influence peddling that affected the League Island question. Senator Frank B. Brandegee of Connecticut resurrected New London port as the proposed location for an iron shipbuilding plant, while powerful Senator Schuyler Colfax of New York allegedly voted for whichever group paid him the most.[28]

In February 1866 Welles made one last effort to acquire League Island for the Navy Department. He asked Assistant Secretary of the Navy Fox to delay his return to private business in Massachusetts in order to visit Philadelphia and report on the island site. Fox met with Delaware Valley leaders and, accompanied by members of the House Naval Affairs Committee, took a steam tug from Southwark to League Island. The Washington delegation found that the Navy had laid up dozens of warships there. "Near the middle of League Island," Fox told Welles, "a single wharf runs out to the deep water of the main ship channel of the Delaware River, and off the extremity of this wharf are moored the iron-clads *New Ironsides*, *Dictator*, *Atlanta*, and the old sailing sloop of war *St. Louis*." Fox saw much more in the Back Channel that separated the island from the mainland. Here "are moored the harbor and river monitors, the light draught monitors, and torpedo boats, also several double-ender and other gunboats." They rested in fresh water to preserve iron hulls, secure from ice damage and close to Pennsylvania coal and iron supplies. Skilled craftsmen and mechanics of the large industrial port city

The old wood and sail sloop of war *St. Louis*, 10 February 1874. The *St. Louis* was housed over in 1973 to serve as the League Island Navy Yard's first receiving ship. Naval Historical Center Photo NH 497; courtesy Rear Amiral Ammen Farenholt.

who resided nearby could on a moment's notice reactivate and ready this reserve fleet for sea duty.[29]

City political and business leaders convinced Fox that unless the U.S. government acted at once it would lose a vital naval resource. "The intervening land will soon be occupied by buildings, when this valuable island must be appropriated for city uses, and will form its most important commercial front," Fox reported. There was in fact little chance that city or private interests would soon develop League Island in the wake of the Pennsylvania Railroad's engineering report that found the island unsuitable for industrial and business development. However, those same interests wanted the Southwark waterfront site of the current navy yard for private development and sold Fox on League Island. The connection was made clear five days after Fox left the city, when the Philadelphia Board of Trade paid for and published Fox's entire highly confidential official report to the secretary of the Navy as a promotional pamphlet: A New England Man, *Advantage of League Island for a Naval Station, Dockyard, and Fresh-Water Basin for Iron Ships*. The "anonymous" pamphlet included both Welles's annual reports to Congress and a study by Chief Engineer of the Navy James W. King (a close friend of Du Pont's) on his recent visit to European dockyards, which revealed how important it was to plan a dockyard in America on a site similar to that provided by League Island.[30]

Fox's report, released to Congress and circulated widely in public as a "New England Man'"s pamphlet, caused further controversy. A "New London Navy Yard Committee" countered with its own pamphlet describing the superiority of the Connecticut site presented by the Stringham report of 1862. As the debate continued in late 1866, a suspicious fire broke out on board *New Ironsides*, the most recognizable and popular Civil War vessel moored at League Island. The fire utterly destroyed the Philadelphia-built ironclad, with only its iron plating, port shutters, boiler parts, and machinery remaining near the low water mark on the Delaware side of Broad Street. The wreckage posed a public hazard to river travel and commerce, provoking demands

Assistant Secretary of the Navy Gustavus Vasa Fox led the lengthy l862–66 fight for government acquisition of League Island for a naval station, dockyard, and fresh water basin for iron ships, writing a promotional pamphlet for the Philadelphia Board of Trade. *USNIP* 19 (September l965): 64.

that the federal government take over the area. A few months after the *New Ironsides* incident, in February 1867, Congress voted to accept title to League Island, and Welles immediately dispatched a board headed by Captain Charles H. Davis to survey the site and drill test boring holes.[31]

The city transferred League Island to the U.S. government on 12 December 1868 for one dollar. The property included the Back Channel that ran at high tide from the Schuylkill River to the Delaware and the "northerly shores" of the channel bordering the city. Philadelphia Mayor Morton McMichael, a Pennsylvania Railroad director, accompanied by a committee of city councilmen delivered the title at a private ceremony held in the Navy Department, attended by president-elect Ulysses S. Grant. Welles eschewed a formal public ceremony because of continuing controversy over the issue. Moreover, the postwar Congress would not appropriate money to develop League Island until "the Navy Yard at Philadelphia [in Southwark] shall be dispensed with and disposed of by the United States as soon as the public convenience will admit."[32]

Naturally, Welles was anxious to close Southwark "as a matter of economy," since "little work is being done at any of the yards." But it was again election year in Philadelphia, where a resurgent Democratic Party threatened the local Republican machine. Philadelphia Republican leader Kelley pressed the Navy Department to expand, not close, and increase, not reduce, business at the old navy yard, thereby creating jobs for party loyalists and voters.[33]

The prospect of developing a government naval facility on League Island improved in early 1869 when President Grant asked wealthy Philadelphia financial backer Adolf E. Borie to become secretary of the navy. A prominent city merchant, society figure, and University of Pennsylvania graduate, Borie appeared the ideal figure finally to conclude the city's six-year crusade to move the naval shore establishment from the busy Southwark neighborhood to the nearly unoccupied riverfront island at the southernmost extremity of Broad Street in Philadelphia. Borie had neither energy nor ability to manage naval affairs in the politically charged Reconstruction era. He abdicated control over his department to battle-toughened Civil War Admiral David Dixon Porter, superintendent of the U.S. Naval Academy in Annapolis and a good friend of former general and now President Grant. Porter consulted

regularly with the president, dominating naval affairs and policies. "Poor Borie was a passive tool," former Navy Secretary Welles observed. "He is now a mere clerk to Vice-Admiral Porter."[34]

Porter opposed development of steam-powered "tin clads," or establishment of an expensive government iron shipbuilding yard. He wanted to avoid postwar Congressional scrutiny over naval spending by saving money on navy yard and warship development. The admiral advised extending sails on older wooden steamers and providing apparatus to lift screw propellers into wooden hulls on cruisers to avoid burning expensive Pennsylvania anthracite coal. He wanted a single navy chief of staff to organize and plan all naval business, hoping to reduce the influence of civilian experts such as Chief of the Bureau of Construction and Repair Lenthall or technological innovators such as Chief of the Bureau of Steam Engineering Isherwood, both advocates for a League Island naval shipyard.[35]

Porter stopped development of a new navy yard in Philadelphia, sharing close friend Commodore Stringham's opposition to U.S. Navy acquisition of League Island. Porter prevented the government from renting out meadows on the island to local farmers for pasturage or crops (a policy encouraged earlier by Welles). Weeds and rotting vegetation choked League Island, reportedly breeding disease that made the low-lying muddy place too unhealthy for engineers and laborers to lay out a navy yard there. At the same time, Porter accepted the advice of Samuel L. Breese, a former naval officer and now Philadelphia port administrator, to expand work on wooden steamers at the old navy yard. This included outfitting the 3,000-ton screw sloop *Pushmataha* (renamed *Cambridge* and then *Congress*), laid down late in the war and launched in July 1868, and *Astoria*, then on the stocks under a ship house on the Southwark riverfront. Launched on 10 June 1869, *Astoria* exemplified Porter's ideal of using sail power to save on burning coal to fire boilers. Naval constructors redesigned *Astoria* (renamed *Omaha*) so that the screw propeller could be disconnected and raised into an enclosed housing in the hull.[36]

Porter's influence declined abruptly when Borie

resigned barely three months into his term of office. New Secretary of the Navy George Maxwell Robeson, a Camden, New Jersey, lawyer and former state attorney general, ignored the admiral. Robeson had his own ties to the president and was more closely connected than Borie to Delaware River business interests that favored a naval shipbuilding facility on League Island. Robeson and former Philadelphia City Council leader and current New Jersey Senator Alexander Gilmore Cattell headed the local Union Party Republicans for Grant organization (called by opponents the Cattell Cabal). Robeson became Cattell's partner in the Delaware Valley world of iron shipbuilding, industry, finance, and railroad development. Cattell presided over the Corn Exchange Bank of Philadelphia (which he founded in 1857), the Philadelphia Board of Trade, and the Pennsylvania Railroad Company, where he sat with Robeson on the board of directors. During the war Cattell had initiated City Council's efforts to develop a navy yard on League Island so that the government would abandon the Southwark site to his own business interests.

Robeson's policies toward League Island exemplified the venality and influence peddling that surrounded politics in Philadelphia and the Delaware Valley during Grant's administrations. Robeson's long tenure as navy secretary (26 June 1869 to 12 March 1877) contributed to the image of this era as one of the darkest in U.S. Navy history, where party loyalty governed policy as never before and generous naval appropriations went to political friends rather than to development of the latest technology in ship construction and ordnance. Consequently, the U.S. Navy fell behind those of other world powers, and the American fleet seemed inadequate to confront even a minor naval crisis such as that caused in 1873 when the Spanish seized passengers from the American-flag vessel *Virginius* and summarily executed them as alleged Cuban rebels.[37]

Robeson was not entirely responsible for the sorry state of naval development. The Reconstruction era Congress that refused to appropriate money for new ship construction or technological innovation funded overhaul of older warships at politically important sites. So Robeson employed his Delaware River political and economic connections to rebuild the postwar navy under the guise of routine maintenance. Between 1869 and 1877, he channeled millions of dollars of navy contracts to the area for reconstruction of warships, purchase of naval supplies, and hiring of a large civilian work force for both the Southwark and League Island navy yards. Contracts for rebuilding warships went to the likes of John Roach's Shipyard and Foundry in Chester, Pennsylvania, the Harlan and Hollingsworth Iron Foundry and Shipbuilding Company of Wilmington, Delaware, and William Cramp and Sons Shipyard and Engine Works in the Kensington section of Philadelphia. The latter became the most technologically advanced industrial shipbuilding plant in the United States.[38]

During the war scare with Spain in 1873 over the *Virginius* affair, Robeson transferred nine ironclad monitors to Delaware River firms, and between 1874 and 1876 he gave Roach another thirteen vessels to rebuild. Delaware River shipyards used iron plate and steam machinery from older ships to reconstruct monitors, and Robeson needed a ready supply of ironclad hulls and steam engine parts for these private industries. When Robeson took office in 1869, he found that League Island held more than a dozen single-turret ironclad coastal monitors, larger double-turret monitors, and the captured Confederate States Navy ironclad *Atlanta*, scheduled for sale to the Haitian government. To preserve this iron and steam fleet, Robeson ordered development on League Island of a fresh water storage basin and iron shipyard with the latest machine shops and foundries necessary to maintain ironclad warships. To this end, in 1869 the Navy Department dispatched Naval Constructor Melvin Simmons to take up permanent residence in one of the half-dozen large frame houses standing on the island when it was transferred to the federal government.[39]

Meanwhile, Commandant George F. Emmons at the Southwark Navy Yard instructed civil engineer Francis C. Prindle to plan permanent improvements to League Island. Prindle had served during the war with former Navy Yard Commandant Du Pont, and his proposal for "a first class dock yard in every respect" reflected one envisaged by Du Pont in 1861 for Philadelphia. "On the plan of the Island, I have indicated in blue lines and adjoined Broad Street on the west, a basin 600 feet by 1200 feet, with four dry docks of 400 feet and two of 350 feet in length, opening into it, and two first class docks, 400 feet long, six ship houses, and two building slips on the river front adjoining and lying to the eastward of Broad Street as contiguous works of improvement." He added deep freshwater basins in the Back Channel for ship and timber storage.[40]

The Prindle blueprint of 1871, revised by a Board of Civil Engineers headed by Chief Engineer W. P. S. Sanger, served as the model for League Island's development for the next 125 years. They divided the 923-acre island into a series of rectangular industrial blocks. Five wide east-west avenues crossed a central north-south thoroughfare, and a series of parallel streets extended on either side, forming an orthogonal grid. Buildings were grouped along a main street (later Broad Street), according to the Navy Department bureau reorganization of 1867 that Secretary Welles had introduced. Buildings for the Bureaus of Construction and Repair, Equipment and Recruiting, Provisions and Clothing, Medicine and Surgery, and Navigation were laid out along the western side. The Bureaus of Steam Engineering, Yards and Docks, and Ordnance established buildings on the eastern side. The careful separation of functions reflected the nearly autonomous bureau system with overlapping tasks, personnel, payrolls, and facilities. Each bureau managed its own little empire, receiving separate budgets and duplicating work.[41]

Prindle moved to League Island in 1871, residing temporarily in House A and then in 1874 at Quarters A, the first building constructed by the navy on the island, so that he could direct improvements, oversee dredging of the Back Channel, and reconstruct the raised causeway (built by Charles Wharton between 1835 and 1841) that connected the southernmost extension of Broad Street to the island's main thoroughfare. Fill from dredging the channel raised the low areas along the Broad Street exten-

sion by some six to eight feet where Prindle laid foundations for buildings. The Navy Department started moving material from Southwark to League Island in 1873, while private contractors tore down wooden buildings at the Southwark plant and shipped the material by barge to League Island for reassembly on stone foundations and piles driven into the wet meadows. Between 1873 and 1877, League Island built the Bureau of Yards and Docks Storehouse and Office (Building No. 1), Boiler and Engine House (Building No. 2), Bureau of Steam Engineering Storehouse and Shop (Building No. 4), Bureau of Construction and Repair Mold Loft (Building No. 7), and Iron Plating Building (No. 3). League Island also contained fifteen temporary wooden "shanties" to house carpenters, craftsmen and inspectors, and boasted a nearly finished ship house and bell tower.[42]

While the Navy Department was developing League Island, Secretary Robeson conducted naval business in Philadelphia at the navy yard in overcrowded Southwark. In 1870–72 Commandant Emmons leased (or bought) more Southwark riverfront acreage for the U.S. government to provide additional space for Civil War surplus material and ships. Secretary Robeson, Congressional visitors, and foreign guests, including Grand Duke Alexis of Russia, stopped by the Southwark yard during the early 1870s. As the Navy Department cut work forces at other East Coast naval establishments and concentrated on development of a Pacific Coast navy yard on Mare Island in California, Robeson poured funds into a Delaware River government facility scheduled for closure.[43]

Expansion at the old yard while a new one was under construction provided jobs for local voters during election time. The nearly defunct yard added 500 civilian employees during each city or state election in Pennsylvania and New Jersey, swelling the work force to 1,400 (four times as many as would be employed annually at the navy yard on League Island over the next thirty years). Hundreds of mechanics and laborers puttered about on sloop-of-war *Brooklyn*, readied for auction the purchased steamer *George H. Stout*, and deactivated monitors

Passaic, *Manhattan*, *Yazoo*, *Wyandotte*, and *Nahant* for storage downstream in the Back Channel. The Southwark crew decommissioned iron-clad gunboat *Terror* and prepared the monitor *Puritan* for passage to John Roach's iron shipyard in Chester, where it would be entirely rebuilt. A steam tug towed the historic frigate *Constitution* alongside the recently launched sloop-of-war *Omaha*. Workers stripped away rigging and masts and floated the frigate on the sectional dry dock for complete reconstruction. Carpenters resumed construction in the large Ship House on *Antietam*, a screw steamer authorized during the war but never built. Wooden frames, boilers, and machinery, some actually condemned as unusable, lay in storage sheds around the yard. To deflect Democratic Party criticism of building another expensive steamer at a moment when the Navy was disposing of identical vessels purchased during the war, Robeson eliminated machinery from the plan and designated *Antietam* as a sailing storeship.[44]

Antietam occupied one of the covered ship houses in Southwark and required few unskilled workers. Repair work declined and the old navy yard laid off 600 laborers, mostly from South Jersey. "Some of them were promised a winter's job, but they found there was no further use for them after the election," the anti-Robeson *Camden Democrat* complained, "this was ungrateful in Robey to say the least." In response, the Camden native found another project to keep his Delaware River friends employed— "rebuilding" sloop-of-war *Quinnebaug*, partly assembled on the stocks at the Gosport Navy Yard in Virginia but untouched for years. Robeson assigned *Quinnebaug* to the navy yard in Southwark in 1872, a ruse that allowed him to bypass the convening of a naval survey board required when a warship needed a major overhaul. Southwark Navy Yard engineer Harman Newell designed eight new boilers and naval constructor Robert W. Steele raised a completely new hull. Southwark launched *Quinnebaug* from the Frigate House on 28 September 1875 and *Antietam* a few weeks later, the last U.S. Navy ships built at the old yard.[45]

As *Antietam*'s graceful wooden hull slipped from the Ship House into the Delaware River and was towed to League Island where carpenters hosed over the deck, the nearby Frigate House was "all torn down." The rest of the old navy yard appeared barren as civilian contract labor completed the transfer of material to League Island. "The lower ship house is razed to the ground and the entire yard presents a scene of desolation," a visitor reported. The Navy Department ordered Commodore George Henry Preble, the Southwark establishment's last commandant, to vacate his office in late November 1875 and move to one on South Broad Street nearer League Island. Engineers cut Federal Street through his old building.[46]

Secretary of the Navy Robeson came unannounced to Philadelphia in early December, walked about the nearly vacant Southwark yard, and joined city tax official Thomas Cochran at the Public Room at the Philadelphia Merchants' Exchange to auction off the riverfront property. Sale of the surplus navy yard was announced as the "largest real estate transaction that ever took place in this city." The U.S. government subdivided the property into choice waterfront lots that the Pennsylvania Railroad expected to obtain at a bargain price, but as the bidding started a representative of the Baltimore & Ohio Railroad, the Pennsylvania's arch rival, offered $500,000 for the entire tract. When bidding reached $1 million, with an offer made by John C. Bullitt for the Pennsylvania Railroad Company, Robeson's personal attorney signaled the auctioneer to close the bidding at once.[47]

A few days later, eighteen Pennsylvania Railroad armed guards occupied the old navy yard before the official closure ceremonies. Philadelphia contractors Nathaniel McKay and John Rice hastily removed remaining government property, but during the transfer navy property was lost, destroyed, or stolen. Patrick Gallagher, for one, was arrested for stripping copper from the rudder of frigate *Constitution*. Finally, on 7 January 1876, Captain of the Yard Clark H. Welles led closure ceremonies at 11:15 A.M. with the striking of the flag at the old Marine Barracks. The Marine Guard presented arms, marched with fife and drum to the old wharf,

The frigate *Constitution* being hauled out for repairs, 1873. Symbolically, the *Constitution* was the last ship repair contract at the Navy Yard that once built sister frigate *United States*. Naval Historical Center Photograph NH 55582; courtesy of Capt. W. P. Robert.

boarded a steam tug, and headed for League Island and new quarters aboard the receiving ship *St. Louis*. Only *Constitution* remained at the old yard, evidence of one hundred years of an American government naval presence along the Southwark water-front. Soon the Navy removed this relic of the past as well, towing the historic frigate across the Delaware River for outfitting at the Wood and Dialogue Shipyard in Camden, New Jersey.[48]

PART TWO

LEAGUE ISLAND

7 ★ LEAGUE ISLAND NAVY YARD

A U.S. NAVAL shore establishment opened on League Island in early 1876 during a period of decline and national indifference to development of the United States Navy. Americans showed little interest in foreign affairs or overseas expansion as they concentrated on reconstructing the Union, settling the West, and industrializing the nation. Economic troubles, labor violence, business excess, and government corruption marked the period. The erratic development of League Island reflected this national malaise. The Delaware River navy yard remained a strange assortment of modern brick industrial buildings, temporary wooden storehouses, workshops, and stone cellar foundations left unfinished by a Congress that refused to appropriate money to place buildings on them. After sunset five small oil lamps lit the site. Piles of gravel, cobblestones, old brick and ship house timbers from the former Southwark Navy Yard, and even furniture lay unattended in the open air around the muddy, low-lying island. Neglected property posed a constant danger to the public, and one female visitor fell to her death while touring the unfinished deck house over store ship *Antietam*, berthed at the island.[1]

League Island, a league or three miles in circumference, covered six hundred acres, most of it kept dry by a system of dikes, grew fields of corn and grazed cattle as it had since colonial times. Several older frame structures housed a lighthouse keeper, gardener, head teamster, and foreman of labor, with their families and several Irish-American female and African-American male servants. The largest house, a former hunting lodge, was occupied by Navy Yard Clerk William J. Manning and his family. A Marine Corps guard of fifteen men and two officers, including young Lieutenant William Phillips Biddle of Philadelphia (destined one day to become Marine Corps commandant), resided on a badly leaking re-

ceiving ship and in temporary barracks near the main gate. The naval constructor, chief engineer, paymaster, and six naval line officers assigned to League Island and most of the hundred permanent civilian employees lived in the southern section of the city, walking or riding in horse-drawn wagons or carriages down Broad Street to the Navy Yard each morning for work. The "Germantown Mail Wagon" visited the place twice daily.[2]

Despite League Island's primitive appearance, a naval establishment took shape there with a physical structure and organization that would become the permanent twentieth-century Philadelphia Navy Yard. A visitor traveling to League Island down Broad Street from the center of Philadelphia during the last quarter of the nineteenth century entered a gate flanked on the east by a two-story frame guard house and on the west by a two-story frame gatehouse. Just below lay Quarters A for the civil engineer and a frame building for the captain of the yard, night watchmen, and inspectors, who supervised dredging the Back Channel and filling low spots along a central dirt and plank road. This 125-foot-wide main roadway (an extension of Broad Street) ran north to south through the middle of the island.[3]

In 1877 engineers built up the old earthen causeway that at low tide connected the island to the city. They excavated an opening and erected a 36-foot-long wooden trussed bridge at the center of the causeway to allow tidal waters to flow through between the Schuylkill and Delaware Rivers to prevent Navy Yard waste from fouling the ship and timber basins in the Back Channel. "The current would not all times cleanse the locality," admitted Navy Yard Commandant Pierce Crosby, "consequently it would make it very disagreeable in the vicinity."[4]

At the southern end of the causeway stood an

David Kennedy watercolor, *Entrance to the Old United States Navy Yard, south Front Street, foot of Federal in 1870*. HSP, Kennedy Watercolors, K: 4-43.

Kennedy watercolor of the Southwark Navy Yard before closure, *The Old U.S. Navy Yard Philadelphia Penna, 1870*. HSP, Kennedy Watercolors, K: 2-114.

Kennedy watercolor, *Gate to League Island Navy Yard Philadelphia, looking up Broad Street towards the City Hall, 1891.* HSP, Kennedy Watercolors, K: 6-72.

The Civil Engineer's Residence (Quarters A), c. 1880. The Engineer's Residence was the first permanent building on League Island, opening in 1874. Note the monitor in the Back Channel at the extreme left. Naval Historical Center Photo NH 75465.

artillery emplacement, raised to defend League Island when labor violence and rioting rocked the area during the Railroad Strike of 1877. Nearby, on the western side of Broad Street, a small Bureau of Medicine and Surgery building—manned in 1880 by Cuban-born nurse Joseph T. Lefadain, an assistant naval surgeon, and "Apothecary"—served as the Navy Yard dispensary and hospital. Below the dispensary stood an unfinished cellar foundation for a planned stone administration building (Building No. 6). A block south of the cellar hole (around present-day Porter Avenue) lay a group of brick and wooden buildings, lining either side of Broad Street and extending to the Delaware riverfront. On the eastern part stood the Bureau of Yards and Docks buildings, stables, and firehouse with a permanent staff of twelve, headed during the early 1880s by civil engineer Mordecai T. Endicott and including two clerks, a master carpenter, draftsman, and a telegraph operator who ran a tiny station connected to the city by overhead wires strung on poles along Broad Street.[5]

The Bureau of Steam Engineering plant lay next to the Bureau of Yards and Docks. "It was built of material brought from the old Navy Yard, [and] is somewhat decayed and very combustible," one commandant observed. Steam Engineering provided office space for the chief engineer, storekeeper, four assistant engineers, foreman of machinists John Rowbotham, and Chief Clerk Henry Vallette. The nearby Bureau of Ordnance building was a fine permanent brick structure that contained the offices of an ordnance officer, clerk, "Gun Car Maker," and gunner in charge of the Naval Magazine located a mile downriver near the inactive U.S. Army post at Fort Mifflin.[6]

Across the main street to the west lay several larger buildings for the Bureau of Construction and Repair. These contained a mold loft, docking apparatus, storehouse, sawmill, and blacksmith shop. The Construction and Repair complement included a master carpenter, eighteen assistants, a timber inspector, and brilliant Naval Constructor Philip Hichborn, whose study of European dock yards later forced the modernization of the entire U.S. navy

yard system. Closer to the waterfront along Delaware Avenue and to the west of the L-shaped ferry dock at the foot of Broad Street stood an iron plating shop (Building No. 3), and further west the 318-foot x 1000-foot Ship House No. 5 that by 1876 was nearly finished with materials taken from the old navy yard and new doors, tin roof, and siding.[7]

The government never used Ship House No. 5 on League Island. Before it became functional, a storm in September 1876 blew down the south end that faced the Delaware River. Another violent northeaster struck the huge building as extensive repairs neared completion. "The tide is one foot higher than ever known," Commandant Crosby observed, with water 10-½ feet "above mean low water mark." The flood tide inundated the island, and Chief Clerk Manning and family, one of five families and fifty-two permanent residents on the island, fled to the mainland. Wind and rain ripped away slated roofs, flashing, and tin siding. The storm destroyed the Bureau of Yards and Docks storehouse; "Ship House No. 5 was crushed by the force of the gale at half-past seven o'clock, and now lies a mass of ruins with scarcely a stick of its frame unbroken."[8]

Another severe storm hit League Island the following month, washing away dikes and exhausting the supply of ballast stone used for "walling and rip-rapping." The second storm convinced the U.S. government not to rebuild Ship House No. 5, and the devastation caused by the storms ushered in a decade of utter neglect and rumors of closure. The Navy Department ignored Crosby's request to surface the plank and dirt road that had washed away during the flood tide. Navy bureaucrats refused to replace the destroyed Ship House with a new sectional dry dock launched as Hull No. 171 in 1877 from John Roach's shipyard in Chester, or to reassemble the old floating dry dock that lay partly submerged in the Back Channel. Crosby received no money to improve drainage around buildings, and at high tide water seeped constantly across floors.[9]

Storm damage and subsequent neglect of the League Island shore facility coincided with increasing Congressional scrutiny of Navy Department business in Philadelphia, particularly Secretary of the

Navy George Robeson's maintenance of two navy yards on the Delaware River. House Naval Affairs Committee Chairman Washington C. Whitthorne and committee members visited Philadelphia and examined log books and steam engineering records for both navy yards. The committee suspected wrongdoing in the simultaneous operation of two competing naval facilities, but never indicted Robeson. In order to avoid further investigation of Republican management of the naval establishment, Robeson's successors William H. Hunt (1881–82) and William E. Chandler (1882–84) reformed Navy Department operations in Philadelphia and at other government naval shore facilities and planned a "New American Navy."[10]

In 1882–83 Secretary of the Navy Chandler convened a Naval Advisory Board to examine the navy yard question and recommend reorganization or possible closure of inefficient facilities, while Naval Constructor Frank Lysander Fernald, returning from a visit to the most modern European dockyards, investigated League Island. Fernald discovered a disheartening situation. He found no dry dock, ship-building way, marine railway, or modern machine tool shop. He observed the local steam engineering force puttering about on the old engines and boilers of the obsolete sloop of war *Ossipee* and rebuilding outdated machinery on nearly useless low-freeboard ironclad monitors *Jason*, *Montauk*, and *Dictator*. Reform-minded naval ordnance officer Edward Simpson, who assumed command of League Island in 1881, admitted that with the existing facilities the Navy Yard could not overhaul monitor *Miantonomah* without assistance from the machinists at John Roach's facility in Chester.[11]

Simpson blamed public neglect for League Island's flaws. The place remained dark at night because the government refused to replenish the supply of oil to light evening lamps. Worse, the Navy Department kept such a small force on the island that

The only known photograph of Ship House No. 5 on League Island, begun in 1873/74, blown down in a violent storm in 1878, and never rebuilt. League Island would have no ship building ways until 1915. NARAM.

all available hands were employed in cleaning horse stables, making mosquito frames for windows, or cutting drainage sluices rather than maintaining an industrial plant. There was no base housing for line officers, so that after closing no one stayed to take command in case of an emergency, particularly a fire. "You are aware that this yard is very isolated, and in case of fire must depend on its own machines," Simpson explained, asking Chandler to bend the "fixed rules" so that a naval officer could reside permanently in a house on League Island to take care of such matters. Instead, the navy secretary ordered Simpson to dismiss the "civil employed force."[12]

The likelihood of closure increased in 1883 when scandal rocked League Island. In response to charges of corruption in the Steam Engineering department, Chandler ordered Simpson to convene a top-level board of inquiry headed by Chief Engineer of the U.S. Navy George W. Melville. The board found foreman John Rowbotham guilty of employing Navy Yard time, labor, and materials to fix his personal carriage and that a second-class machinist in Rowbotham's charge had altered the books to cover up the crime. The Melville board recommended suspension of the machinist "and [of] Chief and Time Clerk H. M. Vallette for falsifying and altering the book of payroll in the Steam Engineering department." The latter charge came as a shock. Vallette was a respected government employee and well-known author of a popular ten-month series of articles in *Potter's American Monthly* on the history of the old Philadelphia Navy Yard in Southwark.[13]

Chandler rescued Simpson from this seemingly tarnished command during the summer of 1884 by sending the reform-minded officer to preside over the Naval Advisory Board in Washington that was studying development of a new steel navy. Captain of the Yard William E. Fitzhugh continued the investigation on League Island, uncovering further "discrepancies" in the pay accounts of receiving ship *St. Louis* and records kept by Paymaster Clerk John Wallace, who disappeared suddenly from the island. Continuing evidence of venality at the Philadelphia naval shore establishment convinced Chandler (who

had earlier tried unsuccessfully to prosecute a ring of thieves caught stealing government property at the Southwark Navy Yard during the Civil War) to close League Island. He ordered the Navy Yard to dispose of supplies and material at public auction in September 1883. But he never closed League Island. Station logs showed no interruption in daily operation or any reduction in work force during Chandler's secretaryship, 1882–84.[14]

Sparing League Island kept the facility near the center of an American naval renaissance that began along the Delaware River. Between 1881 and 1891, Congress funded construction of seventy-eight new naval vessels, most incorporating the latest technology in rolled steel plating, steel rifled ordnance, and triple expansion steam engines. Delaware Valley industry built the first ships of this new American steel navy. John Roach's Delaware River Iron Shipbuilding and Engine Company fabricated the cruisers *Atlanta*, *Boston,* and *Chicago* and the steel-hulled dispatch boat *Dolphin*, while William Cramp & Sons Ship and Engine Building Company in the Kensington section of Philadelphia constructed the steel cruisers *Baltimore*, *Newark*, and *Philadelphia*, gunboat *Yorktown*, and dynamite cruiser *Vesuvius*. The Carnegie, Bethlehem, and Midvale Steel Companies of Pennsylvania rolled steel plate and forged castings for these ships and Pennsylvania mines provided coal to power both the shipbuilding industry and ships of the U.S. Navy.[15]

League Island assumed a secondary role in this late nineteenth-century naval rebirth along the Delaware River because the Navy Yard lacked the modern industrial plant, machine tools, foundry, shipbuilding ways, and dry docks necessary to serve the new steel navy. Commandants Edward E. Potter (1886–88) and Henry B. Seely (1888–91) continually bombarded the Navy Department with reports of the local plant's shortcomings. Without dry docks, the Navy Yard could not inspect the hulls, propellers, or rudders of new vessels brought to League Island for commissioning. Potter reported somewhat sarcastically that the most his navy yard could do for the steel cruiser *Baltimore*, recently launched by the Philadelphia shipbuilding firm of William Cramp up-

Yards and Docks Store House, League Island, c. 1880, one of the half-dozen modern brick buildings finished when the Navy Department decided in l883 to close the League Island Navy Yard. Naval Historical Center Photo NH 75464.

river, was to mount a new flag. Moreover, the Navy Yard had no outfitting pier on the riverfront, forcing captains of the yard to berth new warships in the mud-choked and partly dredged Back Channel. To accommodate outfitting there, old ironclads were moved to moorings near the main ship channel on the Delaware River in front of the Navy Yard, where in 1887 the commercial steamer *Madrid* smashed into the anchored monitors *Montauk* and *Nahant*. Hurriedly, the Navy Yard ordered sections of the old floating dry dock raised from the mud and moved out of the narrow Back Channel, only to have a "boat capsized and four men drowned as old dry dock tanks were towed from back channel."[16]

Secretary of the Navy William C. Whitney appointed League Island Commandant Potter to head a Permanent Improvement Board and develop a plan for a modern industrial facility on the island to serve the new steel navy under construction at private yards along the Delaware River. In late 1888 Potter pro-

posed a $14 million development program, as the Democratic administration of President Grover Cleveland and Navy Secretary Whitney prepared to leave office. Incoming Republican Navy Secretary Benjamin Franklin Tracy rejected the grand design and dispatched his own commission under Commodore Andrew E. K. Benham to League Island in July 1889. The Benham Board supported Potter's plan for development of a dry dock, and consequently Tracy endorsed Whitney's decision to expand the navy yard in Philadelphia and fund the first permanent government dry dock on the Delaware River at League Island.[17]

Civil Engineer Robert E. Peary, "in charge of the Construction of the New Timber Dry Dock" (and later decorated for his exploration of the North Pole), surveyed and staked out Dry Dock No. 1 at League Island in early 1889. Private contractors brought steam derricks, pumps, hoisting engines, dumping cars, and "Hayward Excavators" to the island by barge and digging started in April. Soon the riverfront excavation—which lay fourteen feet below the benchmark at the highest grade of the island—flooded badly. Peary discovered that the dry dock construction required far more pilings, digging, and labor than first estimated, and reported a steady increase in the cost of labor and material. At last, on 4 March 1891, the "U.S. Monitors, *Jason* and *Nahant* were Docked in the forenoon, to test the New Timber Dry Dock." Peary told the board of inspection that much work remained, however, including acquisition of new drainage pumps because the old ones proved inadequate to pump out the dry dock, 500 feet long, 90 feet wide at the sill, and 25½ feet deep. The wooden-hulled steamer *Saratoga* became the first U.S. naval vessel overhauled in Dry Dock No. l on 5 June 1891.[18]

League Island's original dry dock was smaller than the granite graving docks constructed earlier at the Charlestown and Norfolk navy yards or the one under construction at Mare Island in California. Moreover, it could not accept the new battleships under construction nearby that became the backbone of the turn-of-the century U.S. fleet. But Dry

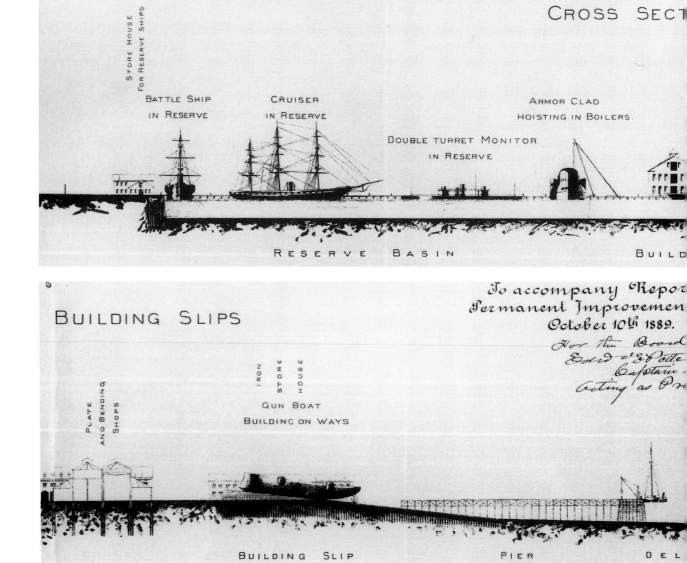

Dock No. 1 accommodated every other type of warship, and its opening in 1891 marked a watershed in development of League Island as a naval shore facility because it meant that the U.S. Navy had committed itself, at last, to establishing a first-class navy yard in Philadelphia.[19]

Improvements to League Island followed introduction of the dry dock. The Navy Yard added concrete sidewalks, paved roadways and pier surfaces, a new sewer and fresh water pumping system, and electric lighting and elevators. During the decade after completion of the dry dock, it built new officers' quarters, store houses, wharves, piers, and a Reserve Basin in the Back Channel. Private contractors filled in the swampy eastern end of League Island with dredge material from Smith and Windmill Islands, removed by the U.S. Army Corps of Engineers from the Delaware River ship channel between Philadelphia and Camden.[20]

The civilian work force on League Island climbed to over three hundred and the government placed laborers under civil service. This required listings of specific positions and hiring based on qualification rather than political connection. A Board of Labor

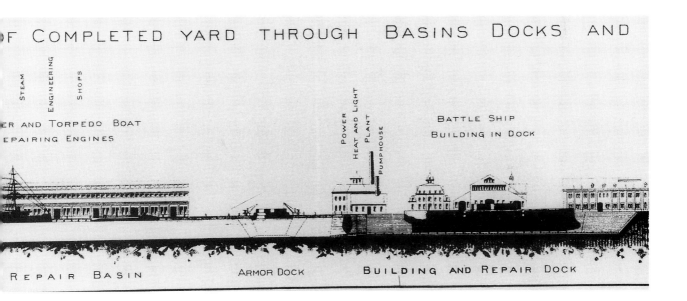

OF COMPLETED YARD THROUGH BASINS DOCKS AND

STEAM ENGINEERING SHOPS

ER AND TORPEDO BOAT
EPAIRING ENGINES

POWER HEAT AND LIGHT PLANT
PUMPHOUSE

BATTLE SHIP
BUILDING IN DOCK

REPAIR BASIN   ARMOR DOCK   BUILDING AND REPAIR DOCK

Navy Yard, League Island, cross section of completed yard through basins, docks, and building slips to accompany Report of Permanent Improvement Board, 10 October 1889. NARAM.

RIVER

Employment, headed by the captain of the yard, published a "List of Trades for Registration, Navy Yard, League Island" in 1892. Schedule A enumerated positions available for traditional unskilled workers, Schedule B listed industrial specializations such as boiler makers, heavy forgers, gas fitters, engine tenders, pipe fitters, riveters, tool makers, and machinists.[21]

Expansion on League Island during the 1890s created a new set of problems. Commandant William Ashe Kirkland (1891–94) complained that officer housing was insufficient for this first-class naval station. "The New Quarters are cottages originally in-

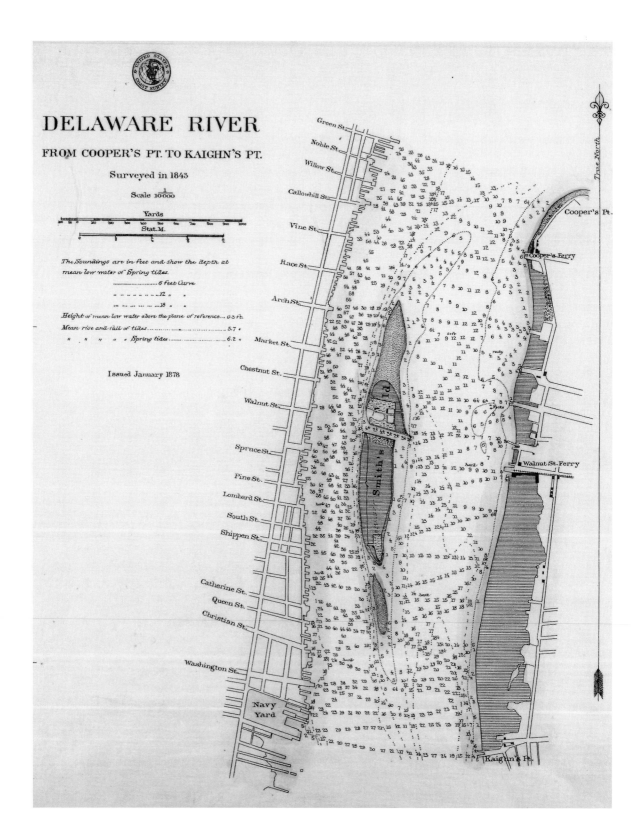

DELAWARE RIVER

FROM COOPER'S PT. TO KAIGHN'S PT.

Surveyed in 1843

Scale $\frac{1}{10000}$

Yards

Stat. M.

The Soundings are in feet and show the depth at
mean low water of Spring tides.

.... 6 feet Curve

" .... 12 " "

" ... ... 18 " "

Height of mean low water above the plane of reference... 0.3 ft.

Mean rise and fall of tides............................5.7 "

" " " " " Spring tides........................6.2 "

Issued January 1878

tended for the use of the Warrant Officers and have no accommodation for servants beyond one small room." He expressed similar disdain for inferior officer housing located in a newly developed working class neighborhood of Philadelphia below Prime Street near the abandoned Southwark Navy Yard. Demonstrating the elitist attitudes of a late nineteenth-century naval aristocracy, Kirkland denigrated the area for its lack of proper social life, particularly society teas and dances for his unmarried daughters. "Neither is the class of inhabitants such as Commandants of Yards are accustomed to mingle with," Kirkland concluded. Captain of the Yard Bartlett Jefferson Cromwell of Georgia agreed. He hated League Island. It "is generally considered exceedingly unhealthy, and for many reasons, one of the most undesirable duties on shore to which an officer could be ordered."[22]

League Island added new officers' quarters and industrial buildings before Kirkland left the Navy Yard command in 1894, and the following year extended electricity to all buildings. But in many ways League Island still was not a first-class navy yard when the armored cruiser *Maine* exploded and sank in Havana harbor in early 1898, bringing the United States to the verge of war with Spain. Commandant Silas Casey and Captain of the Yard Edwin S. Houston mobilized the few resources available on League Island to meet the crisis that became the Spanish-American War. The civilian work force increased from 300 in February to 1,000 by the end of March and 2,000 after declaration of war in April. The Navy Yard docked and overhauled cruisers *Columbia* and *Minneapolis*, part of a South Atlantic squadron operating from League Island since 1893, while Chief Engineer Lewis Wood Robinson, a native of nearby Haddonfield, New Jersey, repaired machinery on the Civil War-vintage monitors *Lehigh* and *Montauk* and experimental harbor defense ram *Katahdin*.

The Army Corps of Engineers removed these islands in the Delaware River between Camden and Philadelphia around 1900 and used the dredge spoils to build up the eastern end of League Island. ISM, Philadelphia.

Ammunition was brought by barge from the Fort Mifflin Naval Magazine to arm these vessels for operations in Cuban waters.[23]

Rumors that a flying squadron of Spanish cruisers was headed for the Chesapeake and Delaware Bays forced Casey to requisition Philadelphia City *Ice Boat No. 3* (commissioned as *Arctic*) and send it downriver to patrol the Delaware breakwater. The commandant sought additional private vessels to guard local waters as part of a district "mosquito fleet" (forerunner of the naval district organization established by the Navy Department in 1903). The threat to the area from Spanish cruisers soon faded, and the Navy Yard turned to purchase and conversion of larger vessels for use as troop transports, supply ships, and colliers—fleet auxiliaries neglected in late nineteenth-century U.S. Navy shipbuilding programs. Between April and May 1898, League Island Naval Constructor John Forsyth Hanscom directed conversion of a dozen auxiliaries, including the 14,900-ton International Navigation Company liner *St. Paul*, the Red D Line steamer *Austin* (commissioned as *Panther*), Delaware River pilot boat *Peoria,* and Philadelphia traction magnate Peter A. Widener's steel yacht *Josephine* (commissioned *Vixen*).[24]

Fighting in Cuban waters lasted but a few months, and mobilization demands on League Island subsided. Dramatic cutbacks in the Navy Yard work force, from a wartime peak of 2,020 in April 1898 to 900 by July when fighting stopped and 600 by the end of the year, testified to the brevity of this little war. The civilian labor force rose to 1,000 again in 1899 as the Philadelphia naval facility laid up warships from the recent conflict and outfitted the purchased steamer *Dixie* for use as a troopship in the Pacific Ocean, where the United States had become entangled in an imperialist war of pacification in the Philippine Islands.[25]

The Philippine insurrection meant continued funding for the Navy Yard expansion started before the war with Spain. League Island received a $1 million appropriation for construction of an Electrical Workshop (1903), Angle Smithery (1903), Block and Cooper Shop (1903), Plate Bending Shop (1905), Foundry

League Island prepared cruiser *Columbia*, built at Cramp's Shipyard upriver in 1892, for duty in Cuban waters during the Spanish-American Cuban War of 1898. NARAM.

(1905), Pattern Shop (1905), Steam Engineering Machine Boiler Shop (1905), and a small boat storage shed. Additional money allowed dredging the Back Channel and completion of officers' quarters, the "Commandant's Office Building," and the first permanent Marine Corps barracks and officers' residence on League Island. The building program brought lucrative contracts to local businesses. John Wanamaker's Department Store in Philadelphia provided furniture for the commandant's office and officers' quarters. The Schuylkill Pressed Brick

Company, Carnegie Steel, Cambria Steel, and William Wharton, Jr. and Company contracted for building supplies and material. The General Electric Company of Schenectady, New York, however, became the largest contractor at the government yard in Philadelphia, establishing over the next fifty years one of the closest Navy Department connections with private industry.[26]

Navy Yard infrastructure improved after the war with Spain. Most streets were paved and curbed. Congress authorized the numbering, lettering, and naming of buildings, quarters, and streets, beginning with the main east-west industrial avenue that was named after Admiral John Woodward Philip, the popular commander of the North Atlantic Squadron during the Spanish War and founder of the movement to build a Sailor's Rest Home in Philadelphia, who died on 30 June 1900. Pier surfaces were "Vulcanized" and trolley tracks laid from the main gate to Philip Avenue. The Navy purchased two trolley cars from the Philadelphia Traction Company to carry passengers into the government establishment. In 1900 a mile-and-a-half broad-gauge railroad spur linked the Navy Yard to the nearby Pennsylvania Railroad system. "The railroad is entirely for the Government use," Civil Engineer Christopher C. Wolcott explained, and was connected "with the Schuylkill River branch of the Penna. R.R. just outside the Yard."[27]

For all the expansion and modernization, however, League Island appeared in 1900 to be still a remnant of the nineteenth-century naval shore establishment, with a mission little changed since the Civil War. Inactive Civil War-vintage monitors cluttered the Back Channel, preventing the berthing of modern battleships. Postwar improvements diverted attention from the Navy Yard's lack of basic facilities for ship alterations, repairs, and new work. Unlike the other major Atlantic coast naval shore establishments in Boston, New York, and Norfolk, League Island had no shipbuilding ways, hauling out slips, concrete dry dock, or heavy cranes. The 40-ton locomotive crane assembled at the yard in 1900 came from the Pittsburgh Bridge Works with the wrong boom and had to be shipped out for recasting to the Edge Moor Works of Delaware. The Navy Yard had to send battleships *Alabama, Maine,* and *Kansas,* launched at local Delaware River shipyards, to distant navy yards for docking trials and inspection.[28]

The dramatic changes in the Delaware Valley society and economy buffeted the Navy Yard, gradually transforming the place into a modern first-class naval shore establishment. By 1915, nearly 20 percent of the two million people residing in Philadelphia and Camden were recent immigrants from Southern and Eastern Europe, particularly Italy,

Russia, Poland, and Austria-Hungary. Many sought work in the shops and factories that made Philadelphia the third ranked wage earning and manufacturing center in the United States during the first decade of the new century. Ethnic, racial, and labor tensions accompanied this transformation of local society and spilled into the quiet, isolated nineteenth-century relic on League Island. A "long period of racial dispute" exploded in riots and fighting. One commandant blamed restless African American workers for setting "mysterious" fires around the place. Cutbacks in seasonal work, combined with violent labor activity among Philadelphia transit workers, led to the most serious threat to Navy Yard order and tranquility, from Italian Americans hired from the southern wards of Philadelphia as temporary unskilled labor. Reportedly, Italian American anarchists planned to kill Captain of the Yard John Bartholomew Collins because he had laid off their countrymen. A death threat signed by the Black Hand Society, an anarchist organization supposed to be behind violent strikes in the textile industries in nearby New Jersey, told Collins that "we will fix your feet." The captain of the yard ordered an increase in the Marine guard on League Island.[29]

To make matters worse, a "spotted-fever" epidemic broke out on board receiving ships *Minneapolis* and *Puritan,* berthed at the yard in 1903 to house the overflow of new recruits and sailors assembled to man the battleships nearing completion at private shipyards along the Delaware River. The Naval Home on Gray's Ferry Road alone received 130 patients. The Navy Yard command ordered that "enlisted men and the officers under thirty-five be required to vacate the *Minneapolis* until all traces of infection from Cerebro-Spinal Meningitis, Mumps and Measles have disappeared from among the men and until the whole ship shall have been completely fumigated with formaldehyde or other approved germicide."[30]

New League Island Commandant Charles Dwight Sigsbee (commander of the ill-fated *Maine* in 1898) ordered more than 600 landsmen apprentices, navy enlisted men, and Marines to a tent city on the easternmost part of League Island that was wet ground raised by dredge spoils. The isolation camp became a stinking, muddy wasteland, where camp personnel consorted with "bum-boat" women. Captain Caspar F. Goodrich of *Puritan* asked that "a stout barbed wire fence of four strands be erected at the eastern limit of the Camp ground to prevent straggling into the jungle with the chances of fouling the soil and of communicating with river pirates etc." In the end, the Navy erected wooden barracks to house this tent community, but the entire situation gave League Island a "bad name."[31]

It seemed as though the Navy Department rated Philadelphia lower than most other shore establishments, dumping older or undesirable officers on League Island. During the first decade of the twentieth century the Navy Yard had twelve different commandants, most nearing the end of largely routine and uninspiring careers, who came to Philadelphia to retire or die at the Naval Home. The Back Channel was only partly dredged, inadequately bulkheaded, and cluttered with single turret monitors, Civil War relics that long ago had been stripped of guns and machinery. In 1904, Secretary of the Navy William Moody ordered disposal at auction of monitors *Nahant*, *Jason*, *Canonicus*, *Lehigh*, and *Montauk*. Two years later, the Navy Yard sold monitors *Mahopac* and *Manhattan*. Once these old monitors had been cleared, dredges hurried to clean out the mud and silt that had accumulated under their anchorages, but an overworked dredge exploded, killing the captain and adding to League Island's reputation as one of the poorest naval shore establishments in the country.[32]

League Island's most glaring deficiency stemmed from an inability to dry dock newer battleships. During the presidential terms of Theodore Roosevelt

and William Howard Taft (1901–13) the navy commissioned an average of two each year as the United States raced to keep pace with the building programs of the world's other naval powers. Private Delaware River shipbuilders constructed one out of every three battleships during this period, but League Island could not dock them, forcing the navy to send the big warships to distant navy yards in Boston, New York, or Norfolk for hull and propeller inspections prior to commissioning back in Philadelphia. Congress had appropriated funds during the war with Spain for a second dry dock of "sufficient depth to accommodate the largest battleship afloat," and construction had started in August 1899. However, architectural alterations, engineering and labor problems, and increased costs delayed completion until July 1907, when the Navy tested Dry Dock No. 2 with its smallest battleship—the 375-foot, 11,540-ton *Kearsarge*. In October the Navy employed Dry Dock No. 2 to overhaul the 14,949-ton pre-dreadnought *Georgia* "for the cruise to the Pacific" as part of Roosevelt's Great White Fleet, assembled in the wake of growing tensions with Japan. But additional alterations to pumping and electrical systems further delayed final commissioning of Dry Dock No. 2 until 1910.[33]

Numerous naval accidents further tarnished the image of League Island. Between 1900 and 1910, League Island commissioned eight battleships, five armored cruisers, a dozen torpedo boat destroyers, and an unarmored cruiser, all built nearby at the Neafie and Levy Ship and Engine Company, William Cramp & Sons Ship and Engine Building Company, or the New York Shipbuilding Company of Camden. The arrival of each new ship to the Navy Yard forced the captain of the yard to move older ships from the

Facing above Civil War monitors *Montauk and Lehigh*, laid up at League Island, 1902. They were sold in 1904 to make room in the Reserve Basin for a squadron of pre-dreadnoughts of the Atlantic Reserve Battleship Fleet. Naval Historical Center Photo NH 45896.

Facing below The U.S. Navy's smallest battleship, the 11,540-ton *Kearsarge*, tested the 715-foot concrete and granite Dry Dock No. 2 in 1907. NARAM.

Delaware River front to the Back Channel, resulting in frequent groundings and collisions. When Cramp delivered armored cruisers *Colorado* and *Pennsylvania* to League Island within a few hours of each other, Navy Yard tugs rushed the troop ship *Dixie*, store ship *Panther,* and torpedo boat destroyer *Chauncey* from the waterfront to make berthing space, and in the process *Chauncey* crashed against the entrance to Dry Dock No. 1. A short time later, the older cruiser *Chicago*, rushing to the Back Channel, cut through and sunk the Navy Yard tug. Troop ship *Prairie* grounded so badly in the Delaware River that it required *Dixie*, five tugboats, and two U.S. Army Corps of Engineers dredges more than a week to free it from the mud. Worse, battleship *Idaho* ran aground as it steamed away from League Island to meet an emergency in the Caribbean.[34]

League Island's obvious strategic and organizational limitations during the first years of the twentieth century again raised the inevitable rumors of closure. Reportedly, the aggressive and reform-minded Secretary of the Navy George von Lengerke Meyer, a Harvard-educated management specialist, planned "to relegate the Philadelphia Navy Yard to the second class," while New York journalists intent on protecting the Brooklyn Navy Yard from closure in 1909 maintained that Meyer's Progressive reforms forecast eventual termination of the inefficient and expendable League Island installation. Since the Brooklyn, Washington, and Norfolk Navy Yards were considered indispensable, "the list of yards in danger would be thus reduced to Portsmouth, League Island, and Charleston, and it can be stated on good authority within the department that the Secretary intends to put them out of commission."[35]

Meyer never considered closing League Island. Instead, the navy secretary authorized the most extensive improvement to the Delaware River navy yard since its permanent relocation to the island in 1876. Meyer was most impressed by the arguments of a group of insurgent naval officers stationed on League Island in 1908–9 who formed a reform association on board *Panther*, berthed at the yard. The *Panther* Associators protested reorganization of

navy yards by Meyer's predecessor Truman Newberry, a midwestern industrial management expert who served as the last of Theodore Roosevelt's six navy secretaries. The Newberry Plan went into effect at the Philadelphia naval shore establishment on 1 February 1909 a few days before he left office, placing command of the entire shipyard organization in the hands of Prussian-born U.S. Naval Constructor Albert G. Stahl. This reorganization removed authority over operations from the navy yard commandant and divorced seagoing officers entirely from the process of overseeing work on the ships. Worse, it deprived the civil engineer of control over public works or deployment of manpower at the shore establishment. This was the cause of League Island's woes, the *Panther* insurgents argued, not poor location or inadequate facilities.[36]

The League Island protest stirred Meyer, in part because it coincided with his own organizational plan. He stopped the implementation of Newberry's scheme until he could convene a top-level naval committee to study a pamphlet issued by the nine League Island insurgents. He also dispatched Assistant Secretary of the Navy Beekman Winthrop to League Island to meet the *Panther* nine, and suspended the Newberry Plan permanently after Winthrop's return from Philadelphia. Most significantly, Meyer soon promoted the *Panther* protest leaders to key administrative posts in the naval bureaucracy. League Island Ordnance Officer Nathan Crook Twining became chief of the Bureau of Ordnance. Engineering Officer Benjamin C. Bryan took charge of the Secretary's Office of Navy Yards, and Josiah S. McKean became Meyer's planning officer for navy yard mobilization and material procurement. Meyer asked League Island Commandant Edwin C. Pendleton to delay retirement and come to Washington to lead the Progressive navy yard reorganization movement, and asked League Island Civil Engineer Homer R. Stanford, who professed personal "loyalty to your administration," to develop a management policy for the naval shore establishment.[37]

Stanford provided Meyer with a definitive statement on the purpose and mission of a navy yard that reflected clearly the views of the *Panther* Association. "The central feature of a navy yard must necessarily be the military control through which the yard is enabled to cooperate with the fleet; next in importance are the two industrial or manufacturing divisions of machinery and hulls which actually perform the work or repair or modification or construction which the vessels of the fleet require," advised Stanford. "All yard facilities and methods should be adjusted to subserve the military control and to facilitate the work of the machinery and hull divisions." Stanford's memorandum contributed to the Meyer reforms in the Navy Department. The Progressive navy secretary advanced the concept of military control through formation of a quasi-general staff of aides for operations, inspection, materiel, and personnel to advise the secretary. Meyer applied this concept at all naval shore establishments by restoring managerial authority over navy yard business to the commandants, public works to the civil engineer, and inspection of ship overhaul and construction to naval officers. He also grafted the organizational system used by private shipbuilding companies onto the government yards by creating a Manufacturing Department with a Hull Division under the naval constructor, a Machinery Division under the chief engineer, and a "Cost [Accounting] and Time Office."[38]

Meyer included League Island prominently in other managerial and technological innovations that characterized the Progressive Era, 1900–1917. The Navy Yard established research and development testing facilities for ship propellers, chemicals, wireless radio transmission, oil burning engines and boilers, and naval aviation. A Board of Scientific Management Methods headed by "unusually able engineering officer" Walter B. Tardy employed League Island to test Frederick W. Taylor's scientific system of management to promote industrial efficiency and productivity through time and motion studies. Meyer also introduced the shop management systems used by the Vickers factories in Britain, with planning, estimating, and evaluating officers who measured progress and performance of navy yard "piece" work or the efficiency of machine tool and assembly

procedure.[39]

Stanford E. Moses, head of the Machinery Division, thought that Progressive reform of the naval shore establishment was an institutional revolution greater than the simultaneous fleet reorganization. "The distribution of work to our navy yards is now receiving systematic handling in the scheduled assignment of the various units of the fleet to established home yards; and in the regular rotation of ships at their respective yards."

Perhaps, but the increasing prominence of the League Island naval shore establishment under the Meyer secretaryship had less to do with Progressive organizational reform than it did with Meyer's growing intimacy with Pennsylvania Senator and Republican Party boss Boies Penrose (credited by President Taft with assuring the party victory in 1908) and Representatives Thomas S. Butler, Michael Donohoe, and J. Hampton Moore of Pennsylvania, leading members of the committees on naval affairs and rivers and harbors. Moore wanted a deep ship channel up the Delaware River to benefit the port of Philadelphia, and he linked river and port development to Navy Yard expansion in order to win support from Meyer. After one visit with Moore, the navy secretary concluded that "the future of the navy yard was great—greater than he or any other Secretary of the Navy had ever dreamed." Meyer visited League Island with Penrose, cousin of current Marine Corps Commandant and Philadelphia native William Phillips Biddle, and in 1909 with Representative Butler to watch his son Smedley Darlington Butler drill a Marine Corps expeditionary force. Meyer told Butler that he expected great things for the Navy Yard on League Island.[40]

With such connections, those great things included an unprecedented expansion of the U.S. Marine Corps Reservation on League Island. The 40-acre reservation on the east side of Broad Street just below the Back Channel underwent steady growth between 1901and 1912, with new Marine Corps barracks, officers' quarters, parade ground, drill field, rifle range, and bandstand. In 1911 Meyer authorized the transfer of the Marine Corps Advance Base Training School to League Island from Newport,

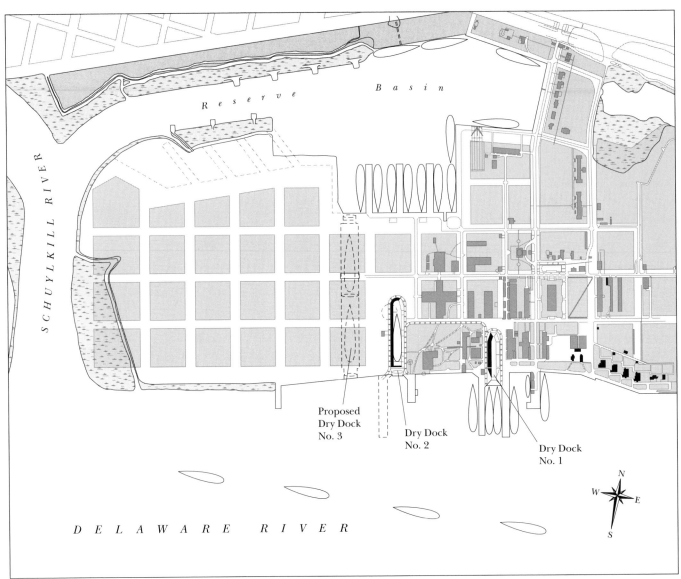

Map 5. 1911. Early twentieth-century naval planners insisted that the Navy Yard would never serve as a first-class yard to support the Atlantic fleet without construction of a 1,700-foot dry dock connecting the Reserve Basin with the riverfront.

Rhode Island, and dispatched a succession of Marine Corps expeditionary forces from the Philadelphia base to the Caribbean to preserve order and protect the approaches to the Panama Canal, then nearing completion. Meyer's patronage extended to the entire League Island facility, leading to funding for construction of Dry Dock No. 3, Pier No. 5, naval enlisted men's barracks, naval prison, and storehouses. The navy secretary pushed development of the Reserve Basin so that League Island could serve as home port for the First Line Reserve Atlantic Fleet, which included a torpedo boat flotilla, armored cruisers *Tennessee* and *Montana*, scout cruisers *Chester*, *Tacoma*, and *Birmingham*, and battleships *Idaho*, *Georgia*, *Kansas*, *Mississippi*, *South Carolina*, *Michigan*, *Connecticut*, and *Alabama*.[41]

Meyer's support for Navy Yard expansion encouraged a generation of Philadelphians to believe that the U.S. government considered their navy yard of great importance to the nation. After Philadelphia Mayor John E. Reyburn introduced the secretary to local society, business, and political leaders at a particularly warm banquet, Meyer promised the gathering that he would make League Island into the greatest government naval facility in the country. In response, Reyburn rushed out and organized a Joint Committee of City Council to promote navy yard development with the Navy Yard Employees Association, South Philadelphia Business Men's Association, and Citizen's Committee for a Greater League Island Navy Yard, the latter headed by leading Philadelphia banker Edward Townsend Stotesbury. The Philadelphia coalition lobbied Congress to support Meyer's call for increased funding of the naval establishment, sponsored another banquet at the Hotel Bellevue-Stratford for members of the House Committee on Naval Affairs, and took the members on a tour of

League Island. Infatuation with Meyer continued after local defeat of Penrose's Republican city machine in the mayoral election of 1911, as the new administration of Progressive Mayor Rudolph Blankenburg made the Republican's "Greater Navy Yard" campaign an integral part of his own Independent Progressive "Greater Philadelphia" movement.[42]

Despite Meyer's constant praise of League Island, though, early twentieth-century naval war planners never conceived of Philadelphia as more than a "subsidiary base" to maintain the Atlantic Reserve Fleet. The Meyer-Taft administration spent more money to develop Brooklyn and Norfolk Navy Yards as home bases for the Atlantic Fleet and thought

Receiving ship *Richmond* housed the Marine Guard on League Island in 1895–1902, until construction of the first Marine Corps barracks in 1901. NARAM.

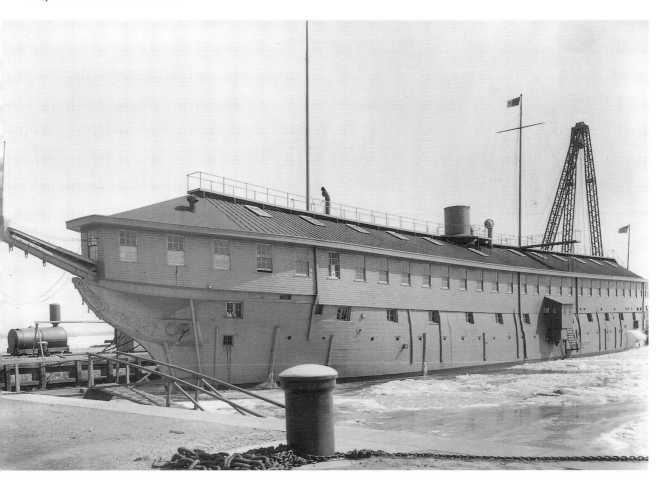

Photograph used in a 1911 campaign by Philadelphia progressive businessmen to promote expansion of League Island as a home base for the battleship fleet. NARAM.

Guantanamo Bay, Pearl Harbor, and Bremerton, Washington, more vital as bases than League Island. Moreover, all the reform, reorganization, and expansion during Meyer's regime had not made League Island into a first class navy yard. There was no shipbuilding facility. In late 1912, as the Republican navy secretary prepared to hand over his office to the recently victorious Democratic Party, the Delaware River naval shore establishment could repair and maintain warships but could construct nothing larger than a coal barge. Incoming Secretary of the Navy Josephus Daniels planned to change this situation so that the Philadelphia Navy Yard on League Island could build its first ship.[43]

8 ☆ NEUTRALITY AND WORLD WAR

PRESIDENT Woodrow Wilson despised the big-navy policies of his Republican predecessors Theodore Roosevelt and William Howard Taft. He opposed creation of a naval general staff or war planning committee and disapproved the recommendations of a Joint Board of the Army and Navy for possible action against Japan during a diplomatic controversy in 1913 over restrictions to Japanese immigration in California. Wilson's Secretary of the Navy Josephus Daniels, a North Carolina newspaperman and liberal Southern Democrat, shared these antinavalist views, particularly opposing a "Prussian" general staff system. But Daniels's main interest lay in democratization and reform of life in the U.S. Navy. In an effort to reduce the gulf between officers and enlisted men, he ordered architects to redesign Marine Corps Barracks No. 3 under construction at the Philadelphia Navy Yard in 1913 so that officers and men shared housing. Daniels banned liquor from navy vessels and facilities and assigned permanent chaplains, established chapels, and opened Young Men's Christian Associations at naval shore facilities as part of his moral reform of the service.[1]

Daniels was not as suspicious of a big navy as was his boss Woodrow Wilson. Daniels's father had been a shipwright and two brothers-in-law served as naval officers. Visits to navy yards and ships of the fleet thrilled Daniels. He dined with top naval officers, flew in primitive naval aircraft, took dives in submarines, and sailed on board mighty battleships during target practice. Daniels favored modernization and expansion of the naval establishment and wanted government navy yards to become self-sufficient industrial facilities. "Within three months after I became Secretary, I became convinced that every Navy Yard should be equipped so that some could build ships of every kind." Moreover, he advocated higher wages and safer working conditions for civilian labor at navy yards in order to make public shipyards competitive with private business that monopolized new warship construction.[2]

Daniels's selection of exuberant New York Democrat Franklin Delano Roosevelt as assistant secretary of the navy indicated a big-navy philosophy as well. Though not as dominant as his cousin Theodore, who ran naval affairs as assistant secretary in 1897, during the Wilson administration FDR exerted tremendous subtle influence on naval expansion. He promoted naval preparedness, war planning, and "enlargement" of navy yards. Roosevelt visited the Philadelphia Navy Yard in late March 1913, and toured League Island with Commandant Albert Weston Grant, who wanted to build a 1,700-foot dry dock system that would connect the Reserve Basin to the Delaware River waterfront.[3]

FDR's visit to the Delaware River navy yard arose partly from efforts to build up the Democratic Party in Philadelphia through contacts with Progressive Democrats George Washington Norris, director of the city Department of Wharves, Docks, and Ferries, and George Howard Earle, III, a future governor of Pennsylvania. Roosevelt's 1913 visit laid the groundwork for a New Deal coalition in the 1930s and contributed more immediately to Secretary of the Navy Daniels's decision to establish the first shipbuilding

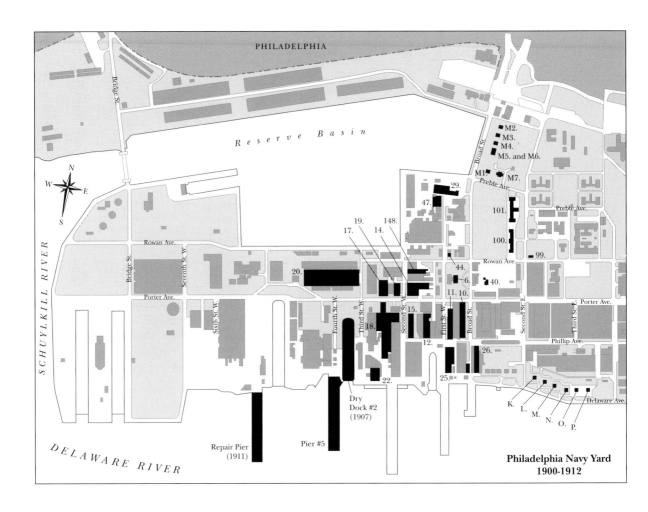

Philadelphia Navy Yard
1900-1912

Building 6.	Yard and Dock Commandant's Office (1901)	Building 99.	Tailor, Barber, Shoemaker's Shop (Marine Barracks) (1901)
Building 9.	Heating Plant for Building 29 (1901)	Building 100.	Marine Barracks (1901)
Building 10.	Workshop for Bureau of Equipment (1903)	Building 101.	Marine Barracks (1911)
Building 11.	Block, Cooper and Spar Shop (1903)	Building 148.	Lumber Storage Shed (1902)
Building 12.	Plate Bending Shop (1905)		
Building 14.	Angle Smithery (1903; 1942)		
Building 15.	Smithery (1903)		
Building 16.	Machine and Electrical Shop (1919)		
Building 17.	Foundry and Coppersmith's Shop (1905)	Quarters K.	Officer's Quarters (1900)
Building 18.	Boiler and Blacksmith Shops, Machine Shop (1908; 1940; 1942)	Quarters L.	Officer's Quarters (1900)
		Quarters M.	Officer's Quarters (1900)
Building 19.	Pattern Shop and Storehouse for Patterns (1905)	Quarters M1.	Marine Officer's Quarters (1912)
Building 20.	Foundry for Sandblast Equipment (1912)	Quarters M2.	Marine Officer's Quarters (1909)
Building 22.	Boiler and Engine House (1907)	Quarters M3.	Marine Officer's Quarters (1909)
Building 25.	Plumbers' Shop (1904)	Quarters M4.	Marine Officer's Quarters (1909)
Building 26.	Ordnance Building (1909)	Quarters M5 and M6.	Marine Officer's Quarters (1912)
Building 29.	Storehouse for Ship's Equipment (1901)		
Building 40.	House for Artesian Well Pumping Machinery (1906)	Quarters M7.	Marine Officer's Apartments (1912)
		Quarters N.	Officer's Quarters (1900)
Building 44.	Disinfecting Plant (1909)	Quarters O.	Officer's Quarters (1905)
Building 47.	Fuel Oil Testing Plant (1912)	Quarters P.	Officer's Quarters (1906)

1900–1916. Between 1900 and 1916 the Philadelphia Navy Yard developed a Marine Corps Reservation, Reserve Basin, and ship repair and overhaul facility, but no shipbuilding or large dry docking facility.

ways on League Island. FDR's attention and Navy Department support for navy yard improvement in Philadelphia encouraged Norris, Philadelphia Mayor Rudolph Blankenburg, and other local urban reformers to expect federal money to construct the long-sought 1,700-foot dry dock. Instead, Daniels favored construction of the dry dock at the Norfolk Navy Yard, designated in Navy General Board studies as the southern home base for the Atlantic Fleet. The General Board advised Daniels, as it had Secretary of the Navy Meyer earlier, that it considered Philadelphia only a "subsidiary base."[4]

Lack of federal interest in a new League Island dry dock agitated the "Greater Philadelphia Movement." Over four hundred Delaware Valley "dry dock boomers" went to Washington in December 1913 to present Philadelphia's case directly to Daniels. The contingent met with Daniels, Vice President Thomas R. Marshall, and Navy Aide for Material Albert G. Winterhalter, an advocate of Norfolk Navy Yard expansion, and left disappointed that the Wilson administration would not support a new dry dock for their Navy Yard. In political retaliation, Representative Robert E. Lee from Delaware County just below the Navy Yard attacked the proposed Norfolk dry dock site before the House Naval Affairs Committee in 1914, killing the scheme and exacerbating a struggle for naval resources between Virginia and Pennsylvania that would not end until final closure of the Philadelphia Navy Yard at the end of the century.[5]

At least the Philadelphia Navy Yard received an appropriation for its first shipbuilding ways two months before World War I exploded in Europe in August 1914. Daniels joined Port Director Norris and Navy Yard Commandant William S. Benson in groundbreaking ceremonies. He told the crowd of 2,000 Navy Yard workers and guests that construction of the facility was an act of peace and neutrality. Norris placed Shipbuilding Ways No. 1 in a different perspective: "The Navy Department is just beginning to realize, the value of the gift which Philadel-

phia made to it 50 years ago in the League Island yard." After the ceremony, Commandant Benson showed Daniels around the yard and hosted a luncheon where the navy secretary met Captain of the Yard Josiah S. McKean, one of the leaders of the *Panther* Association that had opposed Prussian-born Naval Constructor Albert G. Stahl from taking control of League Island Navy Yard management in 1909.[6]

Daniels appreciated Benson and McKean because they rejected the "German idea" of a Prussian general staff system for the U.S. Navy. Georgia-born Benson (one of three Southern officers to command the Philadelphia Navy Yard during the Wilson-Daniels era) seemed more democratic and cautious than aggressive navalist holdovers from the Republican era such as Aide for Operations Bradley Fiske. Benson worked well with Philadelphia Progressives Norris and Blankenburg to develop League Island Park and new working class neighborhoods near the Navy Yard. Daniels liked the idea that Benson, a Roman Catholic, had introduced Protestant and Catholic chapel services at the Navy Yard with a minimum of fanfare, assuring tranquility among the increasing number of South Philadelphia Italian-, Polish-, and Russian-Americans working at the government establishment on League Island. Eventually, Daniels promoted Captain Benson in 1915 over more senior officers as the first chief of naval operations and also brought Captain of the Yard McKean to Washington as one of his closest advisors.[7]

Navy Yard labor and wharf building specialists from the Brooklyn Navy Yard in New York completed Shipbuilding Ways No. 1 on League Island in June 1915, and naval constructors laid the keel for *Transport* No. 1 (named *Henderson* after former Marine Corps Commandant Archibald Henderson). It was appropriate that Philadelphia should build *Henderson*. Over the past decade, League Island had converted and outfitted *Prairie*, *Panther*, and *Dixie* to carry Marine Corps expeditionary troops from the

Ship Building Ways No. 1, under construction in 1914–15. Inset: Philadelphia Port Director George W. Norris (left) with shovel, Commandant Captain William S. Benson, and Secretary of the Navy Josephus Daniels (right) with pick, at dedication, 1914. Naval Historical Center Photo NH 75668.

Advance Base Force in Philadelphia to duty in the Caribbean. By giving the *Henderson* contract to Philadelphia, Daniels continued the former Republican administration policy of expanding the Marine Corps Reservation on League Island. He authorized construction of a third Marine Corps barracks and a storehouse to secure the Advance Base Force outfit that remained in Philadelphia until removal in 1917 to the more spacious Marine Corps Reservation at Quantico, Virginia.[8]

The Philadelphia Navy Yard launched *Henderson* in June 1916, shortly after the Democratic Party renominated Woodrow Wilson as its presidential candidate on a promise to keep the United States out of the war in Europe. Nevertheless, Wilson's policy of neutrality had become increasingly difficult as German submarines and British cruisers intercepted American ships. German U-boats sank the Delaware River-built oil tanker *Gulflight*, the British Cunard liner *Lusitania*, and the French passenger liner *Sussex*, leading to heavy loss of American lives. After the *Sussex* sinking, U.S. Secretary of State Robert Lansing wanted to break diplomatic relations with Germany, but Wilson called instead for a German pledge not to sink ships without warning. The president refused to go beyond the *Sussex* pledge and resisted pressure from former President Theodore Roosevelt, Army General Leonard Wood, and a growing number of preparedness advocates to expand the U.S. naval and military establishments.[9]

Reflecting Wilsonian reluctance to advance preparedness, the Navy in 1916 cut funding for overhaul

Launch of troopship *Henderson*, June 1916, the first ship built on League Island. NARAM.

work on League Island. Consequently "a large re-duction in the working forces at the large East Coast Navy Yard will be unavoidable," Homer Stanford, now chief of the Bureau of Yards and Docks, told Pennsylvania Representative William S. Vare. There was "practically no repair work on hand [in Philadelphia] at present" and he had to lay off Vare's constituents who worked at the government facility on League Island. The Navy Department also refused to purchase heavy rubber boots, wooden-soled shoes, or safety goggles for the remaining la-borers, ship fitters, and painters. "Rest assured at Philadelphia we are a cheap-skate department, very much down at the heel," Public Works Officer Fred-eric R. Harris explained. The work force stood so "very much behind hand [that] our work for the other departments or Bureaus, such as Marine Corps, Magazine, Hospital, Naval Home, etc., is real-ly shamefully neglected." When Navy Yard Comman-dant Benson asked for more money, Assistant Secre-tary Roosevelt replied curtly that "No estimates are to be submitted for additional dry docks, radio sta-tions, coaling plants or fuel oil station."[10]

At least Benson escaped Philadelphia in May 1915 when he came to Washington as the first chief of naval operations. Captain John J. Knapp replaced Benson, but died suddenly a few weeks after taking command; that same week Captain of the Yard Claude B. Price fell ill and was admitted to the Naval Hospital. Meanwhile, Naval Constructor Elliott Snow left for temporary duty with the Bureau of Construc-tion and Repair and Assistant Naval Constructor Richard D. Gatewood vacationed at the New Jersey shore. Seemingly, no one commanded League Island. Chief Boatswain Charles Schonborg had to serve as captain of the yard to deal with a murder case when a supply department laborer shot and killed a fellow worker while on the job in the General Storehouse (Building No. 4).[11]

In late 1915, short on personnel and devoid of leadership, the Philadelphia Navy Yard confronted unusual and difficult tasks. As part of an emerging mission to cooperate with Latin American navies, the local naval shore establishment overhauled an Argentine transport and housed the crew of the Ar-gentine battleship *Moreno*, then under construction across the river at the New York Shipbuilding Corpo-ration in Camden, New Jersey. The "care and preservation gang of the Yard" shuttled the Argen-tinians from their quarters on board the old battle-ships *Indiana* and *Massachusetts* in the Reserve Basin to Camden each day. Twelve newer U.S. bat-tleships crowded League Island as well, including the mighty 20,000-ton, all-big-gun, turbine-powered dreadnought *North Dakota*, which with a 29½-foot draft settled into the mud of the 30-foot-deep Re-serve Basin with engine and boiler troubles. No one seemed to know what to do with this first U.S. fuel oil turbine-powered dreadnought since Congress re-fused to appropriate funds for repair. Nearby, the an-cient monitors *Terror* and *Miantonomoh* were rust-ing away, already stricken from the *Navy Register* and interfering with berthing space for new battle-ships. Such conditions contributed to the collision of battleships *Michigan* and *South Carolina* as they moved around the cluttered and disorganized Navy Yard to find berthing space. A naval board of inquiry found that the accident caused over $8,000 in dam-age to *South Carolina* and ordered a court martial for *Michigan*'s commanding officer.[12]

But political and legal complications accompany-ing belligerent violations of American neutrality dominated Navy Yard operations during the troubled winter of 1915–16. Neutrality officers and radio cen-sors added to the general confusion surrounding an ill-defined neutrality policy as they came and went to monitor river traffic and keep watch on the Ger-man-owned wireless station at Tuckerton on the seacoast above Atlantic City, New Jersey. In Decem-ber 1915 Secretary Daniels appointed Navy Judge Advocate General Robert Lee Russell to command the Navy Yard. Daniels considered Russell some-thing of a "dull" fellow and learned later that he drank too much. Daniels needed Russell's legal background and extensive social and political con-nections. Born in South Carolina and appointed to the Naval Academy from Georgia, Russell was the cousin of important Senator Richard B. Russell, a Georgia Democrat.[13]

Russell and Captain of the Yard John McClane

THE TROPHY.

Atlantic Fleet football champions, 1917. Football games between battleship crews on League Island kept up morale on the eve of American entry into World War I. NARAM.

Luby of Texas (another Southern officer to manage the Philadelphia Navy Yard during the Daniels era) prepared the government facility for the trials of neutrality. FDR helped. In the wake of the *Sussex* crisis, the assistant secretary had visited Atlantic Coast navy yards in early 1916, encouraging the organization of security and preparedness measures. He wanted Philadelphia to submit estimates for construction of an 8-floor storehouse, steel storage shed, steam generating plant, and new receiving station barracks to replace temporary buildings erected in 1903 during the "spotted fever" crisis. Meanwhile, the Wilson administration asked Congress for a naval expansion program that included dreadnoughts and 1,000-foot dry docks for navy yards at Norfolk *and* Philadelphia. To this end, Civil Engineer

Captain Robert L. Russell, c. 1912. Russell, a former JAG, was selected to command the Navy Yard during a difficult period of U.S. neutrality, 1915–17. Naval Historical Center Photo NH 46071.

Clinton D. Thurber arrived on League Island in June 1916 "in connection with the taking of borings for proposed dry dock."[14]

As Congress accepted naval expansion, the Philadelphia Navy Yard pursued preparedness activity. The yard developed additional docking and storage facilities at the Fort Mifflin Naval Magazine to arm torpedo boat destroyers for a neutrality patrol designed to watch for belligerent violation of American neutral rights and inspected merchant vessels for use as naval auxiliaries. Extra-large loads of coal and stores arrived at the Navy Yard so that battleships *Kansas* and *Connecticut* could outfit and raise steam, ready to sail on a moment's notice. Battleships *South Carolina*, *Ohio*, *Wisconsin*, and *Missouri* were moved from the Reserve Basin to riverfront piers for overhaul. Woodwork that would

pose a fire hazard in combat was stripped from *North Dakota* and replaced with metal furniture. The Navy Yard developed a submarine repair station at Pier D in the Reserve Basin and in July 1916 opened a submarine training station in Cape May harbor near the mouth of Delaware Bay. A flotilla of small craft, destroyer *Balch*, and submarine tender *Bushnell* brought naval officers and guests from the Navy Yard down to a Cape May yacht club used as headquarters to dedicate the training station.[15]

Additional work required the hiring of more labor, but the government facility could not compete with "the rate up town for very ordinary help,"— $3.00 a day compared to the $2.24 daily wage at the Navy Yard. "While you are in Washington, will it not be possible to have a word with Mr. Roosevelt and arrange so that we can make formal request for re-

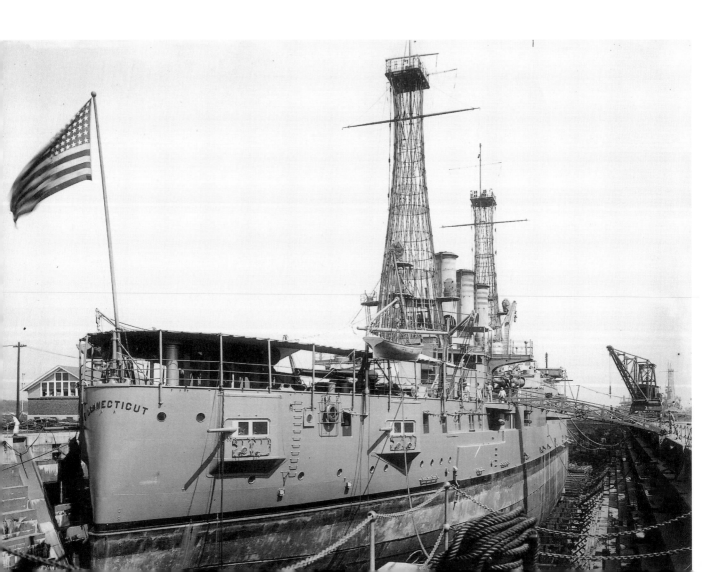

The 663-foot, 23,500-ton German raider *Kronprinz Wilhelm*, 26 March 1917. The former North German Lloyd Lines passenger ship was interned on League Island from October 1916 until American entry into World War I; in June 1917 it was seized and converted into troopship *Von Steuben*, serving until October 1919. Naval Historical Center Photo NH 42420.

vision in wages?" Naval Constructor Elliott Snow asked Civil Engineer Harris. At the same time, Commandant Russell wanted the statutory limitation of an eight-hour day for government work eased for skilled workers. "Telegraphic authority requested to work, not more than twenty machinists and twenty helpers four hours overtime daily until work on *Maine* is completed." FDR resisted such overtime work, and the Navy Department would not issue a revised wage schedule until March 1917. However, the assistant secretary encouraged introduction of night shifts for emergency work, which included replacement of *Michigan*'s eight 12-inch guns and conversion of transport *Dixie* into a destroyer tender and *Melville* into a submarine tender for tiny K-Class submarines assigned in late 1916 to League Island and Cape May harbor.[16]

Naval mobilization planning for Philadelphia matured with Captain Josiah McKean's proposal for wartime expansion of the Navy Yard. The McKean scheme shaped the government compound on League Island for the next eighty years. Drawing on earlier studies, McKean divided League Island into a navy base on the east side and a naval shipyard on

Fourth Division Atlantic Reserve Fleet battleship *Connecticut*, in dry dock on League Island in 1915, was one of the few ships available to patrol the Delaware Bay as German U-boats violated American neutrality and freedom of the seas in the Atlantic. NARAM.

the west side of Broad Street. McKean's design called for creation of a large industrial shipbuilding factory on the undeveloped westernmost portion that extended beyond Dry Dock No. 2 toward the Schuylkill River. The plan featured dry docks, shipbuilding ways, structural steel shops, machine shops, steel storage field, and foundry. The eastern half contained the naval base, storehouses, receiving station, training center, Marine Corps Reservation, and a flying field.[17]

Passage of legislation to greatly increase the Navy in 1916–17 activated the McKean blueprint for League Island that between August 1916 and January 1919 underwent a $25 million expansion. This was twice the $13.7 million spent between 1876 and 1916. Strangely, money for improvement at the Norfolk Navy Yard in the summer of 1916, made a more immediate impact on League Island than did that for Philadelphia. New construction at the Chesapeake Bay naval facility forced the Navy Department to move two large German auxiliary cruisers interned there since 1915 to Philadelphia, which had more berthing space and room for construction of an internment camp to house 700 German sailors and officers. The German auxiliary cruisers *Kronprinz Wilhelm* and *Prinz Eitel Friedrich* escorted by U.S. battleships *Minnesota* and *Vermont* arrived off League Island in early October. A former North German Lloyd liner, the 663-foot, 23,500-ton *Kronprinz* was by far the biggest vessel to enter the Navy Yard so far. The auxiliary cruiser had left New York harbor at the outbreak of war in August 1914, loaded guns and stores, took on additional naval officers off Bermuda, and attacked and sank fourteen Allied merchant ships in the South Atlantic. The 506-foot, 14,180-ton *Prinz Eitel Friedrich*, a former Hamburg-American steamship, also destroyed shipping, including the American-flag merchantman *William P. Frye*.[18]

Philadelphia's large German-American and Irish-American populations cheered the arrival of the Teutonic raiders. But adding hundreds of potentially subversive belligerents, suspected already of sabotage at the Du Pont Powder Works in Wilmington, Delaware, and at area arsenals, railroad bridges, and industrial plants along the river, was either an incredibly stupid or an intentionally provocative act by the neutral and anti-war Wilson administration that had won reelection in November by the narrowest of margins. The Philadelphia Navy Yard, where the raiders berthed in the Reserve Basin, held all kinds of sensitive military targets vulnerable to sabotage. These included an experimental Fuel Oil Testing Laboratory, a new wireless station, the Advance Base training school, and the first excavations of the world's largest naval dry dock. The auxiliary cruisers berthed fifty feet ahead of *Salem*, a scout cruiser used for advanced radio, turbine, and anti-aircraft gun experiments. The hostile German warships lay near *Jacob Jones*, prototype for an innovative new class of fleet destroyers. Moreover, there were twenty-nine other U.S. warships and 3,500 personnel concentrated at the vulnerable island facility.[19]

Secretary of the Navy Daniels admitted that German internees "could have blown up ships and Navy Yard," and rumors circulated that saboteurs had planted explosive mines in the Back Channel, tampered with Dry Dock No. 2, and attempted to set fire to Navy Yard buildings. For the most part, the Germans appeared satisfied to establish contact with the sizable German-American communities in the Delaware Valley. Wealthy Philadelphia German-American import merchant Henry Rohner and city machine company owner Adelbert Fischer were arrested for smuggling chronometers from the German

cruisers to sell in Philadelphia and for allegedly delivering explosives to the warships. "For months the Germans have had the run of the yard and the city of Philadelphia," insisted preparedness crusader and bitter critic of the Wilson administration Henry Reuterdahl. "They lie there, a military menace, and hidden spy system with wire less communication right in the heart of one of our most important navy yards."[20]

Despite Reuterdahl's badgering, the Wilson government refused to move the Germans, maintaining that such an act violated neutrality laws. Former Judge Advocate General Russell allowed the Germans to enter the city, gave outside visitors nearly unlimited access to the internees in the Navy Yard, and permitted crews to roam about the western banks of the Reserve Basin tending their chickens and gardens. But the German announcement on 31 January 1917 that it had reinstituted unrestricted submarine warfare, allowing U-boats to sink all targets without warning (along with the British revelation of the Zimmermann Telegram in March identifying German support for a Mexican attack on the United States in the event of American entry into the war) changed the situation for the German internees on League Island. Russell canceled shore leave, restricted internees to the island, and after the United States severed diplomatic relations with Germany on 3 February 1917, ordered *North Dakota* to train guns on the interned cruisers.[21]

Security precautions intensified. The Navy Yard installed search lights from the battleships along the Back Channel and constructed sentry boxes on League Island Park north of the Reserve Basin. A Marine Corps "armored auto" prowled docks and streets, and armed guards drilled with machine guns at street crossings. Barbed wire fences secured the main power house and guards searched horse-drawn coal and provision wagons entering the main gate from the city. "To prevent spies dressing in women's clothing, or even to guard against female spies, all the women living in the officers' quarters at the yard are given an escort on entering the gates."[22]

Russell increased security further in early 1917,

dispatching the destroyer *Jouett* across the river to guard the battleship *Idaho*, nearing completion at New York Shipbuilding Corporation in Camden. A detachment of Marines patrolled the private ship-yard at night; another left the Navy Yard for Tuckerton to take over the German wireless station and its 800-foot-high transmission tower, which was capable of sending messages directly to Germany. After the Wilson administration decided in February to arm merchantmen, the Navy Yard collected ordnance supplies, and gun mounts and installed tracks and fittings on destroyers for mine laying. The yard

hired additional workers for these tasks, bringing the civilian labor force from 2,600 in mid-1916 to 3,200 by February 1917. In March 1917 the Philadelphia naval shore establishment instituted two 8-hour shifts and a 10-hour day for some skilled workers, advertised job openings with the Labor Board and in local newspapers, and increased wages by 5 to 10 percent. The following month the facility hired its first female industrial employees.[23]

The continued presence of 700 German internees became an embarrassment as League Island prepared for war. The number increased in early February after the diplomatic break, as the German prize crew from the captured British merchant vessel *Appam*, arrived under guard from Norfolk. Mean-

Telephone exchange operations room, 1918. Telephone "girls" were hired just before American entry into World War I. NARAM.

while, the Wilson cabinet vigorously debated policy toward vessels and crews interned in Philadelphia. Secretary of the Treasury William Gibbs McAdoo wanted the United States to seize the large German liners for use by the American merchant marine and lock up the crews in a prison camp. However, Secretary of State Lansing argued that the United States had no legal authority to incarcerate Germans, while Secretary of the Navy Daniels recommended that the government leave the affair to the Navy Yard command. Such a course troubled local officials in Philadelphia. Republican Mayor Thomas B. Smith and Representative Vare, bitter foes of the Wilson administration, demanded that the president expel the Germans from League Island, and turned on Commandant Russell, accusing him of incompetence and drunkenness for allowing the Germans to wander about freely.[24]

Daniels was unhappy with Russell, too, not because of the liberal treatment of the German internees but for the "inexcusable" failure to meet overhaul schedules for battleships. "Under the circumstances the Department cannot but feel that there is something seriously wrong with the conditions at the yard under your command." Nevertheless, Daniels used Russell's Georgia connection to relocate the Philadelphia internees to that state's army camps. Captain of the Yard Luby reported on 26 March 1917 that "411 officers and men of the Interned German Auxiliary Cruiser *Kronprinz Wilhelm* were this day transferred via Penna. Railroad to Internment Camp at Ft. McPherson, Ga., 383 officers and men of the Interned German Auxiliary Cruiser *Prinz Eitel Friedrich*, Steamship *Appam*, and *S.M.S. Geier*, were this day transferred via Penna Railroad to Ft. Oglethorpe, Ga."[25]

With the German internees headed for Georgia, the Wilson administration discussed the disposal of the two large German warships left behind. Once again, McAdoo insisted that the United States seize them for its own merchant fleet, along with three other belligerent steamers interned on the Philadelphia waterfront at the foot of Washington Street. Daniels agreed that the federal government should consider "the use of German warships at Phila," and

after the United States declaration of war on Germany on 6 April he ordered seizure of the interned auxiliary cruisers for conversion into troopships.[26]

Converting the German ships demonstrated that League Island still suffered limitations as a first-class naval establishment. The Navy Yard had never docked a ship as large as *Kronprinz Wilhelm*. Moving the huge German liner from the Reserve Basin to the waterfront for docking caused major problems. Navy Yard tugs *Modoc* and *Samoset* could not free *Kronprinz* when it "grounded on the south side of the entrance to the Reserve Basin in the Schuylkill River." The Navy Yard hired nine tugs from private Philadelphia firms to pull the auxiliary cruiser to the waterfront. Even then, it took days to line up keel blocks and dock the great vessels. The saga continued after docking, as President Wilson, his wife Edith, and Daniels argued over what to call the vessels, settling finally on *Von Steuben* and *De Kalb*, in honor of German fighters for American independence Baron Friedrich von Steuben and Baron Johann DeKalb, who had fought with Washington during the Revolution. *Von Steuben* and *De Kalb* left the Navy Yard in the late summer of 1917 to begin nearly a dozen trips, carrying thousands of troops to the European war zone.[27]

The most urgent need in April 1917 was for antisubmarine patrol and escort vessels. Commander of U.S. Naval Forces Operating in European Waters Admiral William S. Sims demanded antisubmarine craft. "Sims says destroyers are needed more than everything else," Daniels worried. Navy yards up and down the Atlantic Coast overhauled tiny flotillas of older destroyers and orders went out to private shipyards from the Bath Iron Works in Maine to yards on the Gulf of Mexico to rush new destroyer construction. The New York Shipbuilding Corporation alone contracted to build thirty destroyers. The Philadelphia Navy Yard recalled the Atlantic Fleet Reserve Flotilla destroyers *Jouett*, *Beale*, and *Duncan* for installation of mine laying apparatus. Yard tugs towed seven inactive destroyers from the Re-

The *De Kalb* prepares to carry U.S. Marines to France, 12 June 1917. Naval Historical Center Photo NH 54652.

Former German raider *Prinz Eitel Friedrich* being fitted as the troopship *De Kalb*. The search for explosives on the German raiders interned on League Island revealed, instead, the empty beer barrels seen here piled on the pier. Naval Historical Center Photo NH 54657.

serve Basin for overhaul and dispatch on convoy duty by the civilian work force which had reached 4,450 by 1 May 1917. Engineers found the older destroyer *Mayrant* in such poor condition, however, that workers cannibalized torpedo air compressors and other parts to get the equally aged *Ammen* to sea. Meanwhile, the Philadelphia industrial plant overhauled destroyer tender *Dixie* to join Admiral Sims's force in Queenstown, Ireland.[28]

To protect the Delaware after these destroyers left for distant duty, the Navy Yard reactivated the 25-year-old cruiser *Columbia* and five pre-dreadnought battleships for coastal patrol service. But Sims requested cruisers for convoy duty, and the Navy Department dispatched *Columbia* to Queenstown. Cruiser *Chicago*, one of the original steel ships built during the 1880s at Roach's shipyard in Chester, replaced *Columbia* on Delaware Bay patrol, only to be

taken away, as well, for employment as a submarine tender. The battleships left the Delaware for the lower Chesapeake Bay to train crews in gunnery and engineering duty, leaving the approaches to "our main strategical area" completely unguarded.[29]

The Philadelphia Navy Yard command that soon included a reorganized Fourth Naval District Office in charge of local defense searched for replacement vessels. Coast Guard cutter *James Guthrie*, built in 1882, and decrepit lighthouse tender *Woodbine*, armed with a battery of old Naval Militia three-pounders, patrolled the river and bay. When they proved incapable of covering the area, the Navy Yard collected private motor boats, yachts, and fishing trawlers. The idea of a Naval Coastal Defense Reserve Force of private vessels arose in part from Assistant Secretary of the Navy Roosevelt's collaboration with preparedness advocates such as Leonard

Wood, one of the founders of the Plattsburg Military Training Camp Association in New York. "We want the Naval Plattsburg" as well, FDR told the Harvard Club of New York and expected Boston, New York City, and Philadelphia to become centers for such a force. Wood mobilized Philadelphia area leaders at the Union League to organize the naval and military resources of the Delaware Valley. As a result, Philadelphia's social elite connected with New York society circles and formed a Naval Coast Defense Reserve Force along the lines suggested by Wood and FDR.[30]

Edward Townsend Stotesbury, president of J. P. Morgan and Company Bank of New York and Drexel and Company Bank of Philadelphia, and Reading Railroad Company executive John Price Wetherill, Jr., raised $50 million to build private motorboats for the government to use as submarine chasers. Both held a special interest in the local naval establishment. Wetherill's father had orchestrated the original purchase of League Island for the navy in 1862 while a member of Philadelphia City Council, and Stotesbury, the leader of the Greater League Island Navy Yard Citizens' Committee, had worked closely with Norris and Blankenburg to develop League Island Park and the South Broad Street neighborhood approaches to the Navy Yard. Stotesbury's wife Lucretia Cromwell Stotesbury enrolled Philadelphia society women in the naval preparedness movement and from her headquarters at the Merion Cricket Club carried on extensive correspondence with Secretary Daniels about the immoral influence that Philadelphia's "tenderloin district" had on the Navy Yard. Lucretia Stotesbury served as director of Navy Relief during World War I, once narrowly escaping death at the Navy Yard during a loan rally when the scaffold on which she was speaking collapsed, and she influenced the decision in March 1917 to enlist women for the first time in the U.S. Navy. One of the young ladies, Loretta P. Walsh, took the oath as yeoman chief petty officer on the steps of the Naval Home on 21 March 1917. Walsh recruited thirty-seven female yeomen to serve as telephone operators and clerks on League Island.[31]

President Wilson announced on 2 April that he planned to ask Congress for a declaration of war against Germany because of continued violations to American neutrality and freedom of the seas from German U-boats. Four days later the Navy Yard in Philadelphia received a wireless transmission that "The President has signed Act of Congress which declared that a state of war exists between United States and Germany." The radio messages mobilized Stotesbury's Coastal Defense Reserve Force. Fast motor boats and yachts arrived at the Navy Yard for

This untitled photo of a "woman in the U.S. Naval Reserve, World War I" is most likely of Loretta Walsh, the first Yeoman (F) to stand watch on League Island, 1917. NARAM.

Yeomen (F), Philadelphia, 1918. Naval Historical Center Photo NH 53176.

conversion into naval patrol craft. Wetherill brought his motorboat *Little Aie* to League Island for outfitting with machine guns, search lights, and wireless radio set and mast, and headed down river to Lewes, Delaware, to guard anti-submarine nets. Stotesbury's stepson James H. R. Cromwell commanded Stotesbury's powerful motorboat *Nedeva II*. Commissioned as Scout Patrol Boat (S.P. 64), *Nedeva* joined *Shrewsbury* (S.P. 70) and *Absegami* (S.P. 371) on patrol from the Delaware Breakwater to the Corinthian Yacht Club in Essington, Pennsylvania, and back to League Island.[32]

Not every patriotic yachtsman was happy with transfer of his vessel to the U.S. government. Delaware munitions magnate Alfred I. Du Pont complained that the Navy cheated him on his yacht *Alicia*, claiming that the offer of $55,000 for his boat was "manifestly absurd." He blamed the Navy Department bureaucracy: "I confess that I am somewhat at a loss to understand the lack of organization in the Navy Department relative to the taking over of private yachts." On the other hand, the government probably paid too much to the McKeever Brothers and the Coast Fish Oil and Guano Company of Lewes, Delaware, for fishing trawlers used for deep sea patrol duty. Machinery was in poor shape, deck houses had to be removed to mount guns, and considerable ballast had to be added to allow the boats to sail in rough seas. "Got 45 tons of ballast down in the bilges," sighed Commandant's Aid Nelson H. Goss. "God knows that ought to keep them straight."[33]

Frantic mobilization of naval resources in early 1917 led to one of the worst dry dock accidents in Philadelphia Navy Yard history. The tragedy occurred in part because the government facility tried simultaneously to dock three converted mine sweepers, two destroyers, a revenue cutter, a motor patrol boat, and several barges. "At about 5:40 p.m. the *Duncan* and *Allen* fell over to port carrying with them the *Guthrie*," Naval Constructor Elliott Snow discovered. Captain of the Yard Luby reported that an "accident occurred in Dry Dock No. 2 about 5:40 p.m. in which the Destroyers *Duncan* and *Allen* and the Coast Guard Cutter *Guthrie* fell from the keel

blocks after docking, resulting in death of O. C. Hadlock, late electrician, 3rd class U.S.N. of the *Allen*, and injury to seven others." *Guthrie* suffered extensive damage to hull, wireless mast, boat davits, deck house, and rudder, and Snow left the old cutter submerged on her side at the bottom of the dry dock when he refloated the destroyers.[34]

Navy Yard Commandant Benjamin Tappan ordered a full investigation of the *Guthrie* affair. A native of Louisiana and a Spanish-American War hero, Tappan had been recalled from the retired list in April 1917 to replace the discredited Russell. Daniels wanted Tappan to provide moral leadership at a Navy Yard that Lucretia Stotesbury had told him was a sinkhole of immorality and House Naval Affairs Committee Chairman Lemuel P. Padgett of Tennessee had called a place for disreputable street flirts. To free the sixty-one-year-old Tappan from larger regional defense responsibilities so that he could concentrate on the moral and spiritual health of the Navy Yard, Daniels resurrected the local naval district organization, created in 1903 but long dormant. On 27 April, he assigned Captain George F. Cooper as commandant of the Fourth Naval District to coordinate naval business in the Delaware and Chesapeake regions, organize and deploy naval coastal defense reserve forces, and manage the Naval Overseas Transport Service (NOTS).[35]

Tappan provided moral guidance. The task proved difficult. He was greeted in front of the city police station at Fourth Street and Snyder Avenue by rioting sailors trying to free one of their crew mates arrested the night before. Philadelphia church leaders, Mrs. Stotesbury's female yeomen (F), and the Mother's League Against Vice crusaded to clean up the South Philadelphia tenderloin district and provide better housing, recreational facilities, and religious services at the Navy Yard. Meanwhile, Tappan conferred with city police, district attorney, Justice Department officials, and naval staff for the "suppression of vice." He increased Marine patrols in the city and ordered personnel to wear uniforms when off base. Finally, Tappan placed responsibility for controlling the situation on the commanding officer of the Receiving Station, forcing Daniels to order re-

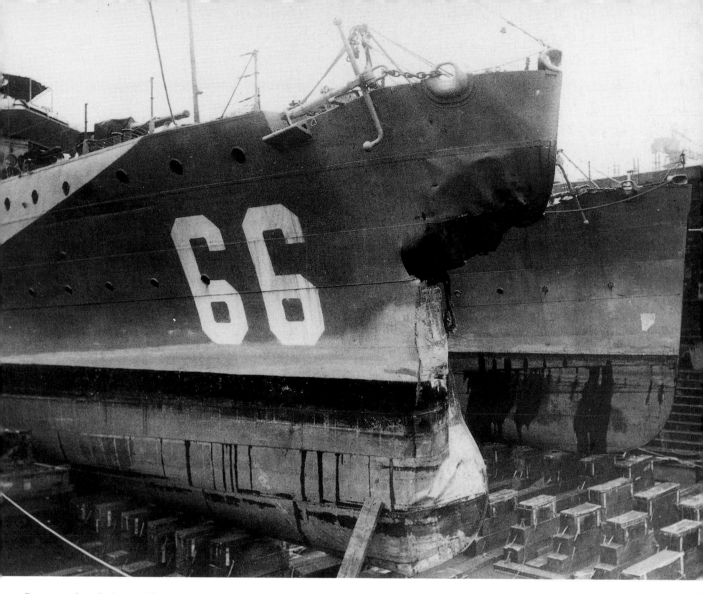

Sampson-class destroyer *Allen* in Dry Dock No. 2 after collision with the *Duncan*, 4 May 1917, the worst dry dock accident in League Island history. National Archives Mid Atlantic Region, Philadelphia, Records of Naval Districts and Shore Establishments, Philadelphia Naval Shipyard, Public Affairs Office, Events, Box 59.

tired Lieutenant Colonel Charles B. Hatch, head of the Marine Corps recruiting office in Philadelphia, to investigate vice and clean up South Philadelphia.[36]

Wartime conditions continued to erode the commandant's control over the naval shore establishment in Philadelphia. Development of a Naval Aircraft Factory on League Island in 1917 further diluted Tappan's authority as Naval Constructor Frederick G. Coburn set up a separate command structure for the new naval aviation facility. League Island had

attracted attention as a flying field for years. The flat eastern part of the island with its wide expanse of riverfront provided ideal space for testing seaplanes, and Marine Corps aviator Alfred A. Cunningham and naval aviator (and acting captain, 1909–11) Henry Croskey Mustin had experimented extensively with gliders and airplanes at least since 1910, when the Aero Club of Pennsylvania built a hangar and flying field there. But it took a large naval aviation appropriation on the eve of American entry into the world war and a visit to the place by Chief of the Bureau of Construction and Repair David W. Taylor to convince the Navy Department to develop a 47-acre aviation facility on League Island. "Decided to build aeroplane factory in Phila.," Daniels noted on 3 August 1917.[37]

Work started during the late summer of 1917 on construction of a hangar, office building, storehouse, and engine and erecting shop. The Naval Aircraft Factory accepted its first construction contract in November for fifty Curtiss H-16 Flying Boats, and by the end of the year employed 700 workers, including a large force of women. Though some officers opposed hiring women, by mid-1918, 890 of the 3,640 employees (nearly 25 percent) were female compared to approximately 5 percent in the rest of the Navy Yard. As the Naval Aircraft Factory grew, it crowded beyond 6th Street East and into the rapidly expanding naval training and receiving station of the navy base. "The Commanding Officer of the Aviation Camp," Tappan reminded Public Works Officer De Witt C. Webb, "should be given an opportunity to express his views before commencing any building on doubtful territory." The commandant gave the go—ahead to construct "Emergency Mobilization Barracks" for 5,000 men only after Marine Corps Aviation Officer Cunningham endorsed the layout.[38]

Construction of a Receiving Station and Training Camp started in May 1917 on an "actual cost plus percentage" contract that assured private contractors a 10 percent profit. The facility included twenty-six units, each to accommodate 192 recruits, two barracks, mess hall, and latrine. Seven small buildings "to contain potato-peeling and dish-washing equipment" were attached to the emergency barracks complex, as were a 200-bed hospital and dispensary with a "venereal treatment room." The camp expanded throughout the war and eventually processed over 7,500 men at a time. To the west of the Receiving Station toward Broad Street, the Public Works Department and outside contractors constructed an eight-story storehouse (no. 83), a seven-story general storehouse (no. 5), an ordnance storage facility for fitting out requisitioned vessels and ships for the Emergency Fleet Corporation at Hog

Island, and three single-story warehouses. Heavily insulated with "Asbestos Protective" coverings, these storehouses provided by the end of the war more than a million square feet of fireproof storage space.[39]

Rapid growth of the naval base on the east side of Broad Street placed tremendous demand on electrical and heating plants. The winter of 1917–18 was particularly frigid, and the heating plant in Building No. 22 could not warm base and shipyard at the same time, severely curtailing work. The Philadelphia Electric Company extended a high tension line to provide temporary power for shops, and the Public Works Office in Philadelphia designed a new heating plant, but Bureau of Yards and Docks engineers thought the plan inadequate. "The logical development would therefore seem to be to put in for Philadelphia, the same plant which we plan to put in for Norfolk, i.e., a pit into which coal cars can dump, a skip hoist to raise the coal to elevated tower in which are located hoppers and coal crushing apparatus, with delivery direct to the elevated track running over the bunkers." Though work began on this new design in late 1917, constant revision to plans and increased labor and material costs prevented completion until after the war.[40]

The Navy Yard dredged, bulkheaded, and built piers for the Reserve Basin during the war, extending berthing space for 125 ships and fifteen boats. A repair base for submarines was developed around Pier D with a marine railway to pull boats out of the water for emergency hull work. The Submarine Base stored and charged batteries, maintained a parts inventory to service fifteen submarines, and quartered boat crews. On the northern side of the Reserve

Facing above Expanded wartime work during 1917–18, such as in this electrical shop, overtaxed League Island's electrical and heating plants and forced the navy to run an emergency power cable from the City of Philadelphia. NARAM.

Facing below Submarines *D-2* and *D-3* (*L-1*, *L-4*, and *L-10* in the background) at the submarine base in the Reserve Basin, 5. March 1919. Naval Historical Center Photo NH 51157.

Basin, private contractors constructed four frame storehouses on government land and two on property rented from the city. Pennsylvania Railroad tracks connected directly to these storehouses, which held 800,000 square feet of naval stores.[41]

Major expansion at the Philadelphia Navy Yard during World War I occurred at "The New Yard," west of Dry Dock No. 2 and extending as far as Pier No. 6 at the foot of 6th Street West. Work had begun before U.S. entry into the war with erection of Shipbuilding Slip No. 1 at the foot of 4th Street West, while appropriations in March 1917 for capital ship construction led to contracts for Shipbuilding Ways Nos. 2 and 3 and 1,000-foot Dry Dock No. 3. Excavation problems and rising wartime labor and material costs increased the price of the dry dock from $3.5 to $6.3 million, delaying completion, and it was only half finished when the war ended. Construction continued in 1919 on the dry dock, structural iron shop, smithery, galvanizing plant, oxyacetylene plant, pattern shop, and heavy machine shop. All opened for business after the war. Shops and foundry featured the latest principles of industrial architecture, with long, single-floor buildings using overhead traveling gantry cranes to rush material along an assembly line to the next manufacturing station. The most visible addition at the shipyard (besides the two 300-foot steel radio towers) was the 350-ton hammerhead crane assembled in 1919 on a rebuilt 1,000-foot outfitting pier adjacent to Dry Dock No. 2. The crane could lift a 14-inch gun turret straight out of a battleship.[42]

Despite extensive improvements during World War I, the Philadelphia Navy Yard laid keels for just three ships and four boats and launched only two before the Armistice of 1918. By comparison, the overcrowded Cramp shipbuilding yard upriver constructed fifty-five warships during the war, and Hog Island Shipyard which opened just below the Navy Yard in 1917 manufactured more than 100 cargo and transport vessels. In the process, Hog Island hired 30,000 workers, New York Ship 25,000, and the newly organized Sun Shipbuilding Company of Chester, Pennsylvania, as many as 25,000; by contrast the Navy Yard never employed more than 12,000 shipyard workers at any one time during the war. Peak employment of 12,400 was reached in 1919 after the war had ended. The lack of new ship construction arose in part from League Island's primary mission to repair damage, overhaul, or outfit with new radio and electrical equipment rather than build ships for the fleet. But even here the Navy Yard displayed serious shortcomings. After a series of failures to meet battleship overhaul schedules "in view of the other urgent work in connection with destroyers, auxiliaries and naval overseas transportation service boats," Chief of Naval Operations Benson expressed his frustration with the poor performance by his former navy yard command. "My soul!" Benson cried, "What is the cause of all that long delay?"[43]

Despite deficiencies, expansion during the war and the immediate postwar period made the Philadelphia Navy Yard a first-class naval industrial shore establishment. Naval Constructor Richard D. Gatewood bragged that it had become "Uncle Sam's Greatest Navy Yard." Its facilities now compared favorably to those in Boston, New York, and Norfolk. League Island contained the most modern industrial plant available in 1919 for building large ships, including a 350-ton hammerhead crane. Indeed, in 1920 the Philadelphia Navy Yard laid the keels for two heavy battle cruisers. The Navy Yard boasted the only government-owned naval aircraft factory, main propeller manufactory, and fuel oil testing lab. It housed a submarine base. The Receiving Station could process 10,000 recruits and the Reserve Basin could berth 140 naval vessels at one time. But as the local government facility emerged as a major component of the U.S. Navy, postwar politics of naval arms limitation threatened to reverse all the progress.[44]

Facing above Foundry building, 27 December 1918. This was the centerpiece of the "New Yard" developed in 1918–19 on the western end of League Island. NARAM.

Facing below Erected in 1919, this 350-ton hammerhead crane on Pier 8 (seen here in 1932 with destroyer tender *Dobbin*, 3 April 1932) was a Delaware Valley landmark until torn down after closure of the Navy Yard in 1996. Naval Historical Center Photo NH 65012; courtesy of Franklin Moran.

9 ☆ BETWEEN WORLD WARS

HE ARMISTICE of 11 November 1918 brought the 11,000 workers on League Island to a virtual standstill. The night shift failed to complete assigned jobs, leaving them to the incoming day shift who also refused to do the work. At the same time, Philadelphia Mayor Thomas B. Smith, a Republican Party critic of the Democratic Wilson administration, demanded that Secretary of the Navy Josephus Daniels release all municipal employees at the Navy Yard to return and "save the city." Even Commandant Charles Francis Hughes, who relieved Benjamin Tappan in late 1918, wanted personnel cutbacks, explaining that the "Commandant fails to see how the large number of work men now enrolled in the yard are to be employed to advantage."[1]

Daniels ignored such advice and during the next few months actually increased the work force on League Island until it peaked in March 1919 at 12,400. There was still much urgent work. The Philadelphia Navy Yard outfitted older battleships to carry troops from the European war zone and overhauled destroyers and cruisers to escort these troopships home. Since the United States failed to ratify the Versailles Treaty ending the war or to join the international League of Nations, Daniels and senior advisors on the Navy General Board warned that the country had to continue building a "Navy second to none," and develop separate fleets for Atlantic and Pacific Oceans. Daniels asked Congress in 1919 for an additional $600 million to build 100 large battleships, battle cruisers, smaller warships, and auxiliaries. At first a tired President Wilson (soon to

be incapacitated by a massive stroke) supported the program, hoping to use the threat of a naval race to force Britain and Japan to discuss naval cutbacks and parity.[2]

In light of these post-Armistice demands, the Navy Department expressed displeasure with the Philadelphia Navy Yard's "most serious falling off in work and the inability of a large and important Navy Yard to meet the requirements of upkeep of the vessels which are assigned to it." League Island appeared in disarray. The captain of the yard failed to schedule routine overhaul for the old battleship *Maine* because battleships *New Hampshire* and *Louisiana*, transport *Camden*, and tug *Kanawha* used up all dry dock space. Emergency repairs took too long on *Minnesota*, which had suffered serious damage in late 1918 when it struck a mine off Fenwick Island Shoal Lighthouse, Delaware. An exasperated Daniels reproached Hughes. "It is expected that the Commandant will take steps to inspire all Navy Yard organizations with the desire and determination to expedite work on vessels at the yard to effect an early completion and return them to service at the earliest possible date."[3]

Prodded by Daniels and bolstered by more than 500 new employees, the Philadelphia Navy Yard returned in early 1919 to near peak wartime activity. During the immediate postwar period, League Island launched more ships than it had during the entire world war. Between April and September 1919, minesweepers *Sandpiper*, *Vireo*, *Warbler*, and *Willet* entered the Delaware River from tiny ship ways erected between Dry Dock No. 2 and Shipbuilding

Way No. 1 at the foot of 4th Street West. On 23 December, the Navy Yard launched its largest ship to date, the 7,000-ton hospital ship *Relief*. Retiring Chief of the Bureau of Medicine Rear Admiral William C. Braisted, Chief of the Bureau of Construction and Repair Rear Admiral David W. Taylor, and Philadelphia Mayor W. Freeland Kendrick attended the launch ceremony. Within moments, shipyard workers reset keel blocks for a 12,000-ton destroyer tender named *Dobbin*, after Secretary of the Navy Daniels's favorite predecessor James C. Dobbin, like Daniels a North Carolinian and "great Democrat."[4]

The Philadelphia Navy Yard received $3.1 million in 1919, more than any other navy yard, from the Congressional appropriation for "Improving and Equipping Navy Yards for Construction of Ships." League Island employed this postwar money to finish Dry Dock No. 3, Shipbuilding Ways No. 2 and No. 3, and an industrial manufacturing plant with heavy machine shop, foundry, structural shop, galvanizing plant, storehouses, and shops. Meanwhile, the postwar shipyard commissioned dozens of new warships completed after the war at private shipyards on the Delaware River. Captain of the Yard Louis Anthony Kaiser took 500 officers and crewmen across the river to the New York Shipbuilding Corporation in Camden, New Jersey, to commission and man the 32,000-ton battleship *Idaho*, and then rushed downriver to Chester to accept the new minesweeper *Quail*. At the same time, Cramp Shipbuilding Company of Kensington, Philadelphia, and New York Shipbuilding Corporation were delivering to the Navy Yard for commissioning the latest 1,200-ton flush-deck *Clemson*-class destroyers completed and launched after the war.[5]

View of the West (or New) Yard from the radio tower, November 1919. The 7,000-ton hospital ship *Relief* can be seen on the stocks; it was launched on 23 December. The larger shipways in the background would in 1920 hold the keels for giant battle cruisers No. 5 and No. 6, later scrapped. The 1,000-foot Dry Dock No. 3, seen under construction between the larger shipways and *Relief*, opened in 1921. NARAM.

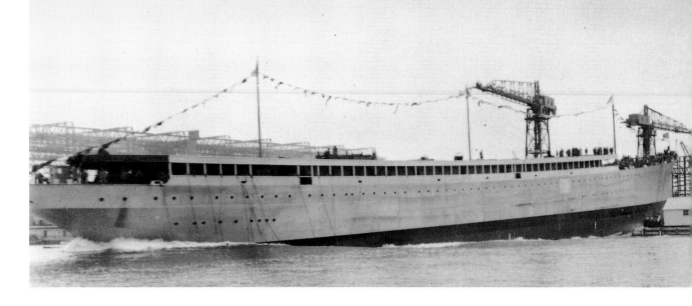

Destroyer tender *Dobbin* enters the Delaware River, 5 May 1921. League Island launched more ships between 1919 and 1921 than during all of World War I. Naval Historical Center Photo NH 54519.

New ship activity diverted attention from the rapid demobilization of forces on League Island during the immediate postwar period. Hundreds of warships of all sizes, from 20,000-ton battleships to 80-ton sub chasers, entered the Philadelphia Navy Yard between 1919 and 1921, until the Reserve Basin berthed 134 inactive ships. Ten 167-ton coastal torpedo boats fresh from patrol duty were the first to return. A board of inspection placed an average $8,000 value on each boat, but Hitner Salvage Corporation of Philadelphia and other local scrap yards purchased them for far less than appraised value.[6]

Next, League Island laid up more than a dozen obsolete battleships. "The *Ohio*, the *Connecticut*, the *New Hampshire*, all older vessels, loomed out of the water like cliffs of stone," observed Philadelphia journalist Christopher Morley, "their two and three high funnels out-topping the squat single stack of the new oil-burners." The Navy Yard decommissioned six on the same day in May 1920. Ship crews and members of the yard work force removed lifelines, davits, and gangways and cleaned ash pans and plumbing. They installed wooden covers on cowls, smokestacks, and hatches. The captain of the yard assigned permanent ship keepers to each vessel to maintain around-the-clock watch. This was a costly process. "When a ship goes out of commis-

sion, the expense of upkeep falls right back on the allotment of the yard," commandant's aide George W. Simpson explained. "The attention of all concerned is especially called to the heavy money loss which can easily result in turning in the large mass of material now on the six ships to go out of commission."[7]

The untidiness of the peacetime naval shore establishment on the Delaware River troubled 53-year-old Navy Yard Commandant Hughes more than did the cost. Though considered a "splendid officer" and destined to become chief of naval operations, Hughes let petty issues interfere with postwar demobilization. He "would go into many details which I felt could be taken care of by others," Captain of the Yard Yates Stirling recalled. Those details included laying up ancient coastal monitors, older cruisers, auxiliaries, seagoing tugs, and 110-foot sub chasers built by the American Car and Foundry Company of Wilmington and Mathis Yacht Building Company of Camden. It meant decommissioning seventy destroyers, ranging in size from prewar 700-ton *Paulding*-class to postwar 1,200-ton *Clemson*-class destroyers. The sheer number of warships laid up in a six-month period from late 1919 to early 1920 made movement into the Reserve Basin difficult for battleships, large transports, or the 9,450-ton hospital ship

Mercy, loaded with patients from the war zone who were being sent to hospitals in Philadelphia. The crowding led to the inevitable collisions and accidents. Forced to berth in the river, the 11,565-ton battleship *Illinois* dragged anchor, drifted across the river, and crashed into the Reading Railroad wharf. The 13,680-ton former armored cruiser *Maryland*, now designated as training ship *Frederick*, wandered into the mouth of the busy Schuylkill River, ran aground, and blocked the channel on the Chester side "about west of League Island Lighthouse."[8]

The arrival of so many warships in so short a time also caused another, more dangerous situation. "All ammunition should be turned in to the Naval Ammunition Depot, Ft. Mifflin, Pa., and all other ordnance material should be turned in to the Supply Officer, Navy Yard, Philadelphia," Chief of the Bureau of Ordnance Ralph Earle told Hughes. Almost daily, ammunition lighter and barge loads of explosives went downriver to Fort Mifflin for storage in the warehouses that had been built there during the world war. "Tug *Somerset* with ammunition lighter #26 loaded with ammunition from the U.S.S. *Strib-*

Destroyers in mothballs, 1923. About one hundred destroyers were stored in the Reserve Basin on League Island after World War I. Naval Historical Center Photo NH 69126.

ling, *Manley*, *Little*, *Sigourney*, *Gregory*, and *Stringham* left the yard for Fort Mifflin," Captain of the Yard Clarence Kempff observed. "Tug #57 towed ammunition lighter #155 loaded with ammunition from the *Illinois* to Fort Mifflin" and "Tug #82 with Ammunition Barge loaded with ammunition from *Kearsarge* left the yard for Fort Mifflin." Storing millions of pounds of explosives so close to a major urban area troubled the superintendent of the Naval Ammunition Depot, who ordered an "unclimbable fence" erected around the facility. The Navy continued to pile explosives at Fort Mifflin until a local investigative reporter discovered in 1929 that Philadelphia sat on a "timebomb."[9]

No one revealed another threat to the Delaware Valley. Hazardous chemicals and toxic materials accumulated during the expanded wartime naval business on the river lay about League Island and nearby riverfront industrial plants. The Navy Department ordered disposal of all such "contaminants," including adamsite, benzol, coronone, and arsenic. For a moment Naval Inspector of Powder Francis J. Comerford thought that some contaminants might be disposed of in local sewer systems, "but complaints have been received from the people who work and live near the outlet of the sewer." Consequently, Ordnance Chief Earle ordered the material sealed in 50-gallon drums and dumped off the New Jersey coast at the "20 fathom curve."[10]

Fearless, a Fourth Naval District section patrol boat, made at least two trips with hazardous and toxic materials down the river, across the bay, and into the Atlantic Ocean. Leakage to drums before reaching the designated deepwater dumping site made "it necessary to sink a drum inside the entrance of Delaware Bay." When *Fearless* arrived off the "20 fathom curve" and tossed the drums overboard, the sealed barrels failed to sink and the crew shot holes in the containers, releasing toxic waste into the air and water. "An incident occurred at sea," the boat's commanding officer reported, "when a small bird landed aboard and fell dead from the effects of the fumes." Apparently, such dumping continued throughout the interwar era. "U.S.S. *Alleghe-ny* left the yard towing Y.C. 256 loaded with con-

demned material for dumping at sea," Yard Captain Frank C. Martin reported in 1936.[11]

In 1919–20, threats lay closer to home, to the health of the postwar League Island population already gripped by a postwar influenza epidemic. Commandant Louis McCoy Nulton, who replaced Hughes in 1920, found rats overrunning the naval shore facility and breeding in refuse from decommissioned warships that spilled into the water or accumulated on piers. Nulton reported that during the first part of 1920 the Navy Yard trapped over 5,000 rats. He wanted more destroyed and issued a "Rat Extermination Order," calling for starvation, trapping, poisoning, rat proofing, and the "use of cats and Ferrets" to control the rat invasion. "Warships going out of commission should be fumigated in order to kill all rats left on board," Nulton commanded. "Each rat killed or trapped should be cremated at once [and] a report of the killing will be made to the Yard Master, through official channels, weekly."[12]

The process of cleaning up the Philadelphia Navy Yard focused on the disposal of sixteen outdated battleships that clogged the Reserve Basin and waterfront piers. Each dumped waste and ammunition on piers or filled already overloaded warehouses with removed equipment. The huge hulls made dredging the basin impossible, and silt built up to over six inches—more than eight inches at the entrance to the Back Channel. Battleship *Indiana* struck bottom at the Preble Avenue berth, even though the ship had a 24-foot draft and should have cleared the berthing space dredged to a 35-foot depth during the war. Meanwhile, submarines *K-2* and *K-8*, with only 13½-foot drafts, struck bottom while entering the Reserve Basin and had to be freed by tugboats.[13]

The incoming Republican administration of Warren G. Harding and Secretary of the Navy Edwin Denby promised to rid the Philadelphia Navy Yard of dirty, expensive battleships as part of a return to "normalcy." Harding promised economy in government, while an isolationist Republican Congress cut naval appropriations by $600 million. This postwar budgetcutting deeply affected Navy Yard work in Philadelphia. Final construction slowed on destroyer

Rear Admiral Louis McCoy Nulton, Philadelphia Navy Yard commandant between September 1920 and June 1923. He found League Island overrun with thousands of rats; he had to decommission 140 warships, dispose of tons of ammunition, and scrap two super battle cruisers on the stocks. Naval Historical Center Photo NH 47533.

tender *Dobbin*, Daniels's favorite ship. The government reduced the work force on League Island from over 9,000 workers at the end of 1920 to 6,000 by September 1921. African American women hired during the war were laid off first, drawing criticism from a Philadelphia veteran. Margaret Hall, one of the laid off workers, was a "Negro women, but American through and through," and should be retained, Marine Corps Sergeant Harry Jackson wrote to Assistant Secretary of the Navy Theodore Roosevelt, Jr.[14]

Layoffs seemed motivated more by budgetary considerations than by racial or gender bias. "Any discharges which the Yard finds it desirable to recommend will be agreeable to the bureau," advised Yards and Docks Chief Taylor. League Island retained all seventeen black servants and maids in the officers' quarters, seven of the twelve white female clerk-typists in the public works department, and

Below League Island employed hundreds of local African Americans during World War I and established the segregated McCoach Recreation Center for Colored People. NARAM.

thirty-six of the seventy-three female stenographers, typists, clerks, and bookkeepers in the supply office, possibly because wages for female clerks were less than those for male counterparts. At the same time, the Navy Yard fired hundreds of male industrial workers, inspectors, clerks, and safety engineers. The Navy Department helped male employees find work only at naval stations on the Great Lakes or at Pearl Harbor, however. "Men transferred to Pearl Harbor will have the same opportunities for promotion as those at any other navy yard or station."[15]

Reduction in force saved some money, but Harding and Secretary of the Navy Denby, a former Detroit automobile company executive, thought it better business to dispose of obsolete warships. The president approved the sale of pre-dreadnoughts *Maine*, *Missouri*, and *Wisconsin* and coastal defense monitors *Tonopah* and *Ozark*, berthed at League Island. Earlier, Daniels had wanted to employ them as he had *Massachusetts*, *Iowa*, and *Indiana* (Coastal Battleships Nos. 1–3) for aerial bombing experiments. The frugal postwar Harding government provided no money for such research and development of new weapons and tactics, and decided that "the *Maine* and *Missouri* will not be used for bombing experiment and that these two vessels will be sold."[16]

The Harding administration sought to save additional money by stopping the postwar naval race that had developed with Britain and Japan after the United States rejected the Versailles Treaty and failed to join the League of Nations. Harding found a way to curb expensive naval competition by sponsoring an international naval arms limitation conference. The president invited representatives from the world's naval powers (Britain, Japan, United States, France, and Italy) to Washington in l921, where U.S. Secretary of State Charles Evans Hughes stunned delegates by proposing that they scrap capital ships. Hughes's dramatic initiative led to the signing of a naval arms limitation treaty in February 1922 that called for scrapping battleships, a measure that gave the United States fleet parity with Britain and superiority over Japan.[17]

The Five-Power Naval Treaty that resulted from the conference deeply affected the Philadelphia Navy Yard, because it forced immediate disposal of capital ships then berthed in the Reserve Basin, including five 16,000-ton *Connecticut*-class battleships. Though the Navy Yard had sold three and prepared five for sale months before the Washington conference, announcement of treaty terms to scrap so many battleships located in Philadelphia started rumors that the government planned not only to sell the U.S. fleet but to dispose of the local navy yard. A 1921 visit of a board under Rear Admiral Hugh Rodman, studying possible closure of naval facilities around the nation, and Denby's announcement in early 1922 that only the New York and Norfolk Navy Yards were untouchable, created "closing rumors" in Philadelphia. These rumors increased after local leaders suggested that the federal government remove its naval facilities downriver to Hog Island, where the wartime Emergency Fleet Corporation had closed its shipbuilding yard, so that city business and industry might develop League Island.[18]

Closure rumors initiated the mandatory pilgrimage to Washington by Delaware Valley political and business leaders to save the Navy Yard. Philadelphia Mayor J. Hampton Moore, U.S. Representative William Vare, and Businessman's Association representative Daniel Gimbel led a delegation to meet Navy Secretary Denby and House Naval Affairs Committee members. The Harding administration kept the Philadelphia Navy Yard active; it probably had no intention of closing the local government facility and circulated rumors as part of the usual postwar retrenchment of the naval shore establishment. This time, naval reorganization included consolidation of the Public Works Department, accounting office, and Hull and Machinery Divisions into one Industrial Department headed by an industrial manager. Divisions of planning, production, and preservation were created, the latter responsible for keeping decommissioned ships in the Reserve Basin ready for Mobilization Day (M-Day) emergencies. "We have endeavored here to try to produce a logical organization which would stand the scrutiny of any business man," Commandant Nulton explained.[19]

Thus reorganized, the Industrial Department began the ship scrapping mission required by the Five-Power Naval Treaty. The process provided enough jobs to slow attrition in the local civilian work force, which had slipped below 5,000 for the first time since U.S. entry into the world war. League Island laborers stripped battleships of fire control towers, revolving parts, range finders, explosive material, and wireless radio sets before sale to the "Hitner Salvage Corporation, Main Office and Docks, 4501 Richmond Street, Bridesburg on the Delaware." The Bureau of Ordnance "required that prior to delivery of these vessels that the turret guns now on board be mutilated by removing breech mechanisms, burning screw box liner and gas check seat beyond repair, and burning large hole through about four feet from the muzzle." In the final stage of scrapping, Nulton dispatched naval inspectors to Hitner Salvage, where they found ships "razed from stem to stern to the level of the water line passing through the torpedo tubes."[20]

Navy Bureau Chiefs David Taylor and Robert S. Griffin worried about the disposal of all these big warships at Philadelphia. They saved battleship *Illinois* by stripping guns in accordance with the scrapping provision of the Washington Treaty and lending

League Island stripped guns from battleships *South Carolina* and *Kansas* before disposal to private Philadelphia companies for scrapping under terms of the Washington Naval Arms Limitation Treaty of 1922. Naval Historical Center Photo NH 69035.

the battleship to the New York State Naval Militia for a training ship, and preserved *Oregon* by giving it to New York as a floating monument. Griffin and Taylor came up with another notion to keep *Kearsarge* by cutting the vessel down to the hull, stripping ordnance, armor, and machinery, broadening the beam, and mounting a 250-ton revolving crane on deck. "The Bureaus expect to be able to make a separate allotment for the work of converting the *Kearsarge* into a crane ship."[21]

Conversion and scrapping of long-obsolete capital ships greatly benefitted the Philadelphia Navy Yard by providing jobs, reducing overhead costs for storage of large decommissioned ships, and clearing the facility of useless relics of the Spanish-American

War era. Such was not the case with the scrapping of Battle Cruisers No. 5 (*Constitution*) and No. 6 (*United States*), laid down in January 1921 on Shipbuilding Ways No. 2 and No. 3. Authorized under the Naval Act of 29 August 1916, these powerful battle cruisers, along with four sister ships under construction in Camden (New Jersey), Newport News (Virginia), and Quincy (Massachusetts) promised to make the United States the world's leading naval power. The 35,300-ton, 874-foot ships boasted 16-inch guns, 18-inch armor, and high-speed turbine engines. For the Philadelphia Navy Yard, these mighty battle cruisers marked the culmination of a $6-million expansion program started by Daniels to create a major government capital ship (heavily armored,

Battleship *Michigan* being stripped of guns in 1923 before disposal to private Delaware River scrap yards. Naval Historical Center Photo NH 76566; courtesy John C. Reilly, Jr.

Battleship *Kearsarge* during conversion into a crane ship, October 1922. The conversion was done to save it from scrapping in 1922. Naval Historical Center Photo NH 520380.

big gun) construction facility on the Delaware River.[22]

Contracts for Battle Cruisers No. 5 and No. 6 stabilized the postwar League Island labor force at 6,000, while private Delaware Valley industries were laying off their entire wartime work forces. The battle cruiser program in Philadelphia gave local contractors postwar naval business. The Duquesne Steel Foundry of Pittsburgh completely retooled its plant to manufacture steel castings and the Buffalo Steam Pump Company and the General Electric Company of Schenectady, New York, built turbines, pumps, and electrical apparatus for Philadelphia's battle cruisers. Regional industrial contractors sent so much material by rail to Philadelphia that the Navy Yard opened an armor plate storage facility downriver at Fort Mifflin and resumed regular ferry runs there that had been canceled after the war.[23]

The battle cruiser contracts stimulated Navy Yard reorganization in 1921 to reflect innovations in private industrial management methods. Naval Constructor William Pierre Robert served as the first industrial manager of League Island, heading the Industrial Department that had combined the old Hull and Machinery Divisions. The restructured Industrial Department rushed ahead on construction of battle cruisers. "All angles, bars and stiffeners have been fabricated, ready for assembly," Robert reported. Hull work on *Constitution* reached 60 percent completion and *United States* over 40 percent. But suddenly, in late 1921, the Harding administration cut off money for new warship construction. "The allotment for work on the Battle Cruisers building at the Philadelphia Navy Yard has been fixed at such a low figure," warned Bureau of Steam Engineering Chief Griffin, "that construction of one will be retarded at least six months, and of the other at least ten months."[24]

Nulton tried to save one battle cruiser, illegally transferring to *Constitution* funds that had been appropriated for construction and repair on other vessels. The League Island commandant also assigned workers to *Constitution* who had contracted to complete destroyer tender *Dobbin*. "There was no real graft involved," former Captain of the Yard Stirling explained. "Often money for one [ship was] juggled for another ship." At the last minute, Nulton learned that the New York Shipbuilding Corporation had saved battle cruiser *Saratoga* on its stocks by convincing the Navy Department to convert it into an aircraft carrier as permitted by the Five-Power Treaty. The Fore River Shipbuilding Company of Quincy, Massachusetts, received a similar deal for *Lexington*, and Nulton sought the same for the battle cruisers lying unfinished in the public shipyard in Philadelphia. In the end, the Navy Department announced "that Cruisers *Constitution* and *United States* are not to be converted into aircraft carriers, but will be scrapped when the Five-Power Treaty of the Limitation of Armaments has been ratified by all of the signatory powers."[25]

The death of the battle cruisers did not end League Island's mission as a scrap yard for capital ships. "Battleship number forty seven USS *Washington* is received this date from the New York Shipbuilding Corporation [and] appropriation chargeable expense moving and mooring scrapping naval vessels," Denby noted. Navy Yard and "outside" tugs towed *Washington* across the river to League Island, where scrappers stripped the big battleship, placed all electrical material "under cover to prevent undue absorption of moisture," and sent 400 tons of material back across the river to New York Shipbuilding so that the private Camden shipbuilder could complete aircraft carrier *Saratoga* and outfit battleships *Colorado* and *West Virginia*. Finally, the Navy Yard towed *Washington*'s stripped hull from League Island to the Chesapeake in 1924, where it sank in "underwater tests."[26]

Disposal of *Washington* was the Philadelphia Navy Yard's last scrapping function under the Naval Arms Limitation Conference of 1922, although a reduced work force of 4,000 later scrapped warships for the London Naval Treaty of 1930. By then, League Island had become a base for enforcement of Prohibition, giving the Navy Yard an active sea mission during the naval doldrums of the mid-1920s.

The Navy Department transferred a flotilla of destroyers, laid up in the Reserve Basin, to the Treasury Department for a "rum patrol" to stop the smuggling of illegal alcoholic beverages into the Delaware bay and river system. Philadelphia homeported and maintained a squadron of 1,000-ton *Wilkes*-class destroyers for the local "rum patrol." At the same time, League Island maintained thirty-seven 400- to 700-ton K-, L-, N-, and O-class coastal submarines. "In a vast drydock, like minnows gasping for breath in a water less hollow," Christopher Morley observed, "lay four diminutive submarines of the K type, their red plates made them look absurdly like gold fish, the diving rudders, like a fish's tail, and the little fins folded pathetically upon their sides toward the bow, increased the likeness."[27]

The Submarine Station, located between Piers D and E on the south side of the Reserve Basin, tested larger submarines, including the 1,100-ton T-3, the first true seagoing U.S. fleet submarine. Also, League Island studied the captured German submarines U-117, UB-148, and U-140, a 2,000-ton cruiser sub equipped with the latest diesel engine technology. Not restricted by the Five-Power Treaty, the Philadelphia Navy Yard developed submarines and experimented with listening devices, storage batteries, torpedo firing apparatus, and diving pumps and engines, periodically using the 30-foot-deep Reserve Basin for test dives. "At 3:15 p.m. Submarine N-5 submerged in Reserve Basin for 15 minutes," and "at 4:45 Submarine N-5 secured to Pier D." In at least one case, a Japanese naval officer quietly watched the experiments.[28]

The Fuel Oil Testing (and Boiler) Laboratory became a center of naval research and development during the late 1920s. Established in 1909, the Philadelphia lab had grown in importance as the Navy built new ships with oil-burning engines and converted coal-fired systems to oil power. At the end of World War I, Bureau of Steam Engineering Chief Griffin had wanted to keep the laboratory busy by expensive upgrades on the troublesome fuel systems on older *Ammen*-class destroyers, decommissioned and laid up in the Reserve Basin. Dramatic cutbacks in naval appropriations, from $1 billion in 1919 to

$320 million by 1924, prevented modernization of fuel oil systems on older destroyers. However, President Calvin Coolidge's Secretary of the Navy, Curtis Dwight Wilbur, secured a supplementary budget in 1925 for converting coal-burning battleships not scrapped under the Washington Treaty to oil power. Wilbur, a Naval Academy classmate of current Philadelphia Navy Yard Commandant Scales, delighted in giving League Island its first major contract since the end of the world war to modernize battleships *Arkansas* (1925–26), *Wyoming* (1926–27), and *Oklahoma* (1927–29). Oil conversion and ordnance modernization of these battleships kept a relatively steady industrial work force of be-

tween 4,500 and 5,000 at the Navy Yard.[29]

The greatest activity at the Philadelphia naval establishment during this era occurred at the Naval Aircraft Factory located on the eastern part of League Island between 4th and 6th Streets East. Chief of the new Navy Bureau of Aeronautics William A. Moffett decided after a visit in August 1921 to expand research and development programs at the four-year-old naval aviation facility. Moffett transferred an aircraft testing laboratory from the Washington Navy Yard to Philadelphia and developed an aircraft propeller research and testing shop there as well. He secured Congressional appropriations for construction of two large hangars, a sea-

In 1921 Journalist Christopher Morley described the tiny submarines in dry dock on League Island as guppies gasping for breath. In the background is the ancient monitor *Miantonomah*, sold for scrapping in 1922. NARAM.

plane "turntable catapult" on Pier No. 1, and recreation facilities. While the rest of the Philadelphia Navy Yard experienced reduction in the work force from 11,000 to 4,000 during the decade, the Naval Aircraft Factory maintained a steady payroll for 1,500–2,000 employees that by 1926 comprised over one-fourth of the entire civilian labor force on League Island. In September 1926, Aeronautics Chief Moffett, Assistant Secretary of the Navy Theodore D. Robinson, and Philadelphia Mayor W. Freeland Kendrick dedicated Mustin Flying Field at the Naval Aircraft Factory, named for Pennsylvania naval aviation pioneer Henry Croskey Mustin (acting captain of the Philadelphia Navy Yard between 1909 and 1911, where he had experimented with early aircraft flight on League Island), killed in 1923 in an aviation accident.[30]

Dedication of Mustin Field accompanied renewed Navy Department interest in improving League Island for the Sesquicentennial Exposition to celebrate the 150th anniversary of the Declaration of Independence. It helped that Secretary of the Navy Wilbur had attended the U.S. Naval Academy (1888) at the same time as two Philadelphia Navy Yard Commandants, Scales (1887) and Thomas Pickett Magruder (1889), who relieved Scales in June 1926. Moreover, the Coolidge administration owed the Philadelphia Republican Party organization for steady political and financial support. Whatever the reasons, in 1926 the Navy Department completely refurbished grounds, streets, and structures and constructed a new recreation center, fence, and attractive Broad Street main gate complex at the Philadelphia Navy Yard.[31]

Wilbur ordered the Naval Academy to send three carloads of ship models, glass display cases, and naval exhibits to Philadelphia for the Sesquicentennial celebration. The Navy Yard spruced up the historic Spanish American War cruiser *Olympia* (laid up in the Reserve Basin since 1919) in order to receive thousands of visitors. Meanwhile, Josiah McKean brought battleship *Wyoming*, new six-inch-gun cruiser *Trenton*, and historic sailing frigate *Constellation* to the Philadelphia Navy Yard for public display. Wilbur visited Magruder on League Island

several times during the Sesquicentennial celebration, though the two soon had a falling out over Navy budgeting and policies. Secretary of State Frank B. Kellogg and dozens of international diplomatic and naval visitors called on Magruder; guests included the Crown Prince of Sweden, the Japanese, Italian, and Spanish ambassadors, and the British consul general in Philadelphia. Brazilian naval officials paid frequent calls attesting to the increased role of the Philadelphia Navy Yard in the development of the Brazilian Navy, including dispatch of League Island's Industrial Manager Julius A. Furer as head of a naval mission to establish a navy yard at Rio de Janeiro.[32]

Eventually the excitement of the Sesquicentennial exposition subsided. Plagued with financial problems that caused a $5 million deficit, the Exposition grounds, located just north of the Navy Yard, closed, and Magruder tried unsuccessfully to acquire the New Jersey pavilion for use as unmarried officers' quarters. In the meantime, *Constellation*, *Wyoming*, and *Trenton* departed. Work stopped abruptly on refurbishing the facility, suggesting that earlier measures had been a public relations campaign. Furthermore, Secretary Wilbur turned his back on Philadelphia, refusing to repair Dry Dock No. 1, which leaked so badly that it could not overhaul ships. Instead, Wilbur threatened to cut 1,000 workers from Philadelphia Navy Yard rolls, and assigned construction of cruiser *Pensacola* to the Brooklyn Navy Yard after promising the contract to Magruder. Wilbur transferred other naval work to Pacific Coast navy yards as the Coolidge administration grew uneasy with reports that Japan had fortified its League of Nations Pacific island mandates.[33]

Magruder had stood at the center of U.S. and international naval attention during the Sesquicentennial exposition, but he now felt abandoned in Philadelphia. He disparaged the shrinking naval budget, cutbacks in personnel, and constant scrapping of capital ships. He feared that the upcoming Geneva Arms Limitations Conference would further destroy the U.S. Navy, expressing the hope that this peace conference end up "a monumental failure."

The main gate on Broad Street before its removal in 1926 to make way for a new gate for the Sesquicentennial Exposition. NARA Washington, D.C., Still Picture Branch.

The main gate as rebuilt for the Exposition, 1927. NARA Washington, D.C., Still Picture Branch.

Dry Dock No. l, 1928. Built in 1891, it leaked so badly during the l920s that it could not dock ships. Commandant Thomas P. Magruder's persistent request for repair contributed to his removal from command. NARA Washington, D.C., Still Picture Branch.

Stirred by Josiah McKean, leader of the 1909 *Panther* revolt and stationed temporarily on League Island during the Sesquicentennial, Magruder lashed out, as McKean had sixteen years before, at the conservative, incompetent naval bureaucracy, administration, and leadership. Magruder claimed that the U.S. Navy was unprepared to defend the nation, publishing in the widely read *Saturday Evening Post* a scathing indictment of Coolidge-Wilbur naval policy and administration.[34]

Magruder's magazine article condemning American naval policies caused an immediate sensation and made front-page news across the country. President Coolidge discussed it at his September 1927 "newspaper conference," while Wilbur confronted constant public questioning about Magruder's charges. Wilbur decided not to court-martial Magruder in order to avoid martyring the outspoken naval hero of the Spanish American and First World Wars, as had the War Department that court-martialed Army aviation enthusiast and decorated war hero Billy Mitchell. But Magruder would not remain silent, continuing his criticism in the Philadelphia press, on the local radio station WIP, and in luncheon speeches before the Rotarians, Knights of Columbus, and other organizations. Worse, he accompanied controversial U.S. Senator-elect William Vare, under suspicion of election fraud, to see the president and navy secretary about the threatened layoff of 1,000 Navy Yard employees.[35]

Delaware Valley interests exploited Magruder, but the commandant was only too willing to defend League Island against other navy yards, advocating closure of the Brooklyn Navy Yard and having all its work transferred to Philadelphia. "The Philadelphia yard is so situated and so well equipped that it could readily be expanded to do the work that would otherwise be done at New York." Magruder testified further, that "I favor the Philadelphia yard because it is far from the coast and is immune from attacks, it is in the very center of our great industrial region, where labor may be readily obtained in time of emergency." Eventually, Wilbur grew tired of the commandant's incessant interference with naval policy-making, and in November 1927 he removed Magruder from command of the Philadelphia naval establishment. The department sent General Board President and former Judge Advocate General Julian Lane Latimer to raise his flag as Commandant of the Navy Yard and Fourth Naval District. The Navy placed Magruder on the inactive list, awaiting orders in Washington, where he testified at a Congressional inquiry into Republican mismanagement of the Navy Department.[36]

Magruder's insurgency contributed to the decline in Navy morale that accompanied arms limitations treaty restrictions and the constant postwar cutbacks in personnel and appropriations. Nevertheless, Magruder raised public concern for the lack of naval preparedness and helped stir an isolationist Congress to spend money on new cruisers to bring the United States up to treaty strength. His activism most likely saved jobs at the Philadelphia Navy Yard and convinced the Navy Department to continue battleship modernization and construction of a treaty cruiser on League Island.[37]

The election of Quaker engineer Herbert Hoover as president in 1928 threatened to reverse gains made during the closing months of the Coolidge-Wilbur administration. Hoover promised efficiency and economy in government, pledged to reduce military and naval spending, announced further limits on armaments through more international treaties, and proclaimed a unilateral moratorium on naval construction. Hoover selected frugal Massachusetts yachtsman and Harvard University treasurer Charles Francis Adams as his secretary of the navy to implement economy in the naval establishment. As with so many of his predecessors, Adams hoped to save money by closing navy yards.[38]

Adams said nothing about closing League Island, however, and the Philadelphia Navy Yard prospered under the Hoover-Adams administration (1929–33). Latimer told the Poor Richard Club of Philadelphia that League Island had become the biggest industrial plant in the city, with $428 million in assets and a

Rear Admiral Thomas P. Magruder, late 1920s. President Coolidge removed war hero Magruder from command of League Island in 1927 for criticizing administration naval policies. Naval Historical Center Photo NH 48052.

$9 million annual civilian payroll. He estimated that the government had a $2 billion investment in the naval shore establishment and that no other industry "ever had so many stockholders, for every taxpayer in the United States is a stockholder in the Navy Yard." The work force increased during the Hoover years from 4,265 in 1930 to 5,000 by 1932, while the Machine Shop ran two shifts to manufacture heavy deck plating for modernization of battleship *New Mexico* and new construction on the 10,000-ton treaty cruiser *Minneapolis*, laid down on 27 June 1931. The Hoover administration developed a lucrative destroyer scrapping business on League Island, netting a profit of $6,000 on each unit scrapped. The ratification of the London Treaty of 1930 gave the Navy Yard more contracts to scrap submarines, destroyers, cruiser *St. Louis*, and 16,000-ton battleship *Florida*.[39]

Hoover's Progressive management style of economy and efficiency, naval arms limitation policies, and the Great Depression itself in fact benefitted the Navy Yard, to the point that one Philadelphia newspaper announced in 1931 that closure was "inconceivable." However, economic troubles and growing unemployment throughout the rest of the country caused voters to choose a New Deal and in 1932 to elect Hoover's Democratic presidential opponent Franklin D. Roosevelt. Roosevelt's election supposedly ended the "Hoover aberration" of not maintaining the U.S. Navy at treaty strength. The dramatic political change made little difference in the operation of the Philadelphia Navy Yard, however. Employment remained the same 5,000 under Roosevelt as it had under Hoover, and actually declined slightly in late 1933, when FDR ended Prohibition and the Navy Yard deactivated seven destroyers from the Coast Guard rum patrol. During the early New Deal era, League Island continued Hoover's ship-scrapping program, between 1933 and 1935 tearing apart thirty-three destroyers and nine submarines and creating a special scrapping team to cut up two destroyers simultaneously. "The Industrial Department completed the scrapping of ex-U.S.S. *Shaw* and ex-U.S.S. *McDougal* in D.D. #2, the ex-U.S.S. *Conyngham* and ex-U.S.S. *Wainwright* placed in D.D. #2

for scrapping by the Industrial Department," Captain of the Yard George B. Landenberger reported.[40]

The scrapping program gradually lost its popularity as the "junk" metal from scrapped vessels at the Philadelphia Navy Yard appeared in Osaka, Japan, reportedly to build weapons for the Japanese conquest of China. Moreover, scrapping became less lucrative for local business, as profit slipped from $6,000 to $2,200 on each scrapped destroyer. Commandant William C. Watts, a Pennsylvania native, blamed the reduction on declining local prices for scrap metal and area labor demands for higher wages. Then labor strife hit League Island as the Industrial Union of Marine and Shipbuilding Workers of America of the Congress of Industrial Organizations (CIO) aggressively sought to unionize the Navy Yard work force as part of a larger organizational campaign that transformed Philadelphia from a conservative American Federation of Labor craft guild and lodge town into an aggressive industrial union city. The Navy Yard command resisted this outside unionization, and in April 1936 Naval Architect Henry C. Robinson, who had worked on League Island since 1909, organized the Philadelphia Navy Yard Development Association with leaders of shop committee organizations at the Naval Aircraft Factory and in the Industrial and Supply Departments. The Association incorporated the traditional craft lodge system of master, leading man, and quarter man, the executive committee of the shop committees, and managers of all civilian offices at the Navy Yard.[41]

The Navy Yard Development Association promoted "a spirit of cooperation and good will between all elements comprising the personnel of the Navy Yard." To boost morale during periods of threatened layoffs, the Association organized social functions, a navy yard band, bowling leagues, and prize money for employees who made "useful safety suggestions" (a forerunner of the Beneficial Suggestions ["Benny Suggs"] program that became popular at the Navy Yard during World War II). The Association published a newsletter to counter union propaganda. Industrial Manager Henry Williams explained that "many erroneous ideas are extant and it is most desirable to have a means of disseminating correct and authentic information as to the activities of the Yard, the prospects for work and the steps being taken to further our interests." The in-house newsletter fostered "stabilization of employment," by warning members to steer clear of outside union agitators and avoid "Communistic activities."[42]

Though the Association vigorously denied charges that it was a "Company Union," the conservative organization of shop managers reported directly "to the Commandant via the Manager." Moreover, Navy bureau chiefs in Washington endorsed the Shop Employees Association, and the secretary of the navy used it as a model for labor-management relations at other navy yards. The most effective method of keeping out Communists and radical unions, however, was to provide sufficient work to prevent layoffs. Unfortunately for this strategy, the FDR administration was slow to deliver new work to the Navy Yard. The first years of the New Deal seemed particularly gloomy ones for League Island. When the U.S. Navy airship *Akron* crashed twenty miles off the New Jersey coast, the Navy Yard launched search planes from Mustin Field to locate possible survivors, including Naval Aircraft Factory patron William A. Moffett. They found none. In August 1933 a violent hurricane struck with a 35-year record tide that inundated League Island and gale force winds that tore down power and telephone lines. Operations ceased entirely for over twenty-four hours. As if an omen, popular League Island officer Homer H. Norton collapsed suddenly and died as he brought the Navy Yard tug *Nokomis* in for berthing.[43]

After a cautious start, the New Deal began to channel monies to League Island through city and state agencies, including funds from the National Industrial Recovery, Federal Emergency Relief, Public Works, and Works Progress Administrations. WPA labor improved roads and railroad tracks, dug steam and water pipe lines, landscaped grounds, and repaired buildings. More than 2,000 WPA workers, including a large number of African Americans, were employed on such projects. WPA labor "removed the unsightly automobile graveyards which clutter up a part of [Broad] Street" near the govern-

ment compound, and built a Naval Hospital at 16th Street and Pattison Avenue on League Island Park, laid out north of the Navy Yard and the Reserve Basin. "All hospitalization and medical treatment at the U.S. Naval Hospital, League Island, Navy Yard, Philadelphia, Pa., will be transferred to the new Naval Hospital as soon as practicable after 12 April and discontinued at the Navy Yard," ordered Commandant Watts.[44]

New Deal money built warships on League Island under the pathbreaking Vinson-Trammell Naval Construction Act of 27 March 1934, which gave FDR the means to begin the greatest naval expansion program in U.S. history. "I hope this is the beginning of a treaty Navy," observed Secretary of the Navy Claude A. Swanson, "a Navy built to the limits of the London Naval Treaty and second to none." The Vinson-Trammell Act and New Deal meant construction at the Philadelphia Navy Yard between 1934, and 1938 of the 1,500-ton destroyers *Cassin*, *Shaw*, and *Rhind*, cruisers *Philadelphia* and *Wichita*, and four innovative 2,000-ton Coast Guard cutters that carried scout aircraft. Contracts came to League Island, after Secretary of the Treasury Henry Morgenthau, Jr., made an extended visit, for *Hamilton*-class cutters *George W. Campbell*, *William J. Duane*, *Samuel D. Ingram*, and *Robert B. Taney*.[45]

Enthusiastic Navy Day celebrations on League Island in 1935 suggested how much New Deal attention revived local morale and interest in American naval expansion. On 28 October, 93,600 visitors flocked to League Island to watch acrobatic and formation flying and dummy parachute drops at Mustin Field and visit the ever popular Spanish American War veteran cruiser *Olympia*. The main attraction was the launch side-by-side of destroyers *Cassin* and *Shaw* and keel laying of cruiser *Wichita*. But Chief Boatswain Edward J. Lysaught thought dry docking of *Shaw* and *Cassin* after the launch drew most public interest. "It required special effort on part of the sentries and others on duty to keep the crowds under control within the immediate vicinity" of Dry Dock No. 2. Such concern seemed justified, when a few weeks later, "John P. MacMullen Mach. 3rd class while going down into Dry Dock #2 lost his balance and fell to the floor of the Dry Dock [and] was instantly killed," explained Captain of the Yard Percy W. Foote.[46]

Navy Day excitement soon waned. After the launching of two 1,500-ton destroyers and completion of four Coast Guard cutters, only *Wichita* remained to keep the Navy Yard busy. In 1936 the Navy Department announced possible layoffs, while the Development Association worried that League Island could not compete for future work with private shipyards, particularly on aircraft carriers *Lexington* and *Enterprise*, assigned to a private yard in Virginia favored by former U.S. Senator (Virginia) and current Secretary of the Navy Claude A. Swanson. At that moment, the Delaware River naval establishment found a patron (as it often did during its long history) in Pennsylvania Governor George Howard Earle, III, wealthy Philadelphian, ardent New Dealer, and FDR friend.[47]

Governor Earle adopted the Philadelphia Navy Yard and visited League Island regularly between 1935 and 1938, arriving at Mustin Field by airplane with much fanfare. Earle brought Congressmen and local civic and business groups to the Navy Yard, and convinced Philadelphia Mayor S. Davis Wilson, a Republican foe of the New Deal, to support distribution of New Deal monies for improvement of League Island. It was no coincidence that Wilson sent the first Philadelphia WPA workers to the Navy Yard on the very day that he and Earle attended memorial services on League Island for aviators Wiley Post and Will Rogers, killed in an aviation accident. The Navy Department showed its appreciation by enrolling Earle in elaborate League Island ceremonies as a lieutenant commander in the U.S. Naval Reserve. The Navy also invited the New Deal governor's wife, rather than the wife of the Philadelphia Republican mayor, to sponsor the city's namesake cruiser *Philadelphia*, launched on 17 November 1936 in front of 12,000 local residents and officials.[48]

The Navy Yard received aggressive new leadership with the arrival of outspoken Louisiana native Wat Tyler Cluverius in June 1937 to assume command of the Fourth Naval District and Philadelphia

Navy Yard. Cluverius entered his new job disparaging the "sentimentalized ballyhoo" of pacifists and isolationists who tried to stop construction of 10,000-ton treaty cruisers and interfered with measures to bring the U.S. Navy up to treaty strength. The Philadelphia Navy Yard commandant watched uneasily as the United States remained neutral while Italians launched aerial gas attacks on Ethiopia; Germany, Italy, and Russia tested new weapons in the Spanish Civil War; and Japan attacked China by air and ground; all within weeks of Cluverius's assumption of command. "The weapons are new—and awesome—as the game is carried into the air and beneath the surface of the sea," Cluverius observed.[49]

Cluverius's new command in Philadelphia contained both air and undersea weapons, including submarines of the larger 800-ton seagoing S-class. The Naval Aircraft Factory developed new types of land-based aircraft and seaplanes, catapult launching systems for warships, and airplane propellers and engines. League Island also researched and developed fuel oil burner parts, ship propellers, "radio direction finders," and machinery for a new type of high-speed motor torpedo boat. But traditional ship types rather than "awesome new weapons" still dominated New Deal-era business at the Navy Yard. One of Cluverius's first tasks was to sell six tiny World War I-vintage *Eagle* boats, berthed in the Reserve Basin among sixty overage destroyers and forty-eight obsolete coastal submarines. He also commissioned the inadequately armed cruisers *Savannah*, *Nashville*, and *Phoenix*, delivered by the New York Shipbuilding Corporation, and in 1937 directed the launch of destroyer *Rhind* on 28 July and treaty cruiser *Wichita* on 17 November.[50]

There were signs of progress toward naval rearmament evident on League Island in 1937. Cluverius joined Captain of the Yard George M. Baum and Industrial Manager Alexander H. Van Keuren at keel laying ceremonies in 1938 for destroyer *Buck* and the 16-inch gun, 35,000-ton battleship *Washington*.

Keel laying for *Washington* was the featured attraction at Navy Day celebrations in October 1938, attended by 65,118 visitors. "At the Welding slab,

Launching of the 10,000-ton treaty cruiser *Philadelphia*, 17 November 1936. During the l930s, League Island also built the *Minneapolis*. Naval Historical Center Photo NH 92060; courtesy of the Naval Historical Foundation, Cdr. Edward S. Moale Collection.

Keel laying ceremonies for the Philadelphia Navy Yard's first battleship, *Washington*, 14 June 1938. The FDR administration contract made League Island one of the world's greatest naval shipyards during World War II. Temple University Urban Archives.

west of Building No. 57 fabrication by electric welding of assemblies for battleship construction may be viewed," the official program announced. Visitors could not really see much of League Island's first battleship, however, as work had only started on hull assemblies and a rope barrier restricted access some distance from Shipbuilding Way No. 3. But at least local residents were allowed to tour League Is-

land. The Navy Department limited foreign visitors "Lt. Commander T. Honguku and Lieutenant Saski, Imperial Japanese Navy" and German naval observers from cruiser *Karlsruhe* to the Naval Aircraft Factory.[51]

The 35,000-ton *Washington* under construction, 7 July 1939. Already made obsolete by the proposed 45,000-ton battleship *New Jersey*, the ship took shape slowly because of labor strikes and steel shortages. NARAM.

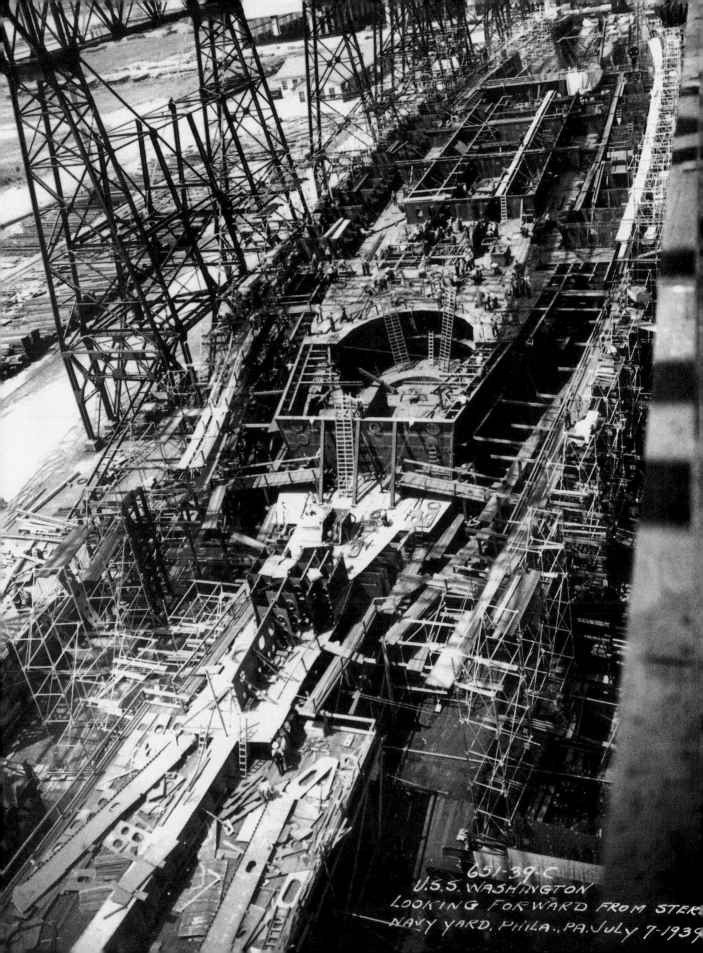

651-39-C
U.S.S. WASHINGTON
LOOKING FORWARD FROM STER
NAVY YARD, PHILA., PA. JULY 7-1939

Featuring the 35,000-ton warship at Navy Day celebrations disguised the fact that this treaty battleship was outclassed already by proposed 45,000-ton vessels. Completing the nearly obsolete warship was even made more difficult because a national labor crisis, including strikes at steel mills, delayed delivery of material to Philadelphia for nearly a year. Also, *Washington*'s construction revealed that the Philadelphia Navy Yard had reached its limit as a capital battleship shipbuilding facility at 35,000 tons. Larger ships could not be fabricated without extension of shipbuilding ways, addition of 20-ton overhead cranes, and expansion of shops to manufacture larger steel assemblies. The electric power plant was outdated, the dry dock facilities inadequate, and the freshwater Reserve Basin so polluted that metal ships berthed there corroded more severely than if they were exposed directly to salt water. The Navy Yard's force of skilled labor proved too small simultaneously to construct a battleship and preserve and maintain the inactive fleet against such pollution.[52]

The Navy Department in 1938 lacked adequate funds to provide the estimated $4 million needed to make League Island truly capable of preparing the fleet for the international crisis that daily worsened with German absorption of the Sudetenland and demands for Danzig and the Polish Corridor. Frustrated by the disintegrating world situation and neglect of his Navy Yard command, Cluverius criticized the lack of American preparedness in a review of R. Ernest Dupuy's book *If War Comes*. "The world is in arms about us, and America, on the side lines is watching with apprehension lest it get out of bounds."[53]

The Atlantic Reserve Fleet of destroyers, submarines, and transports in 1938 await the call of world crisis in the Reserve Basin on League Island. Naval Historical Center Photo NH 60287.

10 ✯ League Island at War

T HE GERMAN invasion of Poland in September 1939, beginning World War II, pushed the Philadelphia Navy Yard toward mobilization. President Roosevelt announced a state of limited national emergency and neutrality patrols for Atlantic coastal waters, while Acting Secretary of the Navy Charles Edison ordered East Coast naval facilities to recommission "Priority I" destroyers that war planners on the Navy General Board wanted ready within thirty days after issuance of Mobilization Day (M-Day) orders. Edison asked Philadelphia to provide eighteen "Priority I" destroyers, causing Commandant Julius C. Townsend to complain that League Island lacked the manpower and facilities required to meet M-Day schedules. Only 5,725 employees worked at the shipyard, little more than during the Great Depression, and the place had few machinists, pipe fitters, heavy derricks, or adequate electrical and machine shops. The local Naval Shore Station Development Board recommended the immediate addition of new crane ways, piers, quay walls, barracks, receiving station, and dispensary for destroyer crews coming to man vessels for the neutrality patrol.[1]

Adolphus E. Watson, who replaced Townsend in January 1940, agreed that lack of funding left the Philadelphia facility short on manpower for M-Day obligations, and wanted more money for a new dry dock and enlarged ship ways to construct the 45,000-ton battleship *New Jersey*, scheduled for construction on League Island. Watson collected a top production team to mobilize destroyers, extend shipbuilding ways, and construct the world's largest

Facing 1939–45. League Island underwent its greatest expansion between 1939 and 1945 with the addition of the world's largest government dry docks, but a 1944 accident near the turbine and boiler laboratory prevented the naval shipyard from ever becoming a nuclear facility.

dry dock. Earl Francis Enright became Production Officer, while Head of the Design Division in the Bureau of Construction and Repair Alan J. Chantry took charge of the Industrial Department. Chantry had served on League Island after World War I as Production Superintendent under the first Industrial Manager, William Pierre Robert. With a master's degree in naval architecture from MIT and a three-year stint as head of the Department of Mathematics and Mechanics at the U.S. Naval Academy, Chantry had the reputation as one of the top U.S. naval constructors. He directed the Industrial Department at the Philadelphia Navy Yard from late 1939 until 1945, retiring at the end of the war. Chantry's partner in 1921 under Robert's tutelage, Shop Superintendent Rufus W. Mathewson, was also brought on board in 1939, as captain of the yard; he served on League Island throughout the war, providing stability and continuity to the dramatically changing wartime navy yard on the Delaware River.[2]

Despite initial grumbling about mobilization, Philadelphia managed to recommission six Priority I destroyers on 25 September, six more on 4 October, and the final six on 16 October. The Navy Yard completed twelve Priority II contracts on 4 and 18 December. In order to ready these neutrality patrol

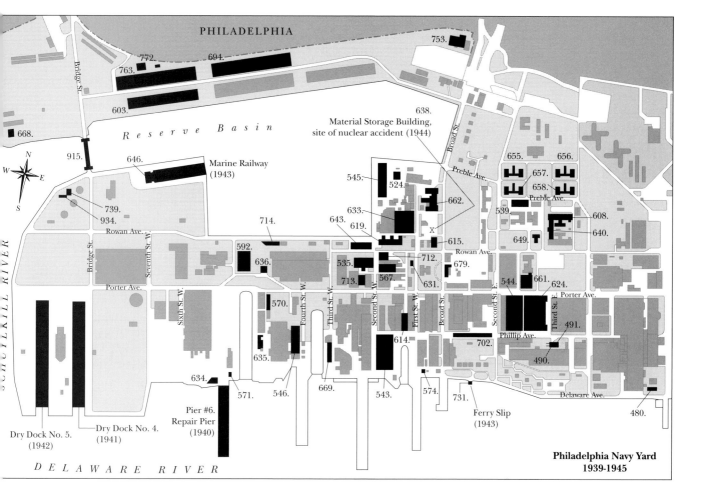

Philadelphia Navy Yard
1939-1945

480. Dispensary (1944)
490. Storehouse (1940)
491. Salvage Storehouse (1940)
524. Heavy Material Boiler Lab Storehouse (1943)
539. USMC QM Storehouse (1937)
542. Storehouse/Reserve Basin (1939)
543. Pipe and Coppersmith Shop (1939)
544. Heavy Materials Storehouse (1939)
545. Storehouse, Paint and Oil (1939)
546. Turret Shop (1939-45)
567. Lumber Storehouse (1934)
570. Pipe Shop (1937)
571. Annealing Furnace (1935)
574. Transformer House (1939)
585. WPA Shop/Building Trades (1935)
592. Material Assembly Shop (1938)
603. Storehouse No. 2 (1941)
608. Receiving Barracks (1941)
614. Accounting/Blueprint Room (1941)
615. Fuel Oil Testing Plant (1912)
619. Fuel Oil Testing Plant (1912)
624. General Storehouse (1941)
633. Naval Boiler Test Lab Office Building (1942)
634. Pier No. 6 Service Building (1941)
635. Service Building/Rigging Shop for Dry Dock No. 3 (1942)
636. Fire House (1941)
637. Radio Transmitter Building (1941)

638. Material Storage Building, site of Nuclear Accident (1944))
640. Receiving Barracks No. 3 (1942)
643. Service Building Pier A (1941)
646. Service Building for Marine Railway/Paint Shop (1942)
649. Chapel (1942)
655-658. Barracks (1942)
661. Recreation Building Annex (1943)
662. Barracks for Crews Ships Overhaul (1942)
668. Incinerator (1942)
669. Service Building DD. No. 2 (1942)
679. Post Office (1943)
694. Storehouse (1943)
702. Bus Terminal (1943)
712. Radio and Sounding Laboratory Radar School (1943)
713. Electrical Shop Extension (1943)
714. Material Segregation Building 1941)
731. Ferry House (1943)
739. Fuel Oil Storage (1944)
753. Locomotive, Crane Repair Shop (1945)
763. Radio, Radar and Shipfitting Material Storehouse (1945)
772. Heating Plant (1945)
915. Vehicular Lift Bridge (1943)
934. Fueling Structure (1943)

craft on time, Mathewson and Chantry cannibalized parts from other decommissioned destroyers in the Reserve Basin, delaying final mobilization of low priority warships. Similar problems occurred at other East Coast navy yards, and design flaws appeared in late model destroyers that made them top-heavy and unstable in high seas. House Naval Affairs Committee Chairman Carl Vinson blamed the Navy bureau system, which had changed little from its founding in 1842 and reorganization of 1862, for delays in mobilization. The influential Georgia representative, a great navy supporter, introduced a measure to abolish the bureau system. The bureau chiefs and Navy General Board opposed such drastic organizational change. Consequently, Ad Interim Secretary of the Navy Edison, who had served as acting secretary since Swanson's death in office, presented a compromise organization that shaped naval shipbuilding and repair at navy yards throughout World War II.[3]

Perhaps not as creative as his scientist-inventor father Thomas Alva Edison, who had headed the Naval Consulting Board during World War I, Charles Edison nevertheless understood the industrial and technical side of the Navy, and as assistant secretary since 1937 had managed the Navy's business for the ailing Swanson. Edison knew that the bureau chiefs would never accept sudden abolition of their bureau fiefdoms, but that some reorganization was essential to ready the shore establishment and the fleet for a world war. Edison favored creation of a central shore establishment office similar to the office of the Chief of Naval Operations, and consolidation of the Bureaus of Construction and Repair and Engineering into a Bureau of Ships. Preparatory to such major reorganization, Edison advised bureau chiefs to merge design divisions, appointing Bureau of Engineering Chief Samuel M. Robinson as coordinator of shipbuilding and Bureau of Construction and Repair Chief Alexander Van Keuren as assistant coordinator.[4]

Lawmakers debated Navy Department reorganization in early 1940 as part of the Roosevelt administration's request for a $1.8 billion defense appropriation. Gradually, isolationist opposition to this huge

defense spending bill eroded as Germany invaded Norway and Denmark in April, new British Prime Minister Winston Churchill asked for American assistance, and FDR urged the nation to move from strict neutrality to "non-belligerency" to allow such aid. Congress passed a $4 billion "two-ocean navy" bill and assented to the exchange of fifty overage U.S. Navy destroyers to Britain in return for the right to develop bases in British possessions in the Western hemisphere. It was the collapse of France in late June 1940, however, that forced the U.S. Navy to consolidate the Bureau of Construction and Repair with the Bureau of Engineering, forming a Bureau of Ships.[5]

Now, at last, Edison had the organization and money to mobilize the Navy. But at this critical moment President Roosevelt decided to replace his loyal and competent navy secretary with Chicago newspaperman Frank Knox. Apparently FDR held the quiet bureaucrat Edison partly responsible for not dealing more aggressively in late 1939 with the flaws in naval organization and mobilization. Moreover, the president wanted Knox and Secretary of War Henry L. Stimson, Republicans who strongly supported administration defense and foreign policies, in a coalition cabinet to ready the nation for war and strengthen FDR's run for an unprecedented third term.[6]

Edison became FDR's hand-picked Democratic candidate for governor of New Jersey, and in this capacity joined former Assistant Secretary of the Navy Lewis C. Compton of New Jersey and Under Secretary of the Navy James V. Forrestal, a Princeton University-educated Wall Street financier, in keel-laying ceremonies at the Philadelphia Navy Yard on 16 September 1940 for the battleship *New Jersey*. As acting secretary, Edison had selected the ship's name after his home state and the League Island construction site to provide jobs for South Jersey. It seemed appropriate that the future New Jersey governor should take an acetylene torch and make the first spot weld on *New Jersey*. In his keel-laying speech, Edison called the giant battleship a living mechanism to "defend human rights" and "for the preservation of peace in our hemisphere."[7]

Battleship *New Jersey*, 9 January 1941. The contract for the 45,000-ton battleship was given to League Island to boost develop-ment of a skilled government work force in the Delaware Valley and provide employment for an area hit hard by unemploy-ment during the Great Depression. NARAM.

President Roosevelt arrived in Philadelphia four days later, ostensibly to receive an honorary degree from the University of Pennsylvania. FDR visited League Island and inspected battleships *New Jersey* and *Washington*, the Naval Aircraft Factory, where he saw a catapult test for a new Grumman fighter aircraft, and top-secret PT Boats 7 and 8. The lengthy tour of the Philadelphia Navy Yard came in the wake of Republican presidential candidate Wendell Willkie's triumphant campaign parade through the city, and FDR used the Navy Yard visit to cement Philadelphia's continued loyalty. Roosevelt motored slowly around the facility in an open car with city Democratic boss John B. Kelley and U.S. Senator Joseph Guffey, followed by a car containing Philadelphians Anthony J. Drexel Biddle and William C. Bullitt, FDR's ambassadors to Nazi-occupied Poland and France. The president promised Philadelphia $4 million for Navy Yard expansion, 11,000 new jobs on League Island, and federal funding for 1,000 low-cost housing units along Penrose Avenue near the Navy Yard. The chants "We want Roosevelt" from thousands of shipyard workers who lined the streets of the government compound as FDR passed by, forecast the president's sweep of city and state in the November elections.[8]

Philadelphia Navy Yard participation in the undeclared Atlantic war against Germany began in earnest after Roosevelt's reelection and the March 1941 passage of the Lend-Lease Act. Navy Secretary Knox arrived on 15 May to commission the Navy Yard's first battleship *Washington*, and stayed late into the evening to tour facilities and discuss the growing crisis over German submarine attacks in the Atlantic which would prompt FDR to proclaim an unlimited national emergency on 27 May. Assistant

President Franklin D. Roosevelt, Democratic city boss and Representative John B. Kelley of Philadelphia, and Senator Joseph Guffey visit League Island Commandant Adolphus E. Watson (in dress whites) a few days before the 1940 election in which Roosevelt sought an unprecedented third presidential term. NARAM.

Secretary of the Navy Ralph Bard arrived a few days later to represent the administration at launching ceremonies for the large minesweeper *Terror*, while Knox returned next day, landing at Mustin Flying Field on his way to the New York Shipbuilding Corporation in Camden to inspect "Battleship X" (*South Dakota*).[9]

All the while, Industrial Manager Chantry rushed expansion of ship ways on League Island to build a second 45,000-ton battleship (*Wisconsin*) and construction of 1,000-foot Dry Dock No. 4. Despite labor walkouts and material shortages, Chantry complet-

Replacing the bow of a battle-damaged British cruiser (barely visible behind the cranes) on League Island. The top secret "Request" program allowed the Roosevelt administration to skirt the Neutrality Laws that prohibited such aid to belligerent forces before the Pearl Harbor attack of December 1941. NARA Washington, D.C., Still Picture Branch.

ed the big dry dock in a record fourteen months, opening for business in October 1941 before U.S. entry into the war. At the same time, Philadelphia outfitted minelayer *Terror* and laid keels for destroyers *Butler* and *Gherardi*, although critical material and manpower shortages tied up work in early 1941 on battleship and destroyer construction. The Navy Yard repaired the Colombian destroyer *Antioquia*, as part of FDR's hemispheric security policy, and readied O-, R-, and S-class submarines, berthed in the Reserve Basin, for transfer to the U.S. submarine base at New London or to British, Canadian, and Dutch naval forces. The neutral Navy Yard performed top secret and probably illegal "Requests" repairs on the belligerent British aircraft carrier *Furious*, battleship *Resolution*, and cruiser *Manchester*.[10]

Clandestine "Requests" contracts typified the Navy Yard's struggles with nonbelligerency in the months before the Pearl Harbor attack in December 1941 brought the U.S. formally into World War II. This troublesome period was reminiscent of that confronted between 1916 and early 1917 by then Navy Yard Commandant Robert Lee Russell. This time, the Navy Department selected sixty-three-year-old Rear Admiral Adolphus E. Watson, a veteran of the Spanish American War, to lead the Philadelphia naval establishment. A fine Virginia gentleman, Watson was overwhelmed by the "great changes" he found at his Northeastern urban industrial command. Instead of naval matters, he had to deal with growing labor strife as various industrial craft union locals walked off the job for higher wages or agitated for union recognition. He confronted charges made by the Dies Congressional Committee Investigating Un-American Activities that Communists had infiltrated the Philadelphia Navy Yard, and fired twenty-three civilian workers who were organizing labor on League Island for alleged Communist Party ties. This act drew an accusation from John Slavin of the Industrial Union of Marine and Shipbuilding Workers that Philadelphia was the "Atlantic Coast's Most Notorious Anti-Labor Establishment."[11]

Watson wrestled as well with city officials over whether they could enter the Navy Yard to arrest shipyard workers for failure to pay the city wage tax—a tax the FDR administration said federal employees must pay. More troublesome, Watson negotiated with Philadelphia Mayor Robert E. Lamberton for condemnation of property near the Navy Yard to erect 1,000 low-rent housing units for federal employees, who were suffering a severe housing shortage as the navy yard work force reached 24,000 by May 1941 and 33,344 by the end of the year. Workers sought housing across the river in New Jersey, competing with New York Ship employees for scarce lodging. So many Navy Yard employees resided across the river (one out of every three) that the government added a second ferry (*League Island*) in March to connect with Red Bank, Gloucester County, New Jersey. Other League Island employees set up trailer camps along Broad Street near the Navy Yard. One gypsy camp at Broad and Bigler Streets drew bitter opposition from local residents who demanded that the city remove it from their quiet neighborhood. After acerbic negotiations, the city decided to close the trailer camp and relocate Navy Yard employees on land condemned from the Girard estate near the mouth of the Schuylkill River. This process required evicting houseowners who had lived for years on this city-managed land.[12]

Security concerns on League Island dominated Watson's command in mid-1941 as the undeclared war intensified and led to the organization of a Sea Frontier System, seizure of French ships in U.S. ports, and declaration on 27 May of an unlimited national emergency. The Bureau of Ships granted the Philadelphia naval establishment $100,000 to develop passive defense at the Navy Yard, which included a security fence, fire alarm system, better lighting around outside buildings, and purchase of police wagons with two-way radios to patrol the island compound. Designated yard security officer, Captain of the Yard Mathewson developed a "passive defense plan" with blackout procedures, an air-raid warning system, and measures to combat chemical attack. "Gas defense officer" Charles F. Hudson held gas attack tests in April 1941, distributing gas masks and organizing teams that included "sniffers," window tapers, and decontamination squads.[13]

The League Island work force rose from 30,000 in 1941 to a peak of 60,000 in 1944. NARAM.

Watson realized in late November 1941 that "the present international situation is such as to call for emphasis on the fact that the emergency may become intensified at extremely short notice." He ordered "preparations to meet any condition which may suddenly confront us"; and that "every responsible senior in the Fourth Naval District make a new analytical study of his responsibility for the security and state of readiness," including the "restudy of War Plans [and] of internal security measures needed (1) to combat unlawful violence or subversive activities under present conditions, (2) to insure the maximum security under war conditions." The commandant's order suggested that the government was fully aware before the Japanese surprise 7 December air strike on Pearl Harbor that U.S. entry was imminent, although the attack fully activated the Navy Yard's passive defenses.[14]

After the Pearl Harbor attack, Mathewson initiated "urgent" passive defense measures on League Island and "blacked out the Yard in conjunction with exercises conducted in the City of Philadelphia and

The Navy Yard pep band tried to boost morale on League Island in the dark days of late 1941. Temple University Urban Archives.

surrounding counties." Storekeeper Thomas Wood recalled that this entailed mostly "the shielding of buildings to prevent lights from being seen in case of enemy night attacks," but in fact Mathewson ordered installation of sand bags and steel window shutters as "protection against incendiary attack and from bomb splinters." He directed "obscurity painting" of buildings, smokestacks, and water towers with dull gray paint, and added more air raid sirens, fire hose connections, and a CO_2 system to fight chemical fires. Security Officer and Civil Engineer Gaylord Church recommended erection of antiaircraft gun platforms and shelters for "roof watcher stations" around Mustin Flying Field, on the tops of buildings, and at the Fort Mifflin Naval Ammunition Depot.[15]

Two weeks after Pearl Harbor, the Bureau of Yards and Docks promised Philadelphia $500,000 to be used "without reference to method" for urgent passive defense projects, and Mathewson initiated a series of air raid battle problems and drills. The first problem drew inspiration from the attack on Pearl Harbor, assuming that an enemy task force of aircraft carriers and destroyers had arrived 300 miles off Delaware Bay "to make a surprise attack on the Atlantic Seaboard." Mathewson set up the problem for his department heads. "Fifty enemy bombers, with fighter escorts, flying in three waves at a high altitude, were reported passing over Lewes, Delaware at 0600," the yard captain stated. "Subsequent reports indicated that the planes were following the course of the Delaware River with the

Philadelphia Navy Yard as their objective."[16]

Stimulated by the shock of a Japanese air strike in the Pacific, these air raid problems and drills were of little value for the Atlantic seaboard. Germany and Italy (Japan's Axis partners that declared war on the United States on 11 December) did not have aircraft carriers, and they posed no immediate threat from the air to the east coast or Delaware River. However, German U-boats off the entrance of Delaware Bay and the New Jersey coast during the first winter of American involvement were a grave danger. War removed neutrality restrictions, and German Commander-in-Chief of U-boats Karl Doenitz launched "Operation Drumbeat" to attack coastal shipping, particularly those ships clearing Delaware Bay headed for the European war zone, or tankers bringing oil from the Gulf of Mexico up the Atlantic seaboard. Doenitz's U-boats struck all along the New Jersey coastline in January 1942, sinking the Norwegian tanker *Veranger* off Atlantic City and a larger merchant ship just off the beach of the Cape May County resort town of Wildwood.[17]

The Navy Department had developed preliminary war plans in 1927 and 1935 for coastal defense with the notion of a sea frontier system based on naval district organization. FDR partly mobilized the sea frontiers in April 1941. Under this scheme, the Delaware and New Jersey coasts became part of the North Atlantic Naval Coastal Frontier. After Pearl Harbor, the commandant of the Fourth Naval District and Philadelphia Navy Yard served as commander, Task Group Delaware, North Atlantic Coastal Frontier, Eastern Sea Frontier. Thus, Watson had to defend the local coast lines from Staten Island, New York, to Cape Henlopen, Delaware, against German U-boat attacks. He had at his disposal a tiny collection of blimps from the Lakehurst Naval Air Station in New Jersey, patrol aircraft and boats stationed at the Section Base in Cape May City near the entrance to Delaware Bay, and several anti-submarine craft berthed at the Navy Yard, 90 miles upriver from the Atlantic Ocean.[18]

Watson commanded few anti-submarine vessels with which to defend Fourth Naval District waters. His destroyers had been recommissioned long ago and sent on neutrality patrol in the Atlantic or transferred to Britain under the Destroyers-for-Bases deal. Some of these Philadelphia destroyers, such as *Greer* (recommissioned on 4 October 1939), had located German U-boats for the British or screened convoys behind the neutral American flag during the undeclared Atlantic naval war of 1940–41. Watson had only *Eagle 56* available for sub chasing because the Navy had sold the remainder of the squadron in 1938 to the Northern Metal Company of Philadelphia. In desperation, the Navy Yard acquired U.S. Coast Guard Tenders *Zinnia* and *Lilac* during the first winter of the world war, but these craft were too small for anti-submarine warfare. The oceangoing tug *Allegheny*, based at Cape May, cruised at the breakwater near the mouth of the Delaware Bay, and Yard Captain Mathewson ordered the tug *Toka* "to make delivery of depth charges to U.S.S. *Allegheny* at Harbor of Refuge." The Philadelphia Navy Yard mobilized private craft owned by wealthy Delaware Valley yachtsmen under a plan similar to that of 1915–16 inspired by then-Assistant Secretary of the Navy and now President Roosevelt. Yachts *Ronaele* and *Scout* "stood in from Corinthian Yacht Club at Essington" in early January 1942, while steam yachts *Cythera*, *Nancy D*, and *Consort IV* arrived the following week. The 1,000-ton *Cythera* (designated as a PY or Patrol Vessel Converted Yacht) rushed to the Chesapeake, where a U-boat sank her off the North Carolina coast, killing Captain Thomas Wright Rudderow of Bryn Mawr, Pennsylvania.[19]

The Navy Yard sent PY conversions across the river to the Mathis Yacht Building Company of North Camden because League Island simply could not make such conversions. Every dry dock was filled with submarines being recommissioned or with contracts to repair battle damage for Allied warships, including a complete reshafting of the British cruiser *Manchester*, torpedoed in the Mediterranean. Moreover, shortages of steel and other critical materials at the government yard, available in some private establishments, tied up work. The government gradually developed a rational priority and scheduling system for allocation of strategic materials to navy yards and other war production industries, and the FDR administration created a War Production Board (WPB) headed by Sears, Roebuck and Company executive Donald Nelson.[20]

Unfortunately, the WPB failed to solve shortages in critical materials and machine tools for the Navy Department, so Under Secretary of the Navy Forrestal appointed his friend and fellow Wall Street organizational expert Ferdinand Eberstadt to reorganize the Army and Navy Munitions Board (the prewar agency for industrial mobilization planning) and, as vice-chairman of the WPB, to introduce Production Requirement and Controlled Materials Plans. Forrestal also established a naval Office of Material and Procurement (OM&P) to improve contracting with private industries. These organizational reforms broke most bottlenecks in Philadelphia as Industrial Manager Chantry defined production needs and material priorities and established schedules for rapid allocation of critical materials. Gradually the gap between material shortages and production needs closed, but in the interim Secretary of the Navy Knox urged that overworked and undersupplied navy yards contract with private outside firms in a "Farm Out" program. The Philadelphia Navy Yard farmed out work to 58 major contractors, most notably Babcock and Wilcox of New York and the Carrier Corporation of Syracuse (New York). Of these secondary contractors, 32 sheet metal fabricators and electrical contractors lay in the Delaware Valley, and 25 in Philadelphia and Wilmington, Delaware.[21]

Delaware Valley war contractors and private Delaware River shipyards drew skilled labor away from the Navy Yard. Major competitors included Sun Ship of Chester, where the government ordered nearly 100 oil tankers, the once bankrupt Cramp Shipyard in Philadelphia, and the huge New York Shipbuilding Corporation, which built 29 warships of over 12,000 tons displacement for the U.S. Navy during World War II. Gradually League Island attracted workers by raising wages, developing apprenticeship training incentives, and providing numerous benefits. The government naval facility also recruited nontraditional labor, including women, who comprised 16 percent, and "non-whites," who comprised

29 percent of the wartime work force on League Island. Still, manpower shortages became so acute that at one point Chantry used Philadelphia radio station WCAU to make an emotional appeal to the patriotism of local workers so that they would come to work on League Island. Workers signed on, but over 13,000 of the 64,000 civilian hires between 1942 and 1945 left the Navy Yard within a few months. Of these, 10 percent joined the armed forces and went to war, 20 percent were dismissed for absenteeism, and a disturbing 40 percent simply quit without recording reasons or providing availability status for war work as required by law.[22]

The Navy Department suggested that such disappearances occurred to avoid military service, but the director of the Third U.S. Civil Service Region in Philadelphia, Charles Demetrius Hertzog, offered another explanation. Hertzog reported that 1,200 "dissatisfied or disgruntled employees" at the Philadelphia Navy Yard had visited his office to complain about unfair promotion policies, abuse by uniformed military personnel, and favoritism toward white male employees. The local civil service czar recommended to Commandant Milo Frederick Draemel, who had replaced Watson in early 1942, that an "inquiry be authorized relative to alleged discriminatory practices in regard to the influence of religious, racial and family connections in employment matters at the Navy Yard."[23]

The Navy Department dealt with such civilian labor issues at all naval shore facilities during World War II through the establishment of personnel divisions and by strengthening traditional shop organizations with better training of supervisory personnel and standardization of promotion procedures. Assistant Secretary of the Navy Ralph Bard created these personnel divisions in August 1942 to coordinate labor-industry relations and two months later convened a labor-management conference in Washington. The larger navy yards at Philadelphia, New York, Boston, Mare Island, and Puget Sound (Washington) each sent eighteen delegates to the first Washington, D.C. naval labor conference. The goal of the meetings was to establish industrial order and "wed all the men and women of the Navy family into

WAVE Jane Moriarity, 20 January 1943. Women held every job on League Island except riggers, including all the chauffeur positions. Temple University Urban Archives.

a harmonious unit." The Philadelphia delegation returned to inform their powerful shop committeemen and industrial clerical group leadership that Washington wanted them to cut waste, save lives, and win the war by ensuring labor stability. Grafting the shop committee system onto the newly organized War Production Committee (WPC) in July 1943 facilitated better coordination of labor and personnel situations. The appointment of female supervisory personnel to the industrial shops helped as well.[24]

Despite these adjustments, the Philadelphia Navy Yard never appeared entirely comfortable with female employees during World War II. "Of course, we are all employing a good many women in lighter jobs, but they are not particularly fitted for work on board ship," one shipyard executive maintained. A female shipfitter recalled that supervisors would not allow women to work on board the battleship *New Jersey* for nearly a year. "They are of dubious value on board ships under construction, especially when the ships are nearing completion," concluded Naval Constructor Charles W. Fisher, Jr., director of the Navy's Industrial Survey Division. Navy Yard officials thought females more accident prone than male workers, as evidenced by Captain of the Yard Mathewson's most graphic Station Log accident report. "Margaret Hilt, age 45 years, Check No 336067 of 237 N. Ramsey Street, Philadelphia Pa. was killed when her skull and neck were crushed by a grinding machine." No accident report for male employees provided such violent detail. Under Secretary of the Navy Forrestal agreed that "due to the physical requirements of many unskilled jobs, it is not feasible in such cases to employ women as replacements."[25]

Forrestal would rather have League Island employ German and Italian prisoners-of-war than patriotic American females for navy yard industrial work. But the use of POWs created bitter opposition from Navy Yard officials. Commander of the Marine Barracks on League Island Albert E. Randall worried about finding enough guards to watch prisoners-of-war while they worked. At the same time, Mathewson explained that enemy prisoners might obtain classified information on ship movement or construction, while Industrial Manager Chantry warned of potential sab-

otage. "The only possible use that could be made of war prisoners would be in the cleaning of grounds, and there are sufficient American prisoners in the Naval Prison available for such assignment." Forrestal kept up the pressure, and toward the end of the war League Island employed nearly 600 German and Italian POWs as cooks, mess men, barracks custodians and in the lumber and scrap salvage yards, jobs traditionally reserved for local African American workers.[26]

Despite reservations, the Philadelphia Navy Yard command preferred female employees, including African American women, to POWs to fill industrial positions at the shipyard and at the Fort Mifflin Naval Ammunition Depot's shell loading plant. In 1942 women comprised 7,000 of 35,380 new hires, in 1943 9,000 of 29,000, and in 1963 6,000 of 24,000. The Naval Aircraft Factory employed the largest number, almost one-quarter of the total work force, compared with 16 percent female employment overall. Approximately 70 percent of the Navy Yard's female employees held clerical, office, or inside shop work. For instance, forty female ship carpenters trained at Dobbins Vocational School to form the "all-girl crew" of the small boat shop. The war drain on male employees opened nearly every job except riggers to females. The Navy Yard lowered height restrictions to allow women to serve as armed guards at entrance gates. League Island hired 500 female welders in 1943 alone, and by early 1945 "girl welders" made up half the deck construction crew for an aircraft carrier. "The girls who invaded the ships about six weeks ago know what to do with a hacksaw and a pair of shears," Navy Yard Personnel Supervisor Elizabeth McNeal Jennings insisted in 1943. "The cold penetrates their six layers of stockings and shirts and slacks and stiffens their fingers and toes; the dirt blackens their faces and eats into their pores; the fumes choke their throats and smarts their eyes, but the girls can take it and they do."[27]

Hiring female and "non-white" employees helped swell the civilian work force at the Philadelphia Navy Yard from 45,820 on 30 June 1942 to an incredible 56,848 by the end of December. The number continued to rise until the civilian work force peaked at

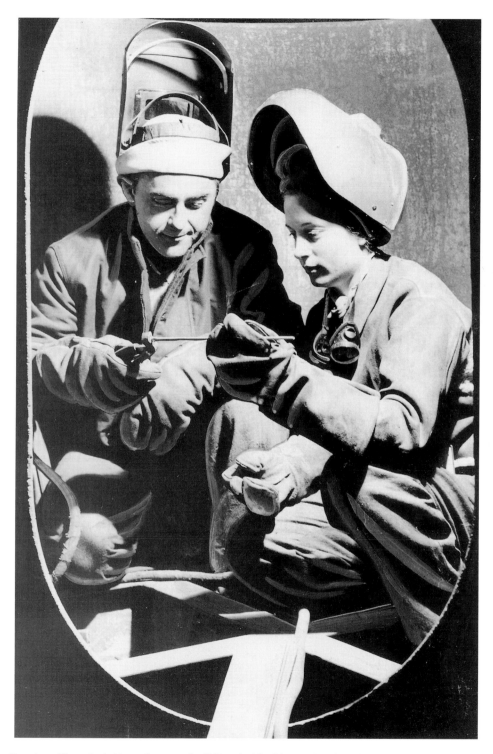

Female welders started to replace men in 1943 on inside ship construction jobs. Temple University Urban Archives.

58,434 on 26 March 1943, as the official station log book recorded 42,382 employed by the Industrial Department, 13,264 at the Naval Aircraft Factory, 1,359 at the Supply Department, and 1,119 "Miscellaneous" employees at the Recruiting Station and Naval Base. Another 330 civilian employees worked on League Island for the Fourth Naval District office.

Fifty thousand workers rushing to and from the Navy Yard each day during the height of the world war placed a tremendous strain on nearby communities. Transportation posed a problem from the beginning, worsening after Pearl Harbor when civilian employees were forbidden to drive private vehicles through the main gate into the Navy Yard. Gasoline rationing, put into effect on League Island on 1 December 1942, further complicated matters, as no supplemental gasoline cards were issued to Navy Yard workers unless they carried a maximum number of passengers in their automobiles.[28]

The *Navy Yard Beacon*, League Island's weekly newspaper published since November 1941, launched a "share your ride" campaign, printing names of drivers needing riders. League Island appointed a rationing officer (reputedly the first at any naval establishment) to determine who might get gasoline and tires, and chartered ten large motor buses to shuttle workers from the parking lots outside the gates to their waterfront dry dock and industrial shop jobs. The Navy Yard added a third ferryboat (*Reliance*) in August 1943 to carry the one-third of the League Island work force who resided across the Delaware River in New Jersey. Still, traffic jams clogged the six motor car lanes and trolley tracks along Broad Street between seven and nine o'clock in the morning and four to six in the evening. The Navy Yard negotiated with the city to restrict parking on Oregon Avenue and McKean Street and to regulate traffic lights during rush hours, based on the successful traffic-control system employed during annual Army-Navy football games at nearby Municipal Stadium (immediately north of the Navy Yard). Establishment of a roadway and east gate leading to the Naval Aircraft Factory, renamed the Naval Air Material Center in 1943, and opening a third entrance on the western end of

League Island, which crossed a new platform lift bridge over the Reserve Basin, gradually improved access to the wartime Philadelphia Navy Yard.[29]

Housing shortages continued to be a critical issue for wartime employees at the Navy Yard as well. The federal government and city of Philadelphia negotiated a deal before Pearl Harbor to construct Passyunk Homes along Penrose Avenue for civilian families and "enlisted men below the first three grades not receiving allowance for Quarters." Passyunk Homes failed to solve the problem. "The housing shortage for war workers in the Philadelphia area is very acute," the *Beacon* announced in October 1942, as it started a housing file that listed rooms, apartments, and houses for rent in the area. Additional low cost housing for Navy Yard workers was built in the city during the war, including Abbotsford Homes along Henry Avenue, Shipyard Homes at Twelfth and Packer, League Island Homes at 36th and Morris, Tacony Homes at Lewis and Tacony, and Oxford Village at Comly and Langdon. In February 1943, 100 dwellings became available for Navy Yard employees in the federal housing projects at Bellmawr Park, New Jersey, seven miles east of the Delaware River.[30]

As the local navy yard sorted out transportation and housing, waves of federal civilian war administrators descended on League Island to encourage industrial production and support for the war through purchase of war bonds. Secretary of the Navy Knox arrived in March 1942 for "a whirlwind inspection," declaring League Island "a magnificent yard" and adding that "the men here in Philadelphia Navy Yard, are just as important as the men aboard the ships." Assistant Secretary Bard visited in May and July and WPB Chief Nelson, Under Secretary of the Navy Forrestal, and Chief of the Office of Material and Procurement Frank M. Folsom inspected Dry Dock No. 5 (dedicated in May 1942) and toured League Island to evaluate production requirements and facilities. WPB Program Vice-Chairman William L. Batt spoke at the launching ceremony for seaplane tender *Currituck*, encouraging the Navy Yard to greater productivity.[31]

Civilian and naval executives boosted production

and morale through promotional programs, the awarding of prizes for "beneficial suggestions," and pennants for "Efficiency" ratings or war bond sales. The Navy Department gave "E" pennants to navy yards exhibiting superior overall efficiency, rate of production, quality of work, and job attendance: the Treasury Department awarded honor flags to navy yards with the highest percentage of employees purchasing war bonds through the government payroll savings plan. The Philadelphia facility won its first efficiency pennant in May 1942, adding stars each consecutive war year. The largest war bond rally in the United States during the war occurred in July 1944 on League Island. Secretary of the Treasury Morgenthau, Commander-in-Chief of the U.S. Fleet Ernest J. King, and Secretary of the Navy Forrestal (who had become secretary after Knox's death on 28 April 1944) arrived to launch the Navy's "Extra Cash War Bond Sale" as part of the Fifth War Loan Campaign. The Philadelphia Navy Yard gave 55,000 employees—whom Bond Coordinator Miles Lilly claimed had already subscribed $1 million in war loans—an unprecedented wartime one-hour lunch break to attend a massive rally held on the Marine Corps parade grounds on the east side of Broad Street. Each war administrator addressed the huge crowd, and then speeches on overseas radio from Commander of U.S. Naval Forces in Europe Harold R. Stark and Commander-in-Chief of U.S. Pacific Fleet Chester W. Nimitz were piped by loudspeakers to the crowd. The entire Philadelphia Navy Yard event was sent on "nationwide hookup of the National Broadcasting Company abroad by short wave."[32]

Local rallies and programs proved more important in boosting morale, promoting unity, and stimulating productivity on League Island. The Philadelphia Navy Yard Development Association, formed during the Great Depression to block outside labor propaganda, organized patriotic wartime rallies. At first these rallies featured the hanging in effigy of

The Philadelphia Navy Yard *Beacon*, which published its first edition in November l941, created the Lester Lukewarm character to encourage workers to full production. NARAM.

"Lester Lukewarm," a cartoon character invented by *Beacon* artists to represent lazy, indolent, or unpatriotic employees. Skits were held as well, condemning "Lucy Lukewarms," female employees allegedly "mooching aimlessly around the yard or washing up on the Government's time." Such activity soon grew tiresome, and the Association formed a Navy Yard Band, inviting Eugene Ormandy, music director of the Philadelphia Orchestra, to conduct the first noontime concert on League Island. The Association also sponsored Esther Williams, Judy Garland, Tyrone Power, John Wayne, and other motion picture stars for lunchtime appearances.[33] An official Incentive Office of the Industrial Department under Frank J. Kelly replaced the semi-official development association in 1943 to organize navy yard entertainment. On Mondays, Wednesdays, and Fridays between

Hollywood stars including Judy Garland entertained lunch-time crowds at the Philadelphia Navy Yard to encourage the sale of war bonds and boost morale. NARAM.

12:05 and 12:25 government employees on League Island watched a "Lunch Time Revue," talent show, beauty pageant, or the "Mid-Day Follies." For a time the Incentive Office showed Mickey Mouse, Porky Pig, or Donald Duck cartoons on screens set up around the Navy Yard, but these proved so popular that workers began to leave the job early to get a place near the screen, and Kelly canceled the movies. Tuesdays, Thursdays, and often Saturdays, workers cheered noontime boxing or wrestling matches held between shop "champions." In the evening, the Development Association continued to promote sports, particularly a "Girl's Bowling League" and an eleven-team Women's Softball League. These games were the most popular wartime event on League Island. "Besides playing a good ball game, the girls attired in natty playing togs, accomplish a good bit more in the way of morale building than is called for on the schedule."[34]

The Philadelphia Navy Yard had become a "completely self-contained" island community of over 50,000 men and women, civilian and naval, dealing with the social and economic issues of any small industrial city on the American home front during the world war. But this community differed from a civilian factory town or industrial community, like nearby Philadelphia, Camden, Chester, Gloucester City, or Wilmington. The Navy Yard's sole reason for existence was to prepare the fleet for total war, providing manpower, supplies, and ships for combat. The Navy Department accentuated this mission in August 1942 by assigning Admiral Nimitz's hard-driving chief of staff Milo Frederick Draemel, who had just reorganized the Pacific Fleet to fight Japan, to command the Philadelphia Navy Yard. Draemel displayed a "comprehensive knowledge, broad vision and executive ability," Secretary Knox observed. He had first showed such organizational ability by greatly tightening telephone and telegraph security during World War I as head of the Code and Signal Section and as commandant of the U.S. Naval Academy, 1937–40. Draemel showed a singleminded purpose to forge the Philadelphia Navy Yard's civilian and military personnel into a fighting machine: "A Navy Yard cannot be run like a [civilian] assembly-line production plant or even like many shipyards," but as a military operation. Draemel suffered no interference from civilian administrators or politicians in Navy Yard management, once telling Philadelphia U.S. Representative Michael J. Bradley that he would not tolerate congressional attempts to reorganize the naval shore establishment during the war.[35]

Pearl Harbor had already defined the war crisis for the Navy Yard in Philadelphia that required Draemel's type of leadership. The Yard had learned on 8 December 1941 that *Cassin* and *Shaw*, two destroyers built on League Island in 1935, were among the eighteen U.S. Navy warships sunk or damaged by the "dastardly attack." Soon the Navy Yard realized that the older battleships *Oklahoma*, *Utah*, and *Pennsylvania*, all modernized on League Island during the 1930s, had been sunk or crippled as well. "We are determined not only to replace the ships which we made, and were lost," Navy Yard officials announced in December 1941, "but to send many more in their place to take up their fight and our fight where the *Cassin* and the *Shaw* had to leave off."[36]

From Pearl Harbor to V-J Day, the Philadelphia Navy Yard built forty-eight new warships, converted forty-one, repaired and overhauled 574, completed and dry docked 650, outfitted over 600 (including all the warships constructed on the Delaware River during the war), and degaussed (protected against magnetic mines) 700 vessels. The Naval Aircraft Factory fabricated 500 aircraft, the Supply Department handled over $100 million a year in stores, the Boiler and Turbine Laboratory trained 14,000 water tenders and fireman, and the Propeller Shop made 5,500 ship propellers. The Receiving Station processed 70,000 Navy recruits (13,000 men and women a year) in this period of the greatest expansion and activity in the history of the Philadelphia Navy Yard.[37]

League Island's first wartime launches occurred on 12 February 1942, when Destroyers *Butler* and *Gherardi* slid down the shipways into the Delaware River. Arguably, the most significant launch in Philadelphia Navy Yard history occurred on the first anniversary of Pearl Harbor 7 December 1942 when League Island launched battleship *New Jersey*. However, *New Jersey*'s launching procedure revealed that the navy yard had more adjusting to do in engineering and industrial planning if it was to contribute efficiently to the war effort. Naval architects and engineers underestimated the drag resistance necessary to safely launch the largest ship yet built on the Delaware River. *New Jersey* raced down the shipways, traveled 1,000 feet beyond the estimated launch distance, and, as if headed to its namesake state, struck the New Jersey shore opposite the Navy Yard. "It sent a huge wave over the Jersey side at National Park and drenched some who were on the beach there," recalled shipfitter James Mutchler. The Navy Yard launching team learned from this exercise and greatly modified the system for the *Iowa*-class battleship *Wisconsin*, under construction nearby on Shipway No. 3, and launched without incident on the second anniversary of Pearl Harbor.[38]

In 1942 the Navy Department changed building priorities from battleships to destroyer escorts for an anti-submarine war in the Atlantic, before moving against Japan in the Pacific with battleships and

Mrs. Charles Edison christening battleship *New Jersey* at the Philadelphia Navy Yard on the first anniversary of Pearl Harbor. Naval Historical Center Photo NH 45485; courtesy Allan J. Drugan.

Bow view of 45,000-ton battleship *Wisconsin* under construction at the Philadelphia Navy Yard in 1943. NARAM.

aircraft carriers. The Navy Department ordered Philadelphia to construct destroyer escorts, before completing the two battleships or starting work on three contracted aircraft carriers. The destroyer escort was smaller, more economical and quicker to build than the standard-sized destroyer. League Island launched six 1,140-ton *Evarts*-class destroyer escorts in July 1942, all earmarked for transfer to Great Britain under Lend Lease. *Anders*, *Engstrom*, and *Drury* (first named *Cockburn*) actually served for a time in the British Navy until returned in 1943. The U.S. Navy retained *Decker*, *Dobler*, and *Doneff* for vital convoy duty, particularly for escorting oil tankers along the U-boat-infested waters of the Atlantic coast. The Navy Yard launched destroyer escorts *Scott* and *Burke* in early 1943, and Under Secretary Forrestal flew into Mustin Field on 29 May to christen the 1,400-ton destroyer escorts *Enright*, *Coolbaugh*, *Darby*, *Blackwood*, *Robinson*, and *Solar*—all built simultaneously in Dry Dock No. 4. Sara Blackwood became the first Navy WAVE (the women's naval reserve auxiliary force created in July 1942) to participate in a navy yard launching ceremony, for *Blackwood*, named after her father, a naval doctor killed in the South West Pacific Theater. The Navy Yard launched *Crosley* and *Cread* in April 1943, *Fowler* and *Spangenberg* in July, *Rudderow* and *Day* on 14 October, and *Ruchamkin* and *Kirwin* in early 1944, the last of twenty-two destroyer escorts built on League Island during World War II.[39]

Larger strategic needs shaped League Island's wartime production schedule. FDR announced an emergency program in April 1942 to build landing craft for North African, European, and later Pacific amphibious operations: consequently the Philadelphia Navy Yard launched the first six of fourteen landing ship tanks (LSTs) on Navy Day, 27 October 1942 and six more in February 1943, all built in one dry dock. The launchings (or more correctly floating ceremonies, since the craft were simply launched by filling the dry dock with water) gave Draemel an opportunity to implement his idea of recognizing local connections to individual craft shop organizations. The wives, daughters, or sisters of employees in Shop 61 (Master Shipwrights/Woodworkers), Shop

17 (Sheet Metal Workers), Shop 11 (Shipfitters), Shop 38 (Outside Machinists), Shop 31 (Inside Machinists) and Shop 53 (Pipe Fitters) sponsored the LSTs in February 1943—two female relatives each from Philadelphia, the Philadelphia suburbs, and Gloucester County in southern New Jersey, representing the residency distribution of Navy Yard workers during the war.[40]

With victory in Europe becoming more certain by 1944, League Island turned to construction of heavy cruisers and aircraft carriers to fight the Pacific War against Japan. On 20 August 1944, in one of the more elaborate and publicized ceremonies at the Philadelphia Navy Yard, Under Secretary of the Navy Bard, the mayors of Chicago and Los Angeles and their wives, and influential Senator Millard E. Tydings of Maryland and his wife christened the 13,000-ton cruisers *Chicago* and *Los Angeles* and *Essex*-class aircraft carrier *Antietam*. The 830-foot, 27,000-ton *Antietam*, the first carrier constructed on League Island, required an unprecedented sixteen tugboats to turn the hull at its berth after launching. The following month, the Navy Yard laid the keel for the 5,000-ton landing ship dock *San Marcos* with Mrs. J. Lloyd Coates of Ardmore, Pennsylvania, who had organized recreational facilities at the Navy Yard during the war, sponsoring the launching in January 1945.[41]

The Navy Yard laid keels for two more aircraft carriers during the war: *Princeton*, launched in July 1945, and *Valley Forge*, completed in November 1945 after V-J Day. Assistant Secretary of the Navy H. Struve Hensel, a Princeton University graduate, spoke at the *Princeton*'s launching ceremony. Alluding to the Revolutionary War Battle of Princeton for which the big carrier was named, Hensel called construction "a pledge to the past." The Navy official also saw the aircraft carrier as "a prophecy for the future" that might prevent the same type of revulsion to war and isolationism that gripped the country after World War I.[42]

Unknown to Hensel, League Island had revealed "a prophecy for the future" already, when in September 1944 a wooden building, storing uranium for the Manhattan Project, exploded killing two and badly

Cruisers *Los Angeles* and *Chicago* (bow visible on the right), launched in August 1944. NARAM.

burning nine. The accident passed virtually unnoticed at the time. Industrial mishaps and deaths had become a frequent occurrence on League Island as thousands of hastily trained employees were rushed into emergency repair or construction jobs. Workers fell into dry docks, became entangled in moving machine parts, or were burned by acetylene torches setting off flammable material or by illegal barrel heaters under oily tarpaulins (called salamanders)

igniting sudden fires. A cigarette tossed into a trashcan caused a fire on the British cruiser *Manchester*. Moreover, there had been worse accidents at the Navy Yard before, including a terrible chemical fire on *Vixen* in Dry Dock No. 1 that killed three and severely burned thirty-one civilian and thirteen naval personnel.[43]

To a careful observer, however, the September 1944 accident near Building No. 638 at the Naval

Tugboats turn the 830-foot, 27,000-ton aircraft carrier *Antietam* after launching on 20 August 1944. *Antietam* was the first of three large attack carriers built on League Island; *Princeton* and *Valley Forge* followed in 1945. NARA Washington, D.C., Still Picture Branch.

Boiler and Turbine Testing Laboratory seemed different from the usual Navy Yard accident. The casualties were U.S. Army privates, first class, said to be technicians from an Army experimental research facility at Oak Ridge, Tennessee. More important, there had been no fire, only intense heat that caused severe radiation burns made by "mineralized explosives . . . known to be more sensitive and of greater power than the standard explosives, such as TNT." Overheated containers of a uranium compound caused the explosion at the "large thermal diffusion plant" designed by brilliant atomic scientist-engineer Philip H. Abelson of Carnegie Institute and Naval Research Laboratory scientists to separate isotope U-235 from uranium ore and produce nuclear fuel for the first atomic bombs. League Island had been selected by the Naval Research Laboratory because the Naval Boiler and Turbine Laboratory at

the Navy Yard generated enough heat (1,000°F) and pressure (1,500 pounds per square inch) to separate the pure isotope.[44]

After the accident, the lab continued to provide 5,000 pounds of uranium for Oak Ridge, most likely used in the atomic bombs dropped on Japan in August 1945. The tiny atomic plant at the Philadelphia Navy Yard operated until September 1945, when it closed and the radiation-contaminated material was reportedly dumped off the New Jersey coast. Abelson stayed at the Navy Yard before returning to the Naval Research Lab in Maryland to work on an atomic propulsion plant for submarines. His research in Philadelphia inspired Hyman G. Rickover and other nuclear submarine pioneers to build the U.S. Navy's first atomic powered submarine *Nautilus*.[45]

Other activity at the Navy Yard during the last

months of 1945 seemed far more mundane, although the Naval Air Material Center experimented with rocket technology. Preparing to retire at the end of the year, Draemel and Chantry confronted the same demobilization, postwar reorganization, and ship scrapping duties as Commandant Hughes and Industrial Manager Robert after World War I. In late 1945, League Island scrapped the partly finished hulls of heavy cruisers *Scranton* and *Norfolk* and 45,000-ton battleship *Illinois*, sold surplus material, and closed rented facilities on the north bank of the Back Channel. Chantry deactivated ships in the Reserve Basin, and Draemel began the process of demobilization and reduction of force. In the midst of planning the most difficult postwar readjustment in

Navy Yard history, Commandant Draemel received orders in November to disestablish the Philadelphia Navy Yard and place all industrial functions in a "Philadelphia Naval Shipyard" and "non-industrial" activities in a "U.S. Naval Base, Philadelphia."[46]

The order came as something of a jolt to the League Island command, prompting Draemel to ask Secretary of the Navy Forrestal to clarify "the uncertainties incident to reorganization." Actually, institutional change had been under consideration throughout the war as part of a struggle between civilian administrators and Commander-in-Chief of the U.S. Fleet and Chief of Naval Operations King over control of naval shore establishment management. The contest became heated at times, and only

This minor visible damage after an explosion on 2 September 1944 in the uranium separation plant attached to the fuel oil boiler laboratory obscured the severe radiation burns that killed two and injured seven others. The accident meant that League Island would never become a nuclear navy facility. Temple University Urban Archives.

FDR's intervention on the side of civilian control and the convening of a "Top Policy" board to draw up a plan for postwar organization that envisaged creation of separate base and shipyard commands eased the bitter bureaucratic infighting. Shortly after Vice President Harry S Truman replaced the deceased FDR as president in April 1945, Forrestal received permission "to proceed with reorganization of the Navy on a staff basis with particular attention to the management and handling of the Navy's industrial establishment—Navy Yard, Ordnance Plants, etc." Consequently, Forrestal rushed reorganization. "The Navy is by the very nature of material forced to take definite action," he told Secretary of War Robert P. Patterson in November, because "ships either have to be laid up or kept in commission; they cannot be left hanging in mid-air."[47]

Forrestal worried that, unless the Navy took immediate steps to reorganize its own house, it would become subject to the unification philosophies of General George C. Marshall, President Truman, and others who expected to centralize the armed services after the war into a single defense department. Moreover, Forrestal wanted to define postwar naval industrial and military institutions, manpower levels, and fleet size at once so that the Navy could plan for what he saw as an inevitable postwar struggle between American democratic capitalism and Soviet Communism. Whatever his institutional vision, on November 1945, Forrestal ordered Philadelphia to establish separate naval shipyard and naval base commands. Thus reconstructed, League Island confronted the inevitable cutbacks in manpower, abandonment of preparedness planning, and disposal of ships that had accompanied every postwar period in Philadelphia since the Revolutionary War.[48]

11 ★ COLD WAR NAVY YARD

LEAGUE ISLAND Production Officer Louis Dreller warned in late 1945 that a return to peacetime presented new problems for what had become one of the largest naval industrial establishments in the world. "Demobilization, reconversion, and tension and uncertainty, which is the aftermath of War is with us," Dreller explained. With engineering degrees from the University of New Hampshire and Columbia University, the naval engineer knew that these postwar tensions arose in part from confusion over how industrial reorganization of the naval establishment would affect further development of wartime advances in weapons technology, including radar, sonar, rockets and guided missiles, jet aircraft, nuclear ship propulsion, and most dramatically "the use of the atomic bomb as a military weapon."[1]

In February 1946, when Chief of Naval Operations Chester Nimitz received a Distinguished Service Award for leading the Pacific Fleet to victory in Japan, he told Mayor Bernard Samuel and a Philadelphia banquet audience that the atomic bomb had not rendered obsolete ships of the fleet such as those in the Reserve Basin at League Island. Nimitz warned, however, that advocates of the atomic bomb and strategic land-based air power doctrine as the first line of postwar national defense sought to diminish the role, mission, and budget of the U.S. Navy. "The sea power built at such tremendous cost of toil and money, is again in danger of neglect." Worse, the Navy might become an appendage of a proposed unified national defense organization fa-

vored by President Truman and Army Chief of Staff Marshall.[2]

As Nimitz suspected, the Navy soon became entangled in a tremendous power struggle between proponents of naval sea-air power, led by Secretary of the Navy Forrestal and senior naval officers, and Army Air Force advocates of strategic air power and a single department of defense. This defense unification struggle, which resulted in passage of the National Security Act of 1947, creating a national military establishment, office of a secretary of defense, autonomous air force, central intelligence agency, and other national security institutions, dominated the postwar naval establishment. President Truman selected Forrestal as his first secretary of defense to secure Navy cooperation in forming this new organization. Forrestal embarked on a troubled two-year crusade to organize the unpopular National Military Establishment and distribute the ever-shrinking defense budget evenly among army, navy, and air force to provide security against the "fanatic Soviet intention to dominate the free world with communisms." Each month this struggle against world communism became more intense, emerging finally into a full-blown Cold War and most likely contributing directly to Forrestal's nervous breakdown and untimely death in 1949.[3]

Signs of such tension appeared in late 1945 at the Philadelphia Navy Yard. Departing wartime Commandant Milo Draemel found "apprehension" everywhere about the "streamlining" of defense organization, disestablishment of the Navy Yard, and seem-

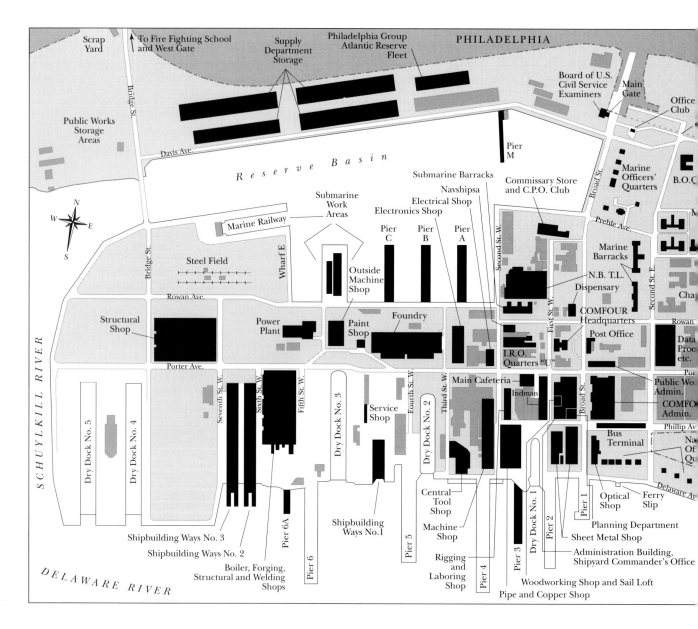

Scrap Yard

To Fire Fighting School and West Gate

Supply Department Storage

Philadelphia Group Atlantic Reserve Fleet

PHILADELPHIA

Board of U.S. Civil Service Examiners

Main Gate

Office Club

Public Works Storage Areas

Bridge St.

Davis Ave.

Reserve Basin

Pier M

Broad St.

Marine Officers' Quarters

B.O.Q

N
W E
S

Submarine Work Areas

Submarine Barracks

Commissary Store and C.P.O. Club

Navshipsa

Electrical Shop

Electronics Shop

Second St. W.

Preble Ave.

Marine Barracks

Marine Railway

Steel Field

Bridge St.

Wharf E

Pier C

Pier B

Pier A

N. B. T. L.

Dispensary

COMFOUR Headquarters

Post Office

Rowan

Data Proc. etc.

Cha

Rowan Ave.

Outside Machine Shop

First St. W.

Porter Ave.

Structural Shop

Power Plant

Paint Shop

Foundry

I.R.O. Quarters "U"

Por

Public Wo Admin.

COMFO Admin.

Phillip Av

Na Of Qu

SCHUYLKILL RIVER

Seventh St. W.

Sixth St. W.

Fifth St. W.

Dry Dock No. 3

Service Shop

Fourth St. W.

Third St. W.

Main Cafeteria

Indman

Bus Terminal

Dry Dock No. 5

Dry Dock No. 4

Pier 6A

Shipbuilding Ways No.1

Central Tool Shop

Machine Shop

Dry Dock No. 2

Dry Dock No. 1

Pier 2

Pier 1

Optical Shop

Ferry Slip

Delaware Av

Planning Department

Sheet Metal Shop

Administration Building, Shipyard Commander's Office

Shipbuilding Ways No. 3

Shipbuilding Ways No. 2

Pier 6

Boiler, Forging, Structural and Welding Shops

Pier 5

Rigging and Laboring Shop

Pier 4

Pier 3

Woodworking Shop and Sail Loft

Pipe and Copper Shop

DELAWARE RIVER

1949–69. The Cold War Navy Yard was capable of building, overhauling, and mothballing in reserve every non-nuclear-powered warship from submarines to the largest conventionally powered attack aircraft carriers.

ingly precipitous creation of the "Naval Shipyard, and Naval Base, Philadelphia, Pennsylvania." The new structures overwhelmed the first commander of the Philadelphia Naval Shipyard, Andrew Irwin McKee, a submarine expert fresh from duty as Nimitz's Pacific Fleet maintenance officer. McKee doubted that he had authority over its local inspection and survey boards, surplus property agencies, and industrial plants responsible for deactivating and disposing of ships and supplies. McKee asked outgoing Captain of the Yard Rufus Mathewson, designated naval base operations officer under the reorganization, where these offices fit into the new scheme. "Are these now part of Production Dept., or Planning or . . .?" Mathewson retired before assisting McKee. Perhaps Industrial Manager Alan Chantry could explain the arrangements, since he had redesigned the civilian industrial and naval-military functions on League Island during the war "specifically and solely to this organization as it related to the Fourth Naval District." Unfortunately, Chantry headed into retirement as well, leaving McKee to face reorganization and industrial demobilization alone.[4]

Former navy yard commander Draemel remained as the only wartime leader on League Island to ease transition to the new system, serving as Fourth Naval District commandant until May 1946. But the reorganization troubled Draemel, too. He asked Forrestal whether the physical plant at the "old Navy Yard" was part of the new naval shipyard organization or a component of the "Naval Base Station"? Anxious to settle such internal departmental reorganization so that he could concentrate on the unification fight, Forrestal defined the proper alignment of functions and offices at the former Philadelphia Navy Yard. "The Secretary of the Navy has specifically directed that the following components located within and contiguous to the geographical area of the Naval Base are district unit," including the naval shipyard, base station, ammunition depot, training school for material preservation, and damage control center. Furthermore, "the Secretary of the Navy has made the U.S. Naval Submarine Supply Center a subordinate component of the U.S.

Philadelphia Naval Shipyard," and placed the naval disciplinary barracks, commissary store, district disbursing office, and Naval Air Material Center under cognizance of the naval base station.[5]

These ambiguous institutional adjustments convulsed League Island for months, but by mid-1946 the Philadelphia Naval Shipyard was aggressively pursuing its postwar fleet deactivation mission to "dispose of the vessels by sale as a hulk for scrapping, by scrapping by the Navy, by transfer to another Government department, by return to the owner, by transfer to foreign power or by placing in an inactive status." League Island deactivated fifty-eight ships in the Reserve Basin as the Philadelphia Group, 16th Fleet U.S. Atlantic Reserve Fleet, including battleships *California*, *Tennessee*, and *South Dakota*, light aircraft carriers *Cabot*, *Monterey*, *Langley*, and *Bataan*, and the nearly completed cruiser *Galveston*. The Naval Shipyard also overhauled ships for transfer to the Maritime Commission, return to private owners, or employment by the Fourth Naval District, Naval Reserve Training Program "to insure ready reserve in case of a future national emergency."[6]

League Island entered the atomic age by stripping armament and installing test equipment on the 27,000-ton battleship *New York* as a target ship for "Operation Crossroads," the atomic bomb test scheduled for July 1946 at the Bikini atoll in the Marshall Islands. At the same time, the Philadelphia Naval Shipyard prepared the captured German battle cruiser *Prinz Eugen*, surrendered at the Potsdam Conference, for inclusion in the 97-ship "guinea pig" task force sent to the Pacific islands for the Bikini test, it also scrapped fifteen 1,200-ton World War I-vintage destroyers and older *Omaha*-class, four-stacker cruisers *Memphis*, *Detroit*, *Richmond*, *Trenton*, and *Concord*. The *Concord* had fired the last shot of the World War II, and its six-inch guns were removed for shipment to a naval museum in Washington, D.C. Scrapping started in early 1946 on cruisers *Scranton* and *Norfolk* and 60,000-ton battleship *Illinois*, all laid down in 1945 and nearly 30 percent complete on the stocks when the war ended. Scrapping *Illinois* continued until January 1947, with surplus material sold and some armor stored under cover.[7]

The ship scrapping program on League Island in 1946 netted the government over $1 million. Meanwhile, League Island disposed of $2.5 million in surplus naval material, supplies, and real estate under the direction of the War Assets Administration. The Navy Yard "Termination Unit" ended the purchasing of new supplies, canceled "farm-out" contracts, and returned property leased from the city, which included the degaussing station on the South Street wharves and storehouses on Delaware Avenue and Lippincott, Wishart, and Venango Streets. League Island disposed of surplus housewares, electrical appliances, paper products, hardware, and plumbing and heating supplies, but threw away valuable canned foodstuffs "needed in European Hunger Areas [and] found discarded by Navy on City Dump." Twenty-six employees were suspended for removing discarded government property from trash cans and scavenging from the DDT-soaked dumps on League Island that were polluting the nearby Delaware and Schuylkill Rivers. At first, the Navy Yard merely suspended employees temporarily for such activity, but after determination in 1947 that such surplus property "cannot legally be salvaged for personal use," arrested and dismissed them for theft.[8]

All the while, the Navy Yard laid off civilian workers and cut naval personnel. Naval officer personnel on board fell from 345 in 1945 to 82 in 1949, and WAVES from 639 to 172. Five thousand civilian workers resigned after V-J Day; the Navy Yard laid off 7,000 mechanical workers, riggers, ship fitters, electricians, welders, and sheet metal workers in late 1945, and 1,000 machinists and helpers, printers, gas burners, welders, electricians, crane men, and unskilled laborers and 600 clerical and professional staffers in February 1946. An additional 85 chauffeurs, locomotive engineers, and firemen, 224 public works employees, and 731 tool makers followed in March, and another 1,094 chippers, caulkers, pipe fitters, riggers, electricians, and shipwrights in April. With the Truman administration setting a peacetime budget ceiling in 1946, League Island released 1,000 workers each week beginning in July until the work force reached the authorized level of 9,000.[9]

Buffeted by the Truman-era peacetime economy in government spending and strict postwar Congressional limitations on naval appropriations, the Philadelphia Navy Yard established a "Reduction-in-Force (RIF) Unit of the Industrial Relations Division" to assist laid-off workers find employment. This agency was remarkably similar to the one that fifty years later would aid League Island employees "riffed" by the Base Realignment and Closure Commission. The RIF Unit decided "who goes and who stays" to complete work on the *Essex*-class aircraft carrier *Valley Forge*, laid down in September 1944 and completed in November 1946. Personnel supervisors gave "Retention Credits" based upon efficiency ratings, veteran status, or length of government service. "For example a Shipyard employee who had been employed in the Frankford Arsenal for 2 years, had worked 8 years in the Naval Shipyard, and had 3 1/2 years of military service would be entitled to a total of 13 1/2 retention credits," explained a RIF Unit spokesman. "Long-service career" employees were placed first in Philadelphia-area business and industrial jobs.[10]

The Navy Yard personnel office developed ties to the U.S. Employment Service on South Broad Street, while the RIF Unit sought job vacancies overseas for shipyard employees, particularly in the Marianas, Pearl Harbor, or other Pacific U.S. naval facilities near postwar Far Eastern trouble spots. The RIF Unit found jobs for female employees, who were veterans of the armed forces and "have emerged from regimental life as charmingly feminine as a woman can be." Assistant Navy Yard Personnel Director Elizabeth McNeal Jennings warned that female employees would not be hired unless they accepted inferior positions with lower pay than those of the men "without talking about them," and then might become eligible "for election to the Shop Committees on the same basis as males."[11]

The civilian work force on League Island plummeted to 8,400 by 1949, while "drastically decreased appropriations" limited Philadelphia to minor overhaul and repair work. "Industrial Managers' work involving ship alteration and overhaul was practically nonexistent in the Fourth Naval District during the six years prior to January 1951," Naval Shipyard Commander Leslie A. Kniskern observed. There was no new naval ship construction for over a decade after *Valley Forge*, and improvements during this interval related to maintenance in the Reserve Basin of the inactive 50-ship 16th Fleet. Constrained by the economy and efficiency programs of the Truman administration, League Island underwent a period of transition, gradually being transformed from an outdated government factory into a modern industrial business with the Naval Industrial Fund, "a corporate type of financial budgeting and accounting control system" modeled after the Hoover Commission of Government Reorganization's recommendation for a comptroller's office in the Department of Defense.[12]

The Naval Shipyard fell under the "coordination control" of the Naval Base with management exercised through the Bureau of Ships. A supervisory staff in planning, production, public works, supply, comptroller, medical and administrative departments, and the heads of industrial relations, management planning, and review departments served as a corporate board of directors. Masters, foremen, and shop supervisory personnel attended management training courses at the Wharton School of Business, University of Pennsylvania. League Island consolidated and centralized material and weight handling functions in the Supply Department of the Naval Base, and crane equipment under the Production Department of the Naval Shipyard. Tool and service shops were "unified" in 1948 under a Central Tool and Service Shop, and the Offices of Naval Architect and Marine Engineer merged into a single office of Ship Design Director.[13]

The Truman administration's obsession with economy in government underlay postwar reorganization and operations on League Island. In 1947, Shipyard Commander Homer N. Wallin implemented an "economy program" and boosted morale by instructing heads of departments to ensure that everyone shared in the Beneficial Suggestion Program, so

successful during the recent war, that offered cash prizes for suggestions designed to save money. In addition, Wallin mobilized the Navy Yard Development Association to assist him in orderly long-range planning and directed the Management, Planning and Review Division to determine "what economies could be effected" through consolidation of offices and functions. "If this Yard is to take its place among the other naval activities that it justly deserves, the maximum of economical measures must be observed."[14]

Despite such economy measures, League Island's work force and the Shipyard's role and mission in support of the fleet declined, reaching a postwar low in early 1949. The nadir coincided with the 1949 appointment of Louis A. Johnson as secretary of defense. Johnson "hated" the Navy, and promptly canceled contracted warships, including the super-aircraft carrier *United States* at the Norfolk Navy Yard and a cruiser promised to the Philadelphia Naval Shipyard. At the same time, Johnson poured money into Air Force development of a long-range B-36 atomic bomber by a corporation in which he held an interest. Such action forced Chief of Naval Operations Louis Emil Denfeld to start a "revolt of the admirals" against Truman administration defense policy and organization.[15]

Rumors circulated in Philadelphia that the Navy planned to close League Island and combine Fourth Naval District offices in Philadelphia with the Fifth Naval District, headquartered in Norfolk, Virginia. Fourth Naval District Commander James L. Kauffman fought to keep his headquarters in Philadelphia, blaming other governmental agencies for plotting to close the district. "I do not believe that any change in the present District boundaries should be made *purely* to agree with areas of other Government departments." Meanwhile, Philadelphia Democratic Representative William J. Green, Jr., and United States Senator Francis J. Myers inquired whether the Democratic administration in Washington intended to close the Philadelphia naval shore establishment. Defense Secretary Johnson evaded the issue, telling Myers that it was all "merely a rumor," while President Truman insisted that he knew nothing about closing the Philadelphia Naval Shipyard.[16]

The Truman administration generated constant fear of base and shipyard closure on League Island, as did budget cutters' decimation of the local military establishment in 1949, which included laying off over 4,000 government employees at the Frankford Arsenal, Marine Corps Supply Depot, Naval Home, Quartermaster General Depot, and Signal Corps Stock Control Office, reducing the federal payroll in Philadelphia by $13 million. Meanwhile, the Department of Defense closed nearby Atlantic City Naval Air Station at Pomona, New Jersey, and announced plans to release an additional 2,000 employees from the Naval Shipyard in Philadelphia. Seemingly, League Island confronting its darkest moment, when the Korean War in June 1950 brought the Philadelphia naval facility back to the center of the American defense establishment.[17]

As part of the United Nations resolution to stop the North Korean invasion of South Korea, Truman ordered American ground troops to Korea and a naval blockade of the Korean coast. The president asked Congress for $10 billion in rearmament appropriations and called for partial mobilization. In July League Island announced 3,700 job openings, recruiting for twenty-one "critical trades" such as electricians, machinists, and draftsmen, and offering a new wage schedule that ranged from $1.51 to $1.72 an hour for skilled machinists to $2.23 for heavy forgers. The employment office at the Main Gate on Broad Street remained open around the clock. The response was disappointing, with local workers preferring the higher wages in area industries, particularly the smaller local companies that saw a 27 percent increase in employment during the Korean crisis. "To attract people to the Shipyard, all recognizable means of advertising were resorted to along with recruiting trips throughout the State of Pennsylvania," Shipyard Commander Kniskern reported. League Island instituted training programs for masters, foremen, and industrial management personnel at the Wharton School, prepared an "Electronics Training Package" with the Philco Corporation, and

offered apprentice drafting and engineering orientation courses in the school district of Philadelphia.[18]

League Island employed 12,500 shipyard workers by the end of 1951 and 14,750 at the height of the Korean "police action" more than at any time during World War I, although well below force levels during the World War II. This modest increase placed a burden on local housing, and the Philadelphia Metal Trades Council urged the city to impose rent controls to benefit shipyard labor and their families. Meanwhile, Kniskern resurrected World War II lunch-hour shows to maintain morale "at a high level," with performances by the Shipyard String Band, war bond rallies, and the Beneficial Suggestion Program.[19]

The Korean War revived passive defense measures on League Island, abandoned after V-J Day. Navy Base and Fourth District Commander Roscoe E. Schuirmann, a former naval intelligence director, tightened security around the installation when the president announced the call-up of reserves in 1950 for the Korean crisis. "The current international situation has increased the problems of sabotage," Kniskern agreed, and instituted sabotage control, closed the government facility to visitors, and installed flood lights around dry docks and steel storage yard. The latter created an "upward reflection" and "sky glow" for enemy bombers that prompted Kniskern to ask the Navy Department for $190,000 to "dim-out" League Island. But the Navy Industrial Relations Office rejected the application, insisting that improvements "in electronic devices" (radar) made dim-outs unnecessary and took money away from the vital training programs for "radiological" defense against an atomic bomb attack.[20]

The Navy Yard instituted some atomic defense training in 1947, and developed "a local disaster plan for radiological, bacteriological and gas defense." With successful Russian testing of an atomic bomb in 1949, the naval establishment on League Island formed a radiological defense unit and designated the Navy Yard's "industrial hygienist" as "radiological health officer." Such measures were necessary, Wallin explained, because "the use of the atomic bomb as a military weapon has introduced a new and insidious hazard to personnel, namely radioactivity." Nevertheless, on the eve of the Korean War in 1950, the Navy Yard had not developed a decontamination program or trained decontamination teams to go on board ship. League Island possessed few "instruments" to detect radioactivity, and the Office of Industrial Relations urged the distribution of pocket cards in industrial shops, describing the proper action to take in the event of an "air burst of atomic bombs." Two months after the United States entered the Korean War, League Island opened a Radiological Decontamination Training Facility in Building No. 681 and distributed manuals and textbooks for the "Ship Decontamination, Radiological Training Program."[21]

Gradually, Philadelphia readied United Nations naval forces for operations in East Asian waters. Ships reactivated included destroyer escorts *Miles*, *Riddle*, *Swearer*, *Wingfield*, *Bright*, and *Cates* and 12,000-ton aircraft carrier *Langley*, renamed *Lafayette* and transferred to France under the Mutual Defense Assistance Act of 1949 to give military aid to America's allies in the Cold War. The Naval Shipyard overhauled submarines *Blower* and *Bumper* for transfer to the Turkish navy, and reactivated nine ships for duty in Korean waters from the Philadelphia Group, Atlantic Reserve Fleet, including cruiser *Macon*, aircraft carriers *Cabot* and *Monterey*, escort carriers *Block Island* and *Siboney*, and destroyers *Warrington* and *Wood*, employing for the first time the standardized planning procedures of the 1951–52 "Cost Reduction Program." Philadelphia modified seaplane tender *Currituck*, built on League Island in 1945, to maintain and refuel heavy seaplane bombers for operations over Korea, and performed snorkel and sonar conversions on submarines *Irex*, *Clamagore*, *Sea Cat*, and *Guavina*. Most submarine conversions during the Korean War went smoothly, earning the yard a Navy commendation, but in April 1952 an explosion and fire in the battery room on *Requin* killed one and severely burned twenty-five.[22]

The Korean War crisis also expanded research and development programs on League Island. The growth of testing, design, and research laboratories had been the bright spot during the postwar era of cutbacks. Five hundred engineers, draftsmen, constructors, and scientists worked on a variety of projects, from completely redesigning the island command center for antisubmarine warfare on the aircraft carrier *Cabot* to a bizarre and dangerous program in 1947 to test the impact of jet engine noise on the human body by enclosing volunteer sailors in a tiny testing chamber where lab technicians "shot a terrific stream of air at them, accompanied by an ear-splitting racket." Most programs proved more rational, notably the Greater Underwater Propulsive Power Program (GUPPY), in which League Island joined the Naval Research Lab in Annapolis to perfect sonar and snorkel installations and create high-speed sonar test target submarines. Philadelphia altered five of the twenty-four *Guppy* II conversions and six of the nineteen snorkel conversions between 1948 and 1952. Also, the Philadelphia Naval Shipyard design staff developed a 25,000-horsepower "water brake" for nuclear-powered submarines for the Atomic Energy Commission and Bureau of Ships.[23]

Philadelphia naval researchers experimented with resins, plastics, stainless steel, and aluminum technology to decrease the weight of warships. "Deep drawing of aluminum for Snorkel and Conning Tower Fairwaters was accomplished by employing Kirksite dies in the 1500-ton press," Shipyard boss Kniskern reported to the Bureau of Ships. League Island tested the aluminum hull of *YP 110* (*PT-8*), built at the Yard in 1940, and fabricated the experimental aluminum hulled *PT-812*. Much of the research in Philadelphia between 1946 and 1952 focused on ship preservation. The Industrial Test Lab developed a system of injected "foamed phenolic resin" to fill voids in laid-up warships to absorb moisture and prevent rust from forming, and test engineers applied a revolutionary "Saran coating" plastic-vinyl compound to the inside of fuel tanks to stop corrosion on fleet submarine oiler *Guavino*. Other projects introduced by Navy Yard researchers during the Korean War era were a "noise-free" bearing developed by the Noise and Vibration Group to allow submarines to run silently and a "Neoprene" coating for propeller shafting to stop metal fatigue and decay.[24]

The naval industrial establishment in Philadelphia regained some of its former importance during the Korean crisis. Kniskern told the rapidly expanding work force on League Island that the United Nations police action in Korea was an "Extremely Critical" situation for the U.S. Navy. The job of reactivating the fleet saved American lives, and "is the First Line of Defense of our American Way of Life." The Navy Department assigned John H. Brown, Jr., commander of the Submarine Force, Pacific Fleet, to run the Fourth Naval District during the Korean intervention. An All-American football player, Brown had brought the Army-Navy football game to Philadelphia while director of the Department of Physical Training at the Naval Academy (1934–36) and had served as the popular officer-in-charge of the Navy Recruiting Station, Philadelphia (1938–40). Brown returned to the city in time to join the June 1951 celebration for the 150th anniversary of the "Grand Old Yard in Philadelphia," and immediately relaxed security regulations in force since the Korean invasion to allow the public to visit aircraft carrier *Palau*, berthed at Pier No. 2 on the riverfront.[25]

The 150th-anniversary celebration was organized by Ship Design Department Head Angelo M. Stefano and Samuel S. Sheller of the Supply Department. They set up window displays at Lit Brothers, Stern's, Wanamaker's, Snellenburg, and other center-city Philadelphia department stores, hired the All-Girl Orchestra for a Convention Hall show, and arranged for the 47-piece Navy Yard String Band led by Jimmy Mullany to parade through the city. Philadelphia Mayor Bernard Samuel declared a "Philadelphia Naval Shipyard Day" and sang with the Navy Yard Glee Club, while popular Under Secretary of the Navy Dan A. Kimball, credited by many for saving the Navy Yard from closure in 1949, addressed a grand banquet at the Bellevue-Stratford Hotel. Secretary of Defense Marshall and President Truman sent greetings. "This pioneering navy yard,"

Truman wrote, "was founded at the dawn of what has been aptly called 'the golden age' of naval achievement."[26]

Secretary of the Navy Francis P. Matthews decided not to attend the gala naval event in Philadelphia. Derided for his support of budget cuts and attempts to dismantle naval aviation, the beleaguered Matthews made one of the more insightful observations about how the Cold War had revived the League Island establishment. "The reputation of the Philadelphia Naval Shipyard, is further enhanced as ships built and serviced here fire their guns and launch their aircraft in the Far East and as other Philadelphia built and maintained vessels stand by in the Mediterranean and the world over, sentinels jealously guarding America's right of freedom." All these Philadelphia sentinels of freedom had been built during World War II. Despite passage of a 350-new ship naval construction program during the Korean crisis (which included an atomic submarine), not a single large warship construction contract arrived at one of the largest government shipyards in the United States, equipped to construct the heaviest warships. Kniskern's successor in 1952, Peter William Haas, Jr., a native of Scranton, Pennsylvania, insisted that this neglect occurred because of "the continued emphasis on shift of work to private shipyards." Certainly this was part of the reason. The departing Kniskern suggested another, hinting that the Philadelphia Navy Yard would never build atomic ships for a nuclear navy. "A disposal area for contaminated material has not been set aside and *due to the location of the Shipyard*, would be a very serious matter."[27]

With a ceasefire in Korea in 1953, warships returned to League Island to be laid up in the freshwater basin. Destroyers *Caperton*, *Dashiell*, *Dortch*, and *Gatling*, escort carrier *Palau*, aircraft carrier *Cabot*, and fifty submarines and auxiliaries arrived in the Reserve Basin. To make space for the returning Korean War fleet, the Navy Department considered scrapping *Olympia*, Admiral George Dewey's historic Spanish-American War flagship that had berthed in the Reserve Basin for the previous thirty-two years. At the same time, the post-Korean War work force gradually fell to below the 9,000 level by 1958, with Naval and Marine Corps personnel on board to a post-World War II low of 79. Meanwhile, League Island weeded out and disposed of World War II-vintage ships that had fought in Korea, until the inactive fleet in the Reserve Basin dropped from fifty-five to thirty-eight ships.[28]

League Island's mission in the post-Korean War era remained overhaul, repair, and modernization of ships to "fight" the Cold War. "Behind the Iron Curtain," Munitions Board Chairman John D. Small cried, "the business of the day is a grim, ruthless buildup of a relentless pressure against us, a pressure designed to engulf us." As the Cold War intensified, the naval facility in Philadelphia modified more submarines, destroyers, and light carriers for anti-submarine warfare (ASW) against what was reported to be a 1,000-sub Soviet Navy. The Naval Shipyard added ASW equipment to the 5,500-ton destroyer Leader *Norfolk* in 1953 (launched across the river by New York Shipbuilding Corporation), and executed *Migraine* III Radar Picket Conversions on diesel submarines *Pompon*, *Rahser*, *Raton*, *Ray*, *Redfin*, and *Rock*. *Migraine* III conversion required

The Spanish American War cruiser *Olympia*, berthed at Preble Street wharf in the Reserve Basin since 1919, being towed to the Keystone Ship Repair company, 19 September 1957, for restoration by the Cruiser Olympia Association. The historic cruiser rests today on the Philadelphia riverfront as part of the ISM. Naval Historical Center Photo NH 81499; courtesy of Capt. Philip Osborn.

cutting through the entire hull on each boat and pulling the two sections apart in order to install a 24-foot prefabricated Combat Information Center.[29]

Philadelphia contributed substantially to the post-Korean War aircraft carrier development program. The Boiler and Turbine Laboratory tested steam catapults for the 60,000-ton aircraft carrier *Forrestal* launched in December 1954 by Newport News Shipbuilding and Drydock Company, and for *Saratoga*, built at the New York Navy Yard. The Philadelphia Naval Shipyard's Foundry and Propeller Shop cast eight 70,000-pound propellers for *Saratoga*. Work on these *Forrestal*-class aircraft carriers ended a postwar absence of new ship construction on League Island, as in March 1958 the Naval Shipyard laid keels in Dry Dock No. 4 for guided-missile frigates *Dahlgren* and *William V. Pratt*. Naval Shipyard Commanders Wallin, Kniskern, Haas, and Joseph E. Flynn had struggled for more than a decade to modernize the local naval shipbuilding plant so that it could accept such new construction. League Island installed a fully integrated data-processing system (UNIVAC II) for inventory control, systems analysis, programming, and guiding the "Engineered Performance Standard" and Electronic Data Processing Department. Shop work became increasingly automated, and shops added "hydraulic operated, optical controlled, dimensional location" machine tools, dynamic balancing equipment, and the "Ampower Lumatrace" optical marking system to lay out and cut ship templates. The Shipyard combined ordnance, missile, and electronics functions into an integrated combat system organization, although the systems management approach would not be fully adopted until 1966.[30]

The guided-missile frigate contracts of 1958 ushered in a new era for League Island. The period from 1958 to 1962 was one of the most productive peacetime eras in the history of the local naval shore establishment, with every dry dock, shipbuilding ways, and pier holding a ship. Assistant Chief for Electrical/Electronic Programs for the Bureau of Ships Joseph E. Rice took command of the Naval Shipyard in 1961 to direct a technically skilled work force of 9,000. During this period, the Navy Yard laid

keels for the amphibious assault ships *Okinawa*, *Guadalcanal*, and *Guam*, each a $37-million contract that brought approximately $15 million in wages and local business to the Delaware Valley. The Shipyard converted cruiser *Galveston* into the first *Talos* surface-to-air-defense armed guided missile cruiser, using plans developed in the Design and Planning Sections. The nuclear-powered cruiser *Long Beach* arrived in early 1961 for installation of Naval Tactical Data and High Capacity Communication and Radar Systems and two 5-inch gun mounts. The Yard started Fleet Rehabilitation and Modernization (FRAM) work in 1959 on submarine tender *Fulton*, and the following year on four destroyers, extending navigation bridge and deckhouse for installation of the Combat Information Center and adding helicopter hangars and sonar domes.[31]

The Navy Department accepted the innovative seven-bladed propeller developed by the Philadelphia Propeller Shop for sixteen nuclear-powered submarines. The Philadelphia establishment installed satellite communication equipment for NASA/DOD on *Kingsport* and tested an aerospace simulator capsule. The Yard commissioned aircraft carrier *Kitty Hawk*, delivered in 1961 by New York Shipbuilding Corporation, and *Dahlgren* and *Pratt*, armed with state-of-the-art surface-to-air *Terrier* missile and anti-submarine rocket (ASROC) defenses. These ships comprised the heart of the national defense, Chief of the Bureau of Ships Ralph K. James announced at the commissioning ceremonies.

Facing above The large transport awaiting decommissioning and the destroyer about to undergo Fleet Rehabilitation and Modernization in front of the dilapidated old Dry Dock No. 1 in 1955 symbolized the declining role of the Philadelphia Navy Yard after World War II. NARAM.

Commissioning ceremonies for the aircraft carrier *Kitty Hawk*, 29 April 1961; the ship was built across the river at New York Shipbuilding Corporation, Camden, New Jersey. The intensifying Cold War and crisis in Southeast Asia revived League Island briefly as a major shipbuilding facility. Naval Historical Center Photo NH 51034.

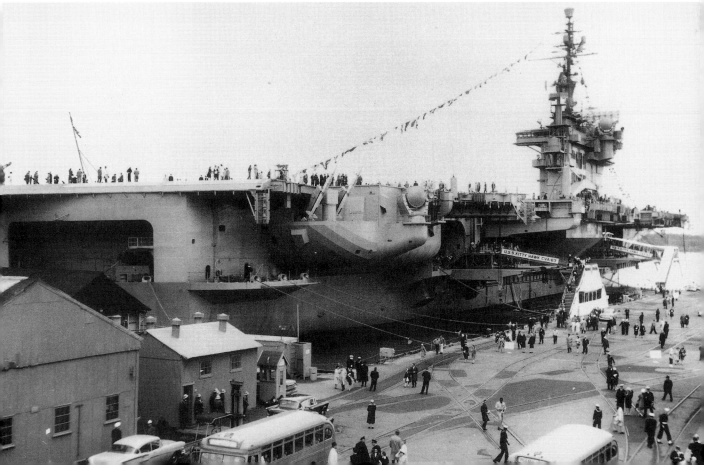

"Nuclear energy, guided missiles, and even spectacular man-in-space have not reduced the importance of sea power for the maintenance of our world position," he assured the League Island crowd, suggesting that the Philadelphia Naval Shipyard would continue to build such ships.[32]

All was not well for the Delaware River naval shore establishment, however. The familiar closure rumors started to circulate again, even as the city celebrated the contracts for the two guided-missile frigates. Philadelphia Mayor Richardson Dilworth feared that cuts in defense spending and proposed defense reorganization by Dwight D. Eisenhower's administration, gravely threatened the local navy yard. Dilworth and Republican Senator Hugh Scott rushed to Washington to meet with Secretary of the Navy Thomas W. Gates, Jr., "a Philadelphian." Gates admitted that the Eisenhower government was studying possible naval base closures and would remove the Submarine Supply Office from Philadelphia and close the Fort Mifflin Naval Ammunition Depot. Moreover, the Navy Department rejected a city bid to scrap warships and turned down a plan to build 500 low-cost houses for government employees funded by the Capehart Act. Dilworth told the city to expect closure of its navy yard on League Island, although Scott was more optimistic. In March 1959 apparently he had been told in confidence by Chief of Naval Operations Arleigh Burke that the Philadelphia Navy Yard would be retained to service and repair nuclear-powered submarines, like those contracted recently for construction across the river at the New York Shipbuilding Corporation.[33]

Dilworth, Scott, and several generations of Philadelphia city and Delaware Valley lawmakers saw the survival of the Philadelphia naval shore establishment in terms of its ability to build nuclear-powered ships. Dilworth worried that, unless League Island became an "Atomic Age Navy facility," the place would become nothing more than a "cedar closet" for old ships. However, Kniskern's earlier warning about League Island's poor location for such work suggested that the 1944 explosion at the atomic plant still weighed heavily against converting the naval shipyard into a nuclear facility. The federal government was forced to cancel simulated mushroom cloud air raid exercises on League Island because officials feared the atomic test would cause panic in the city. To be sure, the Boiler and Turbine Test Laboratory experimented with nuclear ship propulsion systems and Philadelphia laboratories made top-secret nuclear radiation tests on material and equipment. Still, League Island received no license to build or service nuclear-powered vessels. Nevertheless, the nuclear solution continued to be proposed. When Secretary of Defense Robert S. McNamara studied base closure in 1961, city officials turned again to the idea of establishing a nuclear navy yard on League Island to save the government compound. Pennsylvania Representatives William A. Barrett and James E. Van Zandt introduced an amendment to the naval appropriations act to provide money for a nuclear conversion. Senate Armed Service Committee Chairman Richard B. Russell of Georgia removed Van Zandt's amendment, forecasting a campaign by Southern legislators to diminish the importance of the big Northern navy yard to enhance business at their own naval facilities.[34]

Non-nuclear environmental concerns also troubled League Island. "Oil intrusion caused by the rupture of an underwater oil line" that ran through the Navy Yard polluted nearby rivers and cost the government $500,000 to clean up. The city passed a smoke abatement ordinance, forcing League Island to install expensive smoke pollution control systems. The gravest threat to continued operation of the Philadelphia Navy Yard, however, lay in pressure from private shipbuilders and a Congressional mandate that all new ship construction and 35 percent of conversion, alteration, and repair (CAR) work be performed in private shipyards.[35]

Another serious threat to the Philadelphia Navy Yard occurred when Secretary of Defense McNamara, former president of Ford Motor Company and a student of statistical cost performance systems analysis, launched a cost effectiveness study of the national defense establishment to improve efficiency and save taxpayer money. McNamara's Pentagon studies recommended gradual closure of the high-cost Portsmouth and Boston Navy Yards and the naval shipyards in San Francisco and Philadelphia, while venerable House Armed Service Committee Chairman Carl Vinson warned that the government had taken steps already "in the direction of shutting down Philadelphia, Brooklyn and Boston shipyards." At this critical moment, President John F. Kennedy, with ties to fellow Irish-Catholic Democrat and mayor of Philadelphia James H. J. Tate and confronted with a Soviet missile crisis in Cuba, decided to maintain the Philadelphia Naval Shipyard. All five dry docks and two marine railways were "filled as Shipyard Work Picks Up," a Philadelphia newspaper reported.[36]

Kennedy's assassination in November 1963 and new President Lyndon B. Johnson's decision to phase out navy yards resurrected local fears of closure. The continued threat to League Island by Southern Democrat Johnson created strange political alliances in Philadelphia. Local Democratic Party leaders Tate, Representatives William Barrett and James A. Byrne, and Daniel J. Burke, head of the Metal Trades Council, AFL-CIO, representing 7,000 unionized shipyard works, joined forces to save League Island with more conservative Republican Senators Scott and Richard Schweiker and Charles V. Wright, president of the non-union Joint Committee for Yard Development. Tate attacked Johnson administration neglect of Philadelphia, criticized the local Chamber of Commerce for suggesting privatization of the Navy Yard, and castigated New York Shipbuilding Corporation President Edward L. Teale for stealing work from the government yard. Scott and Schweiker went directly to LBJ and asked that the Philadelphia Naval Shipyard build an amphibious force flagship. Soon Secretary of Defense McNamara announced in a radio broadcast to League Island workers that their establishment would remain open, explaining that Philadelphia had twice the space, more dry dock and berthing facilities, and the best industrial layout of any East Coast navy yard, although he considered four navy yards more essential to the national defense—Norfolk, Puget Sound, Long Beach, and Charleston, South Carolina, home of powerful chairman of the House Armed Service Committee Mendel Rivers.[37]

McNamara made Philadelphia a "pilot yard," dispatching a team of engineers from the West Coast to assess the cost of a shipyard modernization program. Investigators from the General Accounting Office and Navy Inspector General's office probed alleged mismanagement and misuse of materials and manpower. As part of a major reorganization of naval shore facilities begun in 1963, League Island incorporated the Naval Material Support Establishment systems commands for air, ships, electronics, ordnance, and supply and a facilities engineering command for navy yards that in 1966 replaced the old bureau system and provided a unified management command structure under the chief of naval operations. In July 1966, the Naval Boiler and Turbine Laboratory, Submarine Antenna Quality Assurance Facility, and Assurance Engineering Field Facility became the Philadelphia Division of the Naval Ship Engineering Center under the Naval Ship Systems Command. In a final McNamara-inspired cost effectiveness measure on League Island, the headquarters of the fourth Naval District was moved to the fifth Naval District headquarters in Norfolk. "This move will not affect the tempo of operations or employment of the Shipyard in any way," assured the in-house Philadelphia Naval Shipyard newspaper.[38]

Boosted by President Johnson's increased defense spending after the Gulf of Tonkin Resolution in 1964 that escalated U.S. involvement in Vietnam, League Island entered one of its most active periods of operations and highest employment levels since the end of World War II. The shipyard work force rose from 8,000 in 1964 to over 13,000 by 1967, with an annual payroll of nearly $90 million. Keels were laid for the amphibious assault ships *Guam* and *New Orleans*, large seagoing landing ship tanks (LST) *Manitowoc*, *Newport*, and *Sumter*, and the amphibious command ship *Blue Ridge*—USMC amphibious warfare ships. They reflected appropriately Philadelphia's role as birthplace in 1775 of the U.S. Marine Corps and home on the Atlantic Coast in the early twentieth century for the Marine Corps Advanced Base. Meanwhile, Philadelphia converted destroyer leaders *Mitscher*, *John Paul Jones*, and *J. S. Mc-*

The twin launchings of amphibious assault ship *New Orleans* and LST *Newport*, 3 February 1968. Temple University Urban Archives.

Cain into guided-missile destroyers, earning the Philadelphia Naval Shipyard commendation for "excellent construction of a communications information center and pilot house mockup."[39]

In 1967 the Philadelphia Navy Yard ran an $800,000 time-cost study on the feasibility of reactivating for duty in Southeast Asia the World War II-era battleship *New Jersey*, laid up in Philadelphia since 1957. Desperately short on ships for the Viet-

nam War, in April 1968 the Navy authorized a $25 million upgrade on communications, fire fighting, and electronic warfare systems, added anti-missile armament and a helicopter platform, and recommissioned the battleship *New Jersey*. Meanwhile, League Island tested new steam catapults by launching deadweight rubber barges into the Delaware River from the decks of the older aircraft carriers *Shangri-La* and *Antietam*, so that such vessels could be modernized to launch heavier jet aircraft for the Southeast Asian conflict, and made alterations to machinery and hulls on older submarines and fast torpedo boats *PT-810* and *PT-811*, earmarked for covert intelligence operations on Vietnam's rivers.[40]

Modifications to these torpedo boats in 1962 were designed to provide stealth intelligence craft for duty on Vietnam's rivers; instead, both were sunk in target practice in 1965. NARAM.

The Philadelphia naval shore establishment supported "Four ships in two docks and one yet to be laid down, plus a large group of conversions," Shipyard Commander James J. Stilwell reported. This activity involved adding new steel handling equipment with high speed overhead cranes, a steel storage facility, and new tracks for 75-ton cranes brought to Philadelphia from the recently closed (June 1966) Brooklyn Navy Yard. Installed at Dry Dock No. 5, in January 1968 the cranes prepared for an 11-month overhaul of the 60,000-ton aircraft carrier *Saratoga*, the largest ship yet to enter the Philadelphia Naval Shipyard. This work seemed to be a "step forward in resurgence of the Yard," but the increased activity during the unpopular war masked serious trouble ahead. Despite new ship construction on League Island, most of the Johnson administration's war budget for Philadelphia went to reactivate older ships such as *New Jersey* or to maintain the existing fleet in action, not to build new ships. Moreover, Congress mandated in 1967 that all new ship construction be made by private yards, so *Blue Ridge*, launched in 1969 and commissioned in 1970, would be the last U.S. naval vessel built by the Philadelphia Navy Yard.[41]

Rising wages and inflated costs for high-tech weapons systems drained the naval budget, which continued to shrink after 1968 from a combination of congressional antiwar sentiment and the channeling of a larger percentage of the defense budget to the Army and Air Force. Hyman Rickover, director of the Division of Naval Reactors in the Department of Energy and Deputy Commander for Nuclear Propulsion in the Naval Ship Systems Command, exerted growing pressure to make the entire fleet nuclear powered. Philadelphia was hurt because it never received a license to build or repair nuclear-powered vessels. Such a license failed to save New York Shipbuilding Corporation in Camden, however, as the government canceled the contract on the partly finished nuclear attack sub *Pogy*, leading to closure of the private Camden shipyard in 1967.[42]

With gradual withdrawal of U.S. forces from Southeast Asia, the U.S. Navy underwent a "tightening-up" process that led during the post-Vietnam War era to one of the darkest periods in U.S. naval history, and cancellation in 1967 of the contract to build a guided-missile frigate on League Island. This marked the turning point in the decline of the Philadelphia Navy Yard. New Jersey Representative Robert Andrews, a long-time defender of the Philadelphia naval establishment, believed that from this point on elements in the naval bureaucracy were determined to close the Philadelphia Navy Yard. Pat D'Amico, head of a Navy Yard Career Transition Center in the 1990s, agreed that about this time "base closure drills [became] a way of life, really upsetting the entire family structure" in the Delaware Valley.[43]

Never before had rumors of closure caused so much anxiety. The realization set in at the end of the Vietnam War that the government might abandon the 200-year-old Philadelphia naval and Marine Corps establishment, and closing rumors began to define daily operations. Every reorganization, labor negotiation, minority training program, or political alliance was placed in the context of base closure. Every annual budget and ship overhaul contract that promised to save the Philadelphia Naval Industrial Fund was evaluated in terms of closure. In the end, the idea that the government planned to abandon League Island permeated every aspect of Navy Yard life, creating between 1969 and 1996 what might be called the culture of closure.

Facing above Commissioning the amphibious assault ship *Sumter*, 14 December 1969. The *Sumter*, along with the amphibious command ship *Blue Ridge*, commissioned in 1970, marked the end of new ship construction at the Philadelphia Navy Yard. Temple University Urban Archives.

Commandant James J. Stillwell showing Philadelphia Representative and later Mayor William Green and other Delaware Valley political and industrial leaders around the propeller shop, 21 May 1965 as part of an ongoing effort to secure support for continued Navy Yard operation. Temple University Urban Archives.

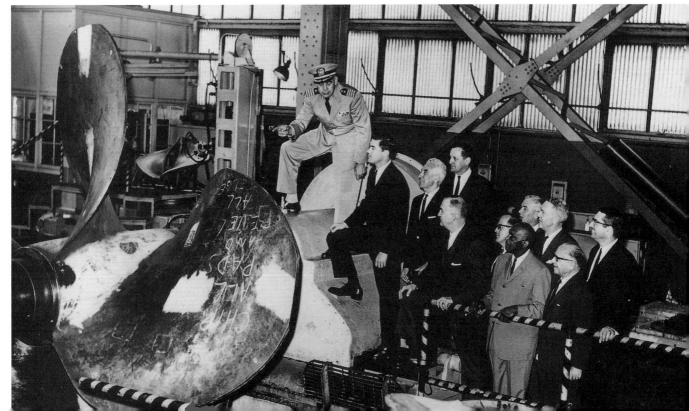

12 ⭑ Culture of Closure

THE "VIETNAM draw-down" decimated the U.S. Navy. Fleet strength fell from nearly 600 ships in 1968 to below 300 by 1978. "Today, the number of ships in the Fleet is the lowest it has been for many years" announced Captain Gerald R. Jones when he assumed command of the Philadelphia Naval Shipyard. Continual cuts during the 1970s meant that the new shipyard commander had to RIF personnel and postpone improvements to facilities. Worse, 800 pipe fitters, pipe coverers, and welders suffered from asbestos poisoning, and League Island underwent expensive and controversial asbestos management and removal programs. "Civil disturbances" gripped the Navy Yard as anti-Vietnam War protestors blocked the main gate and scuffled with workers as they tried to enter. Security against domestic threats increased, and one civilian staffer recalled that "many a night we had to be here until 7 o'clock (to) make plans."[1]

With U.S. Navy morale, discipline, and force levels declining, the strength and size of the Soviet Navy increased. By the middle of the decade the Russian fleet outnumbered the American force in all types of ships except aircraft carriers. Though the U.S. Navy maintained an edge in technology, training, and combat experience, some claimed that by 1976 the Soviet Union had become the world's leading naval power. The Seapower Subcommittee of the House Armed Service Committee held hearings on the "Status of Shipyards" to discover reasons for the growing disparity between the United States and Soviet Union in naval shipbuilding and fleet size. The hearings discovered that older naval shipyards such as the one in Philadelphia could not compete with

private shipyards in the Chesapeake or Gulf Coast managed by Tenneco, Litton Industries, or other modern high-tech industrial conglomerates. The hearings also revealed that the Navy Department considered the Delaware River establishment of less value than warm weather facilities to the south. Not only did cold weather interfere with work on League Island, but the Philadelphia Naval Shipyard failed to meet six of ten basic strategic requirements to maintain the fleet, particularly in its inability to overhaul or repair nuclear-powered ships and submarines.[2]

The Subcommittee's unfavorable comparison of Philadelphia to the Norfolk Navy Yard confirmed New Jersey Representative Robert Andrews's contention that elements in the government and private sector planned to close League Island. "I really believe that there is a pattern here, where people who used to work for the Pentagon, get consulting contracts, and business opportunities of Newport News that would be enhanced if Philadelphia were not there." Representative Joshua Eilberg of Philadelphia called such activity part of a "Tidewater Strategy" by Department of Defense bureaucrats, retired naval officers working for Southern-based defense industries, and their Southern friends in Congress to relocate the entire Atlantic coast naval establishment to southeastern and Gulf coastal regions.[3]

Eilberg promoted Philadelphia to offset the Tidewater Strategy, reading into the Congressional Record how the local navy yard had won the Navy's Zero Defects Award and Cost Reduction Award for completing overhaul and repair ahead of schedule to nineteen ships, including German destroyer *Lutjens*

1960–96. In the years before closure, the Navy Yard berthed everything from *Iowa*-class battleships to amphibious assault ships, aircraft carriers to guided missile destroyers, and submarines and frigates.

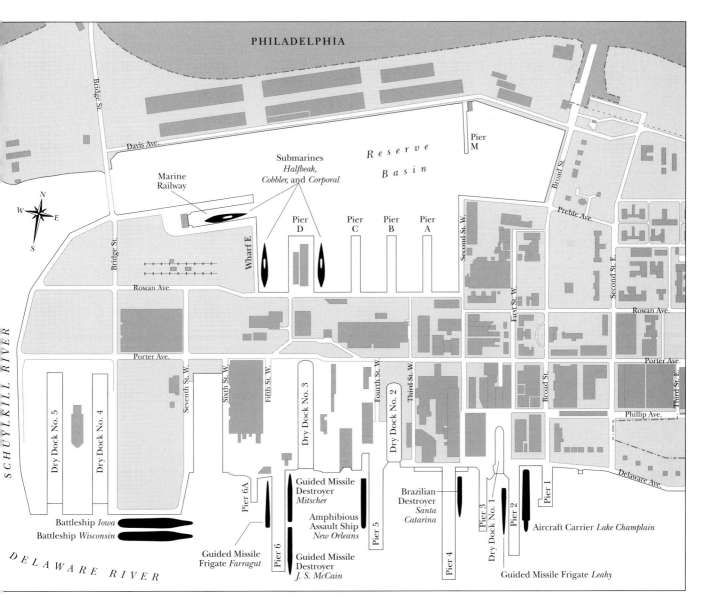

PHILADELPHIA

Bridge St.

Davis Ave.

Marine Railway

Submarines
Halfbeak,
Cobbler, and *Corporal*

Reserve Basin

Pier M

Broad St.

Preble Ave.

N
W   E
S

Bridge St.

Wharf E

Pier D

Pier C

Pier B

Pier A

Second St. W.

Second St. E.

Rowan Ave.

First St. W.

Rowan Ave.

SCHUYLKILL RIVER

Porter Ave.

Seventh St. W.

Sixth St. W.

Fifth St. W.

Dry Dock No. 3

Fourth St. W.

Dry Dock No. 2

Third St. W.

Broad St.

Porter Ave.

Third St. E.

Phillip Ave.

Dry Dock No. 5

Dry Dock No. 4

Pier 6A

Guided Missile Destroyer *Mitscher*

Brazilian Destroyer *Santa Catarina*

Pier 3

Dry Dock No. 1

Pier 2

Pier 1

Delaware Ave.

Battleship *Iowa*
Battleship *Wisconsin*

Pier 6

Guided Missile Frigate *Farragut*

Amphibious Assault Ship *New Orleans*

Pier 5

Guided Missile Destroyer *J. S. McCain*

Pier 4

Aircraft Carrier *Lake Champlain*

Guided Missile Frigate *Leahy*

DELAWARE RIVER

and Iranian warships. Secretary of State Henry Kissinger praised the latter for advancing the Nixon administration's initiative to strengthen the government of the Shah of Iran. "The early completions ac- count for a significant reduction in man days and, subsequently, money expended," Philadelphia Naval Shipyard Commander John B. Berude asserted.[4]

Saving money became most important in 1973 as

Below Ice buildup around a submarine on the Delaware River, 1977. Southern lawmakers and base closure advocates, in an ongoing effort to promote the Norfolk Navy Yard and southern shipbuilding and repair facilities, cited cold weather and ice as one of the limitations of Philadelphia as a major Atlantic coast navy yard. NARAM.

Facing League Island's preparation of overage destroyers for transfer to the Imperial Iranian Navy won special praise from Secretary of State Henry A. Kissinger for implementing the Nixon administration policy of friendship with the Shah of Iran. Temple University Urban Archives.

the federal government closed navy yards in Boston and Hunter's Point, California, naval stations in Newport and Quonset Point, Rhode Island, and other military facilities nationwide. The Department of Defense closed the Naval Air Engineering Center and the Marine Corps Supply Facility at Philadelphia, causing the loss of over 3,000 federal jobs. The Navy Yard was "saved," however, reportedly through the influence of Philadelphia Mayor Frank L. Rizzo and Republican Senators Scott and Schweiker, the latter a member of the Senate Armed Service Committee. More likely, the Defense Department wanted Philadelphia to remain active so that the government controlled the two large dry docks to overhaul the largest conventionally powered aircraft carrier.[5]

In the midst of the post-Vietnam War downsizing and constant closure rumors, League Island quietly converted older warships into state-of-the art, high-tech weapons platforms. This Fleet Rehabilitation and Modernization Program (FRAM) brought the Naval Shipyard longer and more lucrative contracts than traditional routine overhauls. The DLG 6-class modernization program that between 1970 and 1975 installed the latest tactical data systems, computer fire control configurations, and sonar equipment package in destroyer leaders *Farragut*, *Preble*, *Dewey*, *Luce*, and *Coontz*, brought more than $25 million and 12 months of work each to the Philadelphia Naval Shipyard. Even the "Complex Overhaul" of the older guided-missile cruiser *Albany* (launched in 1946 and converted to missiles in 1959) meant ten months of work. League Island also designed and produced innovative propellers for *Trident*-class nuclear-powered missile submarines. To support the new technology, the Navy Yard constructed a machine shop for the Propeller Lab, broke ground in 1971 for the Electronics, Weapons, Precision, and Electrical Facility, and upgraded computer systems, replacing UNIVAC III in 1973 with the Honeywell Multidimensional Information Computer System. To house officers and crews of ships undergoing FRAM, League Island received $10 million in 1974 to construct 350 housing units at the Naval Base.[6]

Meanwhile, the Philadelphia naval shore establishment adjusted gradually to the dramatic social changes in race relations and the status of women in America that accompanied the anti-Vietnam War movement. The Navy Yard introduced an Economic Employment Opportunity policy for minority hiring in 1961 during the Kennedy administration that entailed little more than posting notices on bulletin boards. Shipyard Commander Joseph Rice recruited labor "without regard to race, creed, color, or national origin," while encouraging "minority groups" to overcome their "inertia, without lowering our standards" through more education and training. At this time, African Americans were admitted through "mostly the limited mechanic or helper areas." Advancement seemed perilously slow. In the constant cycle of hiring and layoffs at the government shipyards, African Americans who were the last hired often were the first fired. "'Cause the fact you're Black," an African American Paint Shop employee recalled, "they're going to try to get you out of here some way, and I think I find a little more of the security clearances being pulled."[7]

The Vietnam War increased racial tension in the U.S. Navy and throughout the country, including the 1971 riots in nearby Camden and on board the aircraft carriers *Kitty Hawk* and *Constellation* and fleet oiler *Hassayampa*. Secretary of the Navy John H. Chafee expressed shock at racial animosity and violence in the service. He appointed James E. Johnson, an African American, to the post of assistant secretary of the navy for manpower and reserve affairs, and ordered navy yards to assign a special assistant for equal opportunity and race relations. Philadelphia Naval Shipyard Commander Berude hastened to comply. "I hereby charge all officials of the Philadelphia Naval Shipyard to support the policies stated above, and to ensure that the principles of Equal Employment Opportunity are not just a right and theory, but a fact, and a way of life in this Shipyard."[8]

League Island developed an affirmative action plan in 1973 and recruited minorities more actively. "They came in kind of one at a time and were brought through the apprenticeship program," a shop steward recalled. The Navy Yard introduced Julia Furlong in January 1974 as its first Federal

Women's Program Coordinator, and in 1976 Forrest W. Sellers, an African American graduate of nearby Lincoln and Temple Universities, was hired to direct a program "toward full integration." Under Sellers, League Island organized conferences on Black History, Hispanic Concerns, and Women's Rights, and a service at the Naval Base Chapel to celebrate the birthday of civil rights leader Martin Luther King, Jr. The Naval Sea Systems Command selected Sellers in 1977 as the outstanding civil rights coordinator in the Navy. Still, women comprised less than 10 percent of the League Island work force, and shipfitter and union leader Lorraine Daliessio claimed that only 247 females out of 7,500 employees at the Naval Shipyard worked on ship repair and overhaul.[9]

The election of liberal Southern Democrat James Earl Carter as president in 1976 accelerated civil rights programs and enhanced minority hiring at the navy yards with the appointments of aggressive Equal Economic Opportunity directors. The fact that the new president was a Naval Academy graduate also raised expectations that the inadequate naval budgets of the Nixon years would end. However, Carter introduced an austerity program that reduced the fleet even farther than had that of the Nixon-Ford presidencies. For instance, Carter opposed construction of huge *Nimitz*-class attack carriers, preferring smaller older carriers to defend sea lanes of communication and American strategic interests overseas. Ironically, this limited defense policy benefited the Philadelphia Naval Shipyard, one of only two government facilities on the East coast capable of overhauling and modernizing the older *Forrestal*-class aircraft carriers.[10]

Rumors circulated in early 1978 that Philadelphia might obtain the first Carrier Service Life Extension Program (SLEP) contract for 60,000-ton attack carrier *Saratoga*. News of the coveted SLEP came as Philadelphia was still reeling from the decision in 1977 to close the historic Marine Corps Reservation on League Island and remove the Marine Guard. This was a particularly sensitive issue since the city had given birth to the Marine Corps in 1775 and boasted a nearly continuous 250-year presence, a tradition illustrated graphically the previous year by

the elaborate and costly twelve-acre Marine Corps exhibit and public drills on League Island to commemorate the American bicentennial. SLEP would ease local unhappiness over loss of the Marines.[11]

The Carter administration announced in April 1978 that Philadelphia had won a $500 million SLEP contract for *Saratoga*. This was perhaps the single most important contract in Philadelphia Navy Yard history, opening the way for additional SLEP and extensive overhaul on the even larger conventionally powered aircraft carrier *John F. Kennedy*. The SLEPs employed thousands of Delaware Valley residents for nearly twenty years and brought hundreds of millions of dollars to the region. SLEP reinvigorated the government facility on League Island. "If *Saratoga* did not go to a public shipyard," Senator John H. Heinz observed, "all public shipyards would be slowly phased out of existence."[12]

Heinz and fellow Senator Joseph R. Biden of Delaware fought on the floor of the Senate and behind the scenes to secure Congressional approval of the *Saratoga* contract for Philadelphia. They outmaneuvered opposition led by Virginia Senator John Warner, who lobbied to give the first SLEP to Norfolk. Credit for obtaining the big carrier contract lay in the bipartisan partnership between liberal Democratic city and presidential administrations and more conservative Delaware Valley Republican lawmakers and business interests. Philadelphia Mayor Green's warm support for the Carter government helped, particularly with Vice President Walter F. Mondale, who had failed to live up to campaign promises to save the nearby Frankford Arsenal from closure and hoped to use the SLEP contract to save the Philadelphia Navy Yard from a similar fate. At the same time, Philadelphia Republican Thatcher Longstreth, the Chamber of Commerce, and the tristate consortium known as the Penjerdel Council worked together to win the lucrative SLEP for the Delaware Valley.[13]

At the last minute, a General Accounting Office (GAO) report appeared that criticized public navy yards, particularly Philadelphia, for poor management, waste, and lack of planning. Apparently the report raised doubts in the mind of Secretary of the

Navy W. Graham Claytor, Jr., about the wisdom of sending the important first SLEP to the Philadelphia facility. He inspected League Island under the ruse of visiting the historic cruiser *Olympia*, now berthed as a museum nearby at Penn's Landing on the riverfront. Nevertheless, the Carter government announced that the Delaware Valley would get the contract, not only for the *Saratoga* SLEP, but also for a similar modernization package for *Forrestal*.[14]

Saratoga arrived in Philadelphia from Mayport, Florida, on 30 September 1980 to begin a 28-month, $526-million modernization. Vice President Mondale, Penjerdel executive Longstreth, Philadelphia Mayor Green, and dozens of other area political and business leaders participated in welcoming festivities. More important, the increasingly active Metal Trades Council, AFL-CIO, which had represented union shop locals on League Island since the mid-1960s, signed the "Saratoga Agreement," guaranteeing labor harmony and uninterrupted work. At the same time, the Philadelphia naval shore establishment launched a series of model Equal Economic Opportunity programs to recruit and train female and non-white employees to work on *Saratoga*. These programs included the Philadelphia Regional Introduction of Minorities in Engineering (PRIME), founded in 1973, Programs Reaching Options for Vocational Education (PROVE) "to recruit, counsel, train and place women in non-traditional occupations," and Government and Industry Volunteers for Education (GIVE).[15]

Saratoga was the most extensive and as the first, most difficult Service Life Extension Program project. "We stripped it almost to her skeleton, and we haven't done that to any boat since then," one local shipfitter explained. "Quite as bad as what we opened her up, she hadn't been repaired in a long time; and we stripped her pretty good, down to the bare bones; and replaced her." Mistakes were made. New steam fittings and "Astro Arc" welding on boiler tubes leaked badly and had to be replaced. Quality inspection was flawed, so much so that Secretary of the Navy John F. Lehman warned that similar problems on the upcoming *Forrestal* SLEP would threaten the Yard's future. Lehman criticized work at

other navy yards as well, as part of a strategy to revive the moribund naval establishment and build a 600-ship fleet. Moreover, he attacked private shipyards for cost overrides, contractual deception, and sloppy workmanship. But Lehman's displeasure hurt Philadelphia more, reviving the local culture of closure.[16]

Trouble with Philadelphia's SLEPs continued. Chief of Naval Material Steven White complained that the *Forrestal* modernization had fallen far behind schedule, and found on a visit to League Island that the place was littered with trash and industrial debris. White's inspection tour in August 1984 jeopardized assignment of a third SLEP for aircraft carrier *Independence* and the contract to overhaul older aircraft carrier *Lexington*. Senator Arlen Specter of Philadelphia hurried to League Island in the wake of White's damaging visit and toured *Forrestal*. Specter promised the *Lexington* contract to the local naval facility, and a few months later the old warship arrived for an eight-month, $183.3-million overhaul to replace its wooden flight deck with steel, upgrade firefighting systems, repair old boilers, and repainting so that the ship could serve as a naval aviation training vessel at the Pensacola Naval Air Station. Apparently, Specter convinced Commander-in-Chief of the Atlantic Fleet Wesley L. McDonald to review White's evaluation of the Philadelphia naval shore establishment. After McDonald's visit, Shipyard Commander Thomas U. Seigenthaler reported that "our efforts to improve work site cleanliness" had convinced McDonald "that *Forrestal* was 'head and shoulders' above what he had seen during the production work on *Saratoga*."[17]

Independence arrived in April 1985 for a $620-million modernization that included installation of General Electric turbines, new arresting landing gear, a radar upgrade, *Sparrow* missile defense, and *Phalanx* close-in weapons systems. W. Wilson Goode, the first African American mayor of Philadelphia, local U.S. Representative Thomas Foglietta, and dozens of other Delaware Valley political figures greeted *Independence*. Senators Specter and Heinz toured "Indy" a few days later, reaffirming the growing local political commitment to save the Philadelphia Navy Yard from possible closure. Some credited Foglietta with obtaining $11.3 million to complete Building 1000, the Electronics, Weapons, Precision and Electric Facility authorized in 1971 but unfinished for over a decade. Foglietta dedicated this electronic warfare center on League Island in November 1984. "Philadelphia was a forgotten shipyard for a long, long, time." he observed. "Being a non-nuclear shipyard in a nuclear navy made the yard a step-child for too long." But it was Senator Heinz who emerged as League Island's leading advocate. "All of us in the Congressional delegation," Heinz assured Shipyard Commander William A. Kerr, Jr., "are grateful to you CAPT Kerr, and [to] all these hardworking trades men who have really put the Philadelphia Naval Shipyard at the top of the efficiency list, and we intend to help you keep it that way."[18]

Kerr, assistant chief-of-staff for engineering, naval surface force, U.S. Atlantic fleet, took command on League Island in 1985, and found only one ship undergoing overhaul. RIF notices had just gone out to 263 skilled shipyard employees and the navy yard bus route to Willingboro, New Jersey, had been canceled for lack of riders. Worse, another congressional study decided that the government shipyard in Philadelphia was poorly managed, inefficient, and more expensive to run than private shipyards such as the Newport News Shipbuilding and Drydock Company that was installing robotics, lasers, automatic welding, and computerized machines. "During the past several weeks, the Shipyard has—once again—become the focus of news reports and rumors concerning massive layoffs and possible closure," Kerr announced. "We've been the target of budget cutters for decades, and have always risen to the challenge, deflecting possible closure on several occasions."[19]

The culture of closure obsessed Kerr between 1985 and 1988 as he introduced more modern business and industrial methods to make the government facility competitive with private industry. League Island won a competitive bid against a private shipyard in late 1986 for the $4.5 million overhaul of guided-missile destroyer *Clifton Sprague*.

"This is a whole new way of doing business for us," Kerr explained, "and we need to succeed if we're to survive." Kerr informed the naval shipyard work force that if it wanted to obtain the contract for a fourth SLEP (*Kitty Hawk*), it needed to show more technical discipline, develop quality control, reduce costs through elimination of duplication, and streamline operations. Secretary of the Navy Lehman praised Kerr's "superb improvement in productivity and in the effectiveness of the work force," and called League Island essential to maintain the aircraft carriers that stood at the heart of President Ronald Reagan's 600-ship fleet. "We in the Navy," Lehman asserted, "have no intention of phasing out the Philadelphia Naval Shipyard."[20]

Kerr turned command of the busy, efficient Philadelphia naval establishment over to SLEP Project Officer for the Naval Sea Systems Command Arthur D. Clark in June 1988. Clark announced that *Kitty Hawk* "and her sister SLEP's" promised work on League Island until 1998. "The enormity of her [*Kitty Hawk*] work package is staggering to our imaginations and second only to our country's space program in complexity," Clark marveled. "No other Shipyard, private or public, has ever attempted to execute such a project." There was a dark side to such reliance on Service Life Extension Program contracts, however. "*Kitty Hawk* costs are astoundingly bad and getting worse," Clark confided, and such news added to an urgent concern in Philadelphia as a Commission on Base Realignment and Closure (BRAC), created in 1990 under Title XXIX of the Defense Base Closure and Realignment Act of 1988 closed the local Naval Hospital and placed the Philadelphia Navy Yard "on the Hit List" for future closure.[21]

Few Naval Shipyard workers wanted to believe that this time the government actually planned to close League Island. After all, closure rumors had become a way of life, repeated nearly every year. A 1990 survey of male and female employees, some with more than forty years of employment on League Island, showed great faith in the ability of Senator Heinz and area lawmakers and businesses to save the Philadelphia Navy Yard again, just as they always had before. Only one veteran shop worker realized that there were no large jobs scheduled for Philadelphia after the last SLEP. As Shipyard Commander Clark concluded in early 1990, "I cannot guarantee that this Yard will remain open."[22]

In the midst of emerging pessimism about the future of the Navy Yard, aircraft carrier *Constellation* arrived in April 1990 to begin a 29-month SLEP. Clark instituted improved SLEP planning through "Zone Technology" to "sequence" work packages in one area at a time and a "cost/schedule control system" to reduce overhead and labor costs. Moreover, he improved working conditions and health benefits for "on-call" personnel hired for each SLEP and laid off before the next contract arrived. Clark insisted that all work be carried out under the strictest local, state, and federal environmental regulations in order to make the Navy Yard "a model industry and as a good neighbor in the Delaware Valley."[23]

The Persian Gulf crisis following Iraqi invasion of Kuwait in 1990 brought new work to League Island. The Philadelphia Naval Shipyard activated the aviation logistics support ship *Wright* and transports *Scan*, *Pride*, and *Lake*, mothballed in the Reserve Basin for the past thirteen years, for Operations Desert Shield and Desert Storm. Clark rushed modernization work on *Kitty Hawk* so that the big attack carrier could join Gulf War operations, while League Island held its first commissioning ceremonies in twenty years for guided-missile cruiser *Gettysburg*, built at the Bath Ironworks in Maine and sent to Philadelphia for installation of the Aegis electronics warfare system, engineered by the Navy Combat Systems Engineering Development Site in nearby Moorestown, New Jersey. It was a process reminiscent of those bygone golden shipbuilding days on the Delaware River when freshly launched vessels from Cramp or New York Ship were delivered to League Island for final outfitting, inspection, and commissioning.[24]

Life at the Navy Yard in 1990 bore little resemblance to those earlier days, however. The social fabric of the place had undergone a quiet revolution over two decades, mirroring the diversity present in American society. The shipyard work force was 25 percent African American and 9 percent Hispanic. Nearly half of the monthly Navy Yard awards for excellence went to minority groups and employees. An African American was named Navy Yard "Quality Employee of the Year for 1991." Females comprised only 5 percent of the Shipyard work force, but held key industrial positions including crane operators, painters, shipfitters, and sheet metal workers. Women managed the industrial electronics control, safety engineering, hazardous waste management, and personnel offices and directed the critical Career Transition Center at the Navy Base. Females held military commands on League Island, including Captain Kathleen Dugan at the Naval Station and Rear Admiral Louise C. Wilmot at the Philadelphia Naval Base in 1993, the first female officer to run any U.S. naval base.[25]

Increasing diversity at the Navy Yard undoubtedly helped Pennsylvania Representative John P. Murtha, Pennsylvania Senators Heinz and Specter, and New Jersey Senators Frank R. Lautenberg and Bill Bradley secure for Philadelphia in 1990 the SLEP for the nearly 80,000-ton aircraft carrier *John F. Kennedy*. But this "pet project of the delegation from Pennsylvania" drew immediate criticism from opponents of the George Bush administration's greatly increased defense spending for Operation Desert Storm and failure adequately to address an economic crisis that accompanied a savings and loan scandal. Critics asserted that the *Kennedy* contract was "nothing but a wasteful piece of pork barrel spending." Partly in response, the Navy Department downsized the contract to an "extensive overhaul" rather than the more costly and lengthy SLEP. As a result, Shipyard Commander Clark predicted that *Kennedy* would be the "last availability assigned to the Shipyard."[26]

Philadelphia had been on a Defense Department "hit list" along with more than 100 other military and naval facilities since the base realignment and closure movement had begun in 1988, but meetings early in 1991 by the Navy Department Base Structure Committee forecast imminent closure of the Philadelphia naval establishment. There was some Navy Department support to keep League Island.

Commander of the Naval Sea Systems Command John Heckman and Deputy Commander of Industrial and Facility Management John S. Claman wanted to maintain an active naval presence in Philadelphia; "We recommend that Philadelphia Naval Shipyard be drawn down to a small size activity in the mid 90's as workload declines in order to provide a government controlled CV dry dock site and ship repair capability for the north east." Instead, the Navy Base Structure Committee recommended closure to the Commission on Base Realignment and Closure. BRAC accepted the Navy Department recommendation to close the naval base and naval shipyard in July 1991, but to keep active "the Naval Foundry and Propeller Center, Naval Inactive Ships Maintenance Facility: Reserve Basin, Dry Dock wharf and Piers 4 and 6, and Naval Surface Warfare Center, Carderock Division-Ship Systems Engineering Station" as adjunct functions of the Norfolk Naval Shipyard and 5th Naval District command. Secretary of Defense Richard B. Cheney forwarded this recommendation to President Bush on 1 July and on 10 July the president sent Congress his decision to accept closure.[27]

In the past, powerful Pennsylvania Senators Scott, Schweiker, or Heinz had deflected threats of closure. This time, however, Heinz, a close friend and political supporter of President Bush, could not lead the fight. The longtime advocate for the Philadelphia Navy Yard died in an airplane accident just eleven days before BRAC recommended closure. Heinz's replacement Harris Wofford tried to help, becoming a "member of the firm of Specter, Wofford, Bradley, Lautenberg, and our House colleagues" to introduce a joint resolution to Congress disapproving the BRAC recommendation and bringing a lawsuit against the Navy Department. Congress referred the Delaware Valley resolution to the powerful Senate Armed Service Committee, which was dominated by former Navy Secretary Warner of Virginia and Trent Lott of Mississippi, who represented regions that would benefit from increased naval business if Philadelphia should close.[28]

Left to right, Pennsylvania Democratic Representative Thomas J. Foglietta, Republican Senator Arlen Specter, Secretary of the Navy John Lehman, New Jersey Democratic Senator Frank Lautenberg, Philadelphia Mayor W. Wilson Goode (at podium), and Pennsylvania Republican Senator John J. Heinz try to save the Navy Yard from closure. Heinz's death in an airplane accident removed League Island's most powerful friend and contributed to closure. NARAM.

Though suggesting that other business was more important, the Committee agreed to hear short statements from Senators Wofford and Specter. A novice in the base closure wars, Wofford simply complained that the whole procedure had been unfair. Specter made a more impassioned defense of Philadelphia's interests, attacking the flawed and biased Navy Department process used to reach the closure decision. The former Philadelphia district attorney contended that the Navy Base Structure Committee knew that Philadelphia was "a good shipyard," probably the most cost-effective, efficient, and best managed government naval facility, and located ideally to maintain the AEGIS-system, the basic ship weapons array for the 1990s. The Senate Armed Services Committee listened politely, and heard additional testimony from BRAC Chairman James A. Courter, a former New Jersey Representative, who professed great sympathy and understanding for the impact of closure on the Delaware Valley's society and economy. Courter insisted that BRAC had acted impartially and fairly in recommending closure of League Island, basing its decision on budgetary restrictions, congressional mandate to give repair work to private shipyards, and the Delaware River government shipyard's inability to handle the proposed all-nuclear-powered aircraft carrier fleet.[29]

Congress refused to keep League Island, although the U.S. Court of Appeals for the Third Circuit in Philadelphia allowed Specter's lawsuit against the Navy Department to go forward, raising hopes once more that the Navy Yard would be saved. Specter argued the case of *Dalton v. Specter* before the U.S. Supreme Court in March 1994. In the end, the Justices ruled that the court had no jurisdiction to intervene in the governmental process of base closure. Facing the inevitable, Clark retired after "doing everything in my power to relieve the stress generated by Shipyard/Base Closure," and his successor, Jon C. Bergner, a former production officer at the plant, moved ahead with closure planning. Bergner disposed of ships, placing on sale for scrapping twelve of the thirty-six ships of the Inactive Fleet still in Philadelphia, including amphibious attack ship *Iwo Jima*, guided missile cruisers *Wainwright*

and *Edward McConnell*, frigate *Voge*, and eight obsolete guided missile-destroyers. The old cruiser *Salem*, berthed for two decades in the Reserve Basin next to the Main Gate on Broad Street, went as a floating museum to Fall River, Massachusetts.[30]

Bergner ordered League Island cleaned up for civilian use. Twelve hazardous chemical sites were identified, one supposedly containing buried cylinders of chlorine "hauled to the refuse pit at the west end of the yard" during World War II. Current sites were bad enough, containing PCBs (polychlorinated biphenyls), toxic additives used to cool oil in electrical transformers. The sites also held waste oil, dredge spoil filled with lead and arsenic, chemical ash, blasting grit, and asbestos piping insulation stripped from ships and buildings during the past decade and dumped on the island. "Oil and Hazardous Substances are extensive throughout the Naval Base," pollution control officer Mark Stephens admitted.[31]

At the same time, local governments considered what closure would mean to the Delaware Valley economy. The Philadelphia Industrial Development Corporation studied the impact of closure for Pennsylvania, while the Pennsylvania Economy League made a similar study for New Jersey. They predicted the direct loss of 12,050 Navy Yard and 36,400 collateral jobs, and the end of $113 million in annual local and state tax revenues, particularly the Philadelphia city wage tax of over $40 million. With the most to lose, Philadelphia took a leadership role in responding to eventual closure. Mayor Edward G. Rendell formed a 28-member Commission on Defense Conversion (CDC) to cooperate with the Department of Defense Office of Economic Adjustment and to apply for the property under the Base Closure Community Redevelopment and Homeless Assistance Act of 1994. The CDC issued a "Philadelphia Naval Base Conversion Initiative" concept paper that outlined "a scenario that does not rely on continuing Navy and federal subsidy," and "is to develop a private shipyard at the Naval Base." The mayor's commission hired architects Kohn Pederson Fox, economic consultants Coopers and Lybrand, and environmental engineers Roy F. Weston and Urban Engineers, Inc.

to develop a "Community Reuse Plan" for the League Island naval base and shipyard. The city submitted this study to the Navy Department in late 1994 as the basis for a closure transition and property transfer plan, and the Naval Facilities Engineering Command made a cultural resource survey and developed a "Base Cleanup Plan."[32]

The City of Philadelphia and Department of the Navy held a "Master Lease Signing Ceremony" on 22 November 1994 at the "Naval Station Philadelphia on League Island." The Philadelphia Authority for Industrial Development signed on behalf of the city and acted as landlord for the property, shops, and associated equipment. The Navy retained ownership for "emergent use" until the Base Realignment and Closure Commission of 1995 forced surrender of ownership. However, the Navy continued to operate the Propeller Manufacturing Center, Inactive Ship Maintenance Facility, and Carderock Division of the Naval Surface Warfare Center as components of the Philadelphia Naval Ship Systems Engineering Station (NAVSSES), which employed over 2,000 with an annual operating budget of $315 million. Assistant Secretary of the Navy for Conversion and Redevelopment William J. Cassidy, Jr., insisted that this process was a model for future public-private partnerships in other communities facing BRAC closures, although the Philadelphia CDC thought that the entire complex transfer arrangement had retarded private development on League Island.[33]

The closure and conversion process cost local, state, and federal governments well over $400 million. New Jersey, Pennsylvania, and the city of Philadelphia contributed smaller grants to assist in conversion to private use. The bulk of closure expense fell on the federal government. Mayor Rendell's intimacy with the President William J. Clinton administration assured massive federal support for conversion to civilian usage. Clinton dispatched Secretary of Labor Robert Reich and other administration officials to the city with millions of dollars in funds and visited Rendell in September 1995 with a $3.5 million grant to open a road to the eastern part of League Island for development of an "East End Commerce Park." Shortly thereafter, Vice President

Albert Gore arrived with $30 million in grants for job retraining and $50 million for shipyard infrastructure and technology improvement.[34]

Conversion of League Island to private business became an expensive proposition for local government as well. Reportedly, the Meyer Werft Shipbuilding firm of Germany wanted $167 million in loans from city and state as incentives to open a small shipyard on the western end of League Island. Meanwhile, Delaware County Representative Curt Weldon suggested that a $50 million "start up fund" was necessary to ready the old naval shipyard to scrap ships of the former Soviet Navy, stripped of their nuclear reactors and now useless with the break-up of the Soviet Union and end of the Cold War. The multi-national Kvaerner Construction and Engineering Company of Norway, Europe's largest shipbuilder, reportedly wanted $200–$400 million in subsidies and incentives from local government in order to locate a ship construction facility in Philadelphia.[35]

The Philadelphia Navy Yard, renamed the Philadelphia Naval Business Center in 1994, attracted some private business. Westinghouse Electric Corporation leased part of Building 592 to test and assemble turbine generators for nuclear-powered submarines for the "fleet of the 21st century." Garvey Precision Machine of Willingboro, New Jersey, moved into the machine shop in Building 16 to manufacture hydraulic maritime machinery, and Metro Machine, part of the huge public-private shipbuilding combination in Norfolk, Virginia, rented dry dock space to sandblast, paint, and overhaul *Detroit* and other fast transports. Moreover, the city signed a lease with Kvaerner in late 1997, an event that Mayor Rendell hailed as the dawn of a new era that "will make the city whole again."[36]

Despite Rendell's exuberance, it seemed unlikely that Kvaerner or the Philadelphia Naval Business Center structure could ever replace the Philadelphia Navy Yard to "make the city whole again." A navy yard on the Delaware River had been an integral part of American institutional history and naval tradition for over 200 years. Nothing could replace that. The last Station Log entry, on 27 September 1996, like those entered daily on League Island for the past 120 years, dramatically illustrated the meaning of closure. "Continued the watch and log from 26 Sep 96; Conditions as follow: Shipyard Commander is in Qtrs K, USS *Butte* is berthed at drydock 4, Various Units of the inactive Fleet are berthed through out the harbor," the final entry recorded. "Ship-

yard commander arrives, observed morning colors, commenced closure ceremony, shipyard commander orders the watch secured and Philadelphia Naval Shipyard closed; secured the watch, Philadelphia Naval Shipyard is closed, no further entries this Log."[37]

The battleship *New Jersey* arrives home to the closed Navy Yard in 1999, prior to removal to Camden across the river. *Camden Courier Post*.

ABBREVIATIONS AND SHORT TITLES

These are used in both the notes and the illustration captions.

American State Papers. American State Papers, Documents, Legislative and Executive, of the Congress of the United States: Naval Affairs. 4 vols. Washington, D.C.: Gales and Seaton, 1832–61.

APS. American Philosophical Society.

Barbary Wars Naval Documents. U.S. Office of Naval Records and Library, *Naval Documents Related to the United States Wars with the Barbary Powers: Naval Operations Including Diplomatic Background from 1785 Through 1807.* 6 vols. Washington, D.C.: U.S. Government Printing Office, 1939–44.

Beacon. Philadelphia Naval Shipyard Beacon.

Cadwalader Military Papers. General George Cadwalader Military Papers, HSP.

Central Subject Files. Central Subject Files, PNSR.

Civil Engineer Miscellaneous Correspondence. Civil Engineer, Public Works Office, Miscellaneous Correspondence, PNSR.

Commandants Letters. Letters from Commandants, Navy Yard Philadelphia, PNSR.

Commissioners Letters. Letters from Board of Navy Commissioners to Commandants, Philadelphia Navy Yard, 1827–35, PNSR.

Council/Committee Minutes. Minutes of the Provincial Council of Pennsylvania and of the Committee (Council) of Safety of the Province of Pennsylvania. *Pennsylvania Colonial Records.*

Daniels Diaries. Josephus Daniels, *The Cabinet Diaries of Josephus Daniels, 1913-1921*, ed. E.

David Cronon. Lincoln: University of Nebraska Press, 1963.

Defense Committee Minute Book. Minute Book of the Committee of the Defense of the Delaware, 1814–15. HSP.

Du Pont Civil War Letters. Samuel Francis Du Pont, a Selection from His Civil War Letters, ed. John D. Hayes, 3 vols. Ithaca, N.Y.: Cornell University Press for Eleutherian Mills Historical Library, 1969.

DANFS. U.S. Naval History Division, *Dictionary of American Naval Fighting Ships*, 8 vols. Washington, D.C.: U.S. Government Printing Office, 1959–.

Deed Book. Philadelphia County Deed Books, Department of Records, PCA. Microfilm, HSP.

Delegates Letters. Letters of Delegates to Congress, 1774-1789, ed. Paul H. Smith. Washington, D.C.: Library of Congress, 1976–.

Franklin Papers. The Papers of Benjamin Franklin ed. Leonard W. Labaree (vols. 15–25 ed. William B. Willcox). 25 vols. New Haven, Conn.: Yale University Press, 1959–.

Grice Papers. Joseph Grice Papers, Society Miscellaneous Collection, HSP.

Hamilton Papers. The Papers of Alexander Hamilton, ed. Harold C. Syrett with Jacob C. Cooke. 27 vols. New York: Columbia University Press, 1961–87.

HSP. Historical Society of Pennsylvania, Philadelphia.

Humphreys Papers. Humphreys Papers, HSP.

Humphreys Shipyard Papers. Humphreys Shipyard

Papers, ISML.

ISM, ISML. Independence Seaport Museum/Library, Philadelphia.

Laurens Papers. The Papers of Henry Laurens, ed. Philip M. Hamer (vols. 5–10 ed. George C. Rogers, Jr. and D. R. Chesnutt). 10 vols. Columbia: University of South Carolina Press, 1970.

Mounted Clippings. Philadelphia Navy Yard Mounted Clipping Collection, Urban Archives, Temple University Library, Philadelphia.

NARAM. National Archives, Mid Atlantic Region, Philadelphia

Navy Yard Collection, Princeton. Philadelphia Navy Yard Collection, Princeton University Library, Princeton, N.J.

NDAR. U.S. Navy Department, *Naval Documents of the American Revolution*, 10 vols., ed. William Bell Clark. Washington, D.C.: Naval Historical Center, 1964–96.

Official Records. U.S. Office of Naval Records and Library, *Official Records of the Union and Confederate Navies in the War of the Rebellion*. Ser. I, 27 vols.; ser. II, 3 vols. Washington, D.C.: U.S. Government Printing Office, 1894–1922.

Oral History Project. Philadelphia Naval Shipyard Oral History Project, ISML.

Pennsylvania Colonial Records. Colonial Records of Pennsylvania: Minutes of the Provincial Council of Pennsylvania from the Organization to the Termination of the Proprietary Government. 16 vols. Harrisburg: Theo Fenn, 1851–53.

PCA. Philadelphia City Archives.

PGSP. Publications of the Genealogical Society of Pennsylvania.

PMHB. Pennsylvania Magazine of History and Biography.

PNSR. Philadelphia Naval Shipyard Records, Records of the Fourth Naval District, NARAM.

PNSY History. Philadelphia Naval Shipyard History, collection of ten boxes transferred to PNSR in 1996.

Potter's American Monthly. Potter's American

Monthly, bound volume of collected articles with marginal notes, PNSY History, box 2.

Public Works Office Miscellaneous Correspondence. Public Works Office, Miscellaneous Correspondence, 1899–1921, PNSR.

Public Ledger. Philadelphia Public Ledger and Daily Transcript.

Quasi-War Naval Documents. Naval Documents Related to the Quasi-War Between the United States and France. Naval Operations . . . February 1797 December 1801, 7 vols. Washington, D.C.: U.S. Government Printing Office, 1935–38.

Ship Files. Ship Files, PNSR.

Source Document. Historic Preservation Studio, *Source Document—Philadelphia Navy Yard*. Philadelphia: University of Pennsylvania Program in Historic Preservation, 1993.

Special Committee Minutes. "Special Committee on Extension of the Navy Yard," Joint Committee Minutes, Councils, Committees, Joint Special Minutes, 1860–61, PCA. *Journal of the Select Council of the City of Philadelphia*. Philadelphia: Crissey and Markley, Printers, Goldsmith Hall, 1862.

Station Log. Station Log, PNSR.

Tax List. Philadelphia Tax Record Book. Microfilm, HSP.

USNIP. U.S. Naval Institute Proceedings.

Washington's Writings. The Writings of George Washington from the Original Manuscript Sources, 1745-1799, ed. John C. Fitzpatrick. 39 vols. Washington, D.C.: U.S. Government Printing Office, 1931–44.

Welles Diary. Diary of Gideon Welles: Secretary of the Navy Under Lincoln and Johnson, ed. Howard K. Beale. 3 vols. New York: W.W. Norton, 1960.

Wharton-Willing Papers. Wharton-Willing Papers, Gratz Collection, HSP.

Will Book. Will Books, Department of Records, PCA.

Yards and Docks Letters. Bureau of Yards and Docks Correspondence, PNSR.

NOTES

INTRODUCTION

1. W. T. Cluverius, Discussion, "On Inspection Duty at Navy Yards," *USNIP* 142 (1912): 743; Captain Alfred Thayer Mahan, *Naval Strategy Compared and Contrasted with the Principles and Practice of Military Operations on Land: Lecture Delivered at U.S. Naval War College, Newport, R.I., Between the Years 1887 and 1911* (Boston: Little, Brown, 1915), 195; Julius Augustus Furer, *Administration of the Navy Department in World War II* (Washington, D.C.: Department of the Navy, 1959), 1.

2. The best historical works on navy yards are the British study by Jonathan G. Coad, *The Royal Dockyards, 1760-1850: Architecture and Engineering Works of the Sailing Navy* (Aldershot: Scolar Press of Gowen Publishing, 1989) and the series of short essays in Paolo E. Coletta, ed., *United States Navy and Marine Corps Bases, Domestic* (Westport, Conn.: Greenwood Press, 1985), particularly on Pensacola by George F. Pearce and Norfolk by Peter C. Stewart. The best "in-house" history remains Frederick R. Black, *Charlestown Navy Yard, 1890-1973*, 2 vols. (Boston: Boston National Historical Park, National Park Service, U.S. Department of the Interior, 1988). Good "in-house" histories of the Philadelphia Navy Yard include Navy Yard Clerk Henry M. Vallette's series of ten articles "History and Reminiscences of the Philadelphia Navy Yard" that ran from January to October 1876 in *Potter's American Monthly* 6–7 (Philadelphia); Chief Clerk A[mandus] R. Ritter, "A Brief History of the Navy Yard from Its Inception to December 31, 1920" (Philadelphia: Philadelphia Navy Yard Department of Yards and Docks, 1921), typescript copies in Free Library of Philadelphia and NAMAR; S. E. Zubrow, "The History of Philadelphia Navy Yard: An Official History of the Navy Yard from 1798 to 1945, with Particular Emphasis on the Navy Yard's Role in World War II," *United States Naval Administration in World War II, Commandant, Fourth Naval District*, 6 vols. (Washington, D.C.: Naval History Division, Navy Department, 1946), 3, Part 4; Vice Admiral James Laurence Kauffman, *Philadelphia's Navy Yards, 1801-1948* (New York: Newcomen Society, 1948); and Arthur Menzies Johnson, "The Genesis of a Navy Yard," *USNIP* 81 (September 1955): 992–1011.

3. Captain John Hood, "Naval Policy as It Relates to Shore Establishment, and the Maintenance of the Fleet," *USNIP* 150 (1914): 319–44 (332–33); Mahan, *Naval Strategy*, 170, 195; David Budlong Tyler, *The Bay and River Delaware: A Pictorial History* (Cambridge, Md.: Cornell Maritime Press, 1955).

4. Harry M. Tinkcom. "The Revolutionary City, 1765–1783," and Richard G. Miller, "The Federal City, 1783–1800," in Russell F. Weigley, ed., *Philadelphia: A 300-Year History* (New York and London: W.W. Norton, 1982), 109–207; Thomas R. Heinrich, *Ships for the Seven Seas: Philadelphia Shipbuilding in the Age of Industrial Capitalism* (Baltimore and London: Johns Hopkins University Press, 1997).

5. A New England Man [Gustavus V. Fox], *Advantages of League Island for a Naval Station* (Philadelphia: Sherman & Company, 1866), 29; Gail E. Farr and Brett F. Bostwick, *John Lenthall, Naval Architect: A Guide to Plans and Drawings of American Naval and Merchant Vessels, 1790-1874* (Philadelphia: Philadelphia Maritime Museum, 1991); Heinrich, *Ships for the Seven Seas*, 23.

6. Hood, "Naval Policy as It Relates to the Shore Establishment," 334; William F. Trimble, *Wings for the Navy: A History of the Naval Aircraft Factory, 1917-1956* (Annapolis, Md.: Naval Institute Press, 1990).

7. A. E. Watson, Commandants Order No. 9/42, 28 February 1942, File A3-1, box 21, Central Subject Files; Karl Abraham, "Navy Yard Has Its Own Atom Secret," Philadelphia *Evening Bulletin*, 27 August 1967.

8. Closing Rumors, 1964–68, and 1969– Files, box 173A, Mounted Clippings.

9. Oral History Transcripts.

CHAPTER 1. ORIGIN OF A NAVY YARD

1. Carl E. Swanson, *Predators and Prizes: American Privateering and Imperial Warfare, 1739-1748* (Columbia: University of South Carolina Press, 1991), 123–25; Joseph A. Goldenberg, *Shipbuilding in Colonial America* (Charlottesville: University Press of Virginia for the Mariners Museum, Newport News, 1976), 108–14; J. Thomas Scharf and Thompson Westcott, *History of Philadelphia, 1609-1884* (Philadelphia: L.H. Everts, 1884), 208–10; Peter C. Stewart, "Norfolk, Va., Naval Shipyard, 1767–," in Paolo E. Coletta, ed., *United States Navy and Marine Corps Bases, Domestic* (Westport, Conn.: Green-

wood Press, 1985), 387.

2. Willing to Messrs. Austin and Laurens, 5 December 1755, *Laurens Papers*, 2: 23; William Franklin to Printer of the Citizen, 16 September 1757, *Franklin Papers*, 7: 262; Charles Lyon Chandler, "Early Shipbuilding in Pennsylvania, 1683–1812," in *Philadelphia: Port of History, 1609–1837* (Philadelphia: Philadelphia Maritime Museum, 1976), 10–13; Swanson, *Predators*, 51; Ronald Schultz, *The Republic of Labor: Philadelphia Artisans and the Politics of Class, 1720–1830* (New York: Oxford University Press, 1993); Frederick B. Tolles, *Meeting House and Counting House: Quaker Merchants of Colonial Philadelphia* (Chapel Hill: University of North Carolina Press, 1948).

3. Benjamin Smith, Christopher Gadsden, and Henry Laurens, Commissioners of Fortifications to William Allen, 27 October 1757; Willing and Morris to Messrs. Austen and Laurens, 13 July 1757; Allen and Turner to Messrs. Smith, Gadsden & Laurens, 24 December 1757, *Laurens Papers*, 2: 537–38, 535, 540–42; Provincial Commission to William Denny, 25 January 1757, *Franklin Papers*, 7: 102–5; "Acco't Letters of Marque—1762," *Pennsylvania Archives*, second series, Mathew S. Quay, pub. (Harrisburg: B.F. Meyers, state printer, 1876), 11: 668; Arthur Cecil Bining, *Pennsylvania Iron Manufacture in the Eighteenth Century* (Harrisburg: Pennsylvania Historical and Museum Commission, 1973), 44, 53, 131–32; Marion V. Brewington, "The Designs of Our First Frigates," *American Neptune* 8 (January 1948): 16n.34; Schultz, *Republic of Labor*, 28–30; Chandler, "Early Shipbuilding," 30; Swanson, *Predators*, 93; Scharf and Westcott, *Philadelphia*, 1: 208–14.

4. "Ship *Hero* dimen. 95.6 keel 32.6 beam 10.6 hold," Record Book of Shipbuilder Samuel Humphreys, Dimensions of Ships, 3b, Humphreys Shipyard Papers, ISML; Vote of the Continental Marine Committee, December 16, 1775, *NDAR*, 3: 130; Brewington, "First Frigates," 16; Scharf and Westcott, *Philadelphia*, 1: 256–57. This argument contradicts James C. Bradford's claim that "colonial antecedents for an American Navy did not exist," "Navies of the American Revolution," in Kenneth J. Hagan, ed., *In Peace and War: Interpretations of American Naval History, 1775–1984* (Westport, Conn.: Greenwood Press, 1984), 3.

5. Henry D. Paxson, *Sketch and Map of a Trip from Philadelphia to Tinicum Island, Delaware County, Pennsylvania* (Philadelphia: George Buchanan 1926), 30–45; Margaret B. Tinkcom, "Southwark a River Community: Its Shape and Substance," *Proceedings of the American Philosophical Society* 114 (August 1970): 327–42.

6. John F. Watson and Willis P. Hazard, *Annals of Philadelphia, and Pennsylvania in the Olden Time: being a Collection of Memoirs, Anecdotes, and Incidents, of the City and Its Inhabitants* (Philadelphia: Edwin S. Stuart, 1899), 1: 71–72, 146–55; John L. Cotter,

Daniel G. Roberts, and Michael Parrington, *The Buried Past: An Archaeological History of Philadelphia* (Philadelphia: University of Pennsylvania Press, 1992), 220–33; Chandler, "Early Shipbuilding," 6, 29–30.

7. "Schemes of the First and Second Philadelphia Lottery," 12 December 1747 and 2 June 1748; James Logan to Franklin, 3 December 1747; "Notes on the Association," 28 April 1748; "Proposal to the Associators," 21 March 1748, *Franklin Papers*, 3: 288–89, 219–24, 312–14; Scharf and Westcott, *Philadelphia*, 1: 215.

8. "Plain Truth, 1747," *Franklin Papers*, 3: 197; Watson and Hazard, *Annals of Philadelphia*, 1: 148; Tinkcom, "Southwark," 327–42.

9. Matthew Clarkson and Mary Biddle, eds., *To the Mayor Recorder Alderman Common Council and Freemen of Philadelphia this plan of the Improved Part of the City surveyed and laid down by the late Nicholas Scull, Esqr, Surveyor General of the Province of Pennsylvania, November 1st 1762* (map); see also Southwark Tax List (1774), 436–48.

10. T. Wharton to Fisher, 31 March 1762, Penna. Series. Provincial Congress Papers, case I, box 13, Gratz Collection, HSP; see also Southwark Tax List (1774), 452–62, Scharf and Westcott, *Philadelphia*, 1: 256–57; Tinkcom, "Southwark," 327–42.

11. Deeds, Francis Holton and his wife Mary to John Wharton, 1 September 1759; James Reynolds and his wife Mary to John Wharton, 15 March 1768, Deed Book D 11: 198–200, 194–96; Edward Gottair (Gotear) and his wife Mary to John Wharton, 12 March 1770, Deed Book I 16: 349–51; John Wharton, Will 92, 9 December 1797, Will Book Y: 232–34; "James, Thomas, or John Wharton for London, The Ship *Madeira* Packet," advertisement, *Pennsylvania Chronicle* (May 23–30, 1768); Southwark Tax List (1774), 446–49; Ann H. Wharton, "The Wharton Family," *PMHB* 1 (1876): 324–29.

12. Deed, Jacob Lewis and his wife Sarah to James Penrose, 21 September 1761; Isaac Penrose to James Penrose, 29 September 1769; Deeds, Anthony Duché to James Penrose, 1 March 1768, Thomas Philips to James Penrose, 30 March 1770, Deed Book I 11: 34–35; I 8: 550–52; I 11: 33–34, I 7: 217–18; Southwark Tax List (1774), 444; Josiah Granville Leach, *History of the Penrose Family of Philadelphia* (Philadelphia: Drexel Biddle, 1903), 21–23, 38–40; Chandler, "Early Shipbuilding," 30.

13. See Joshua Humphreys Ledger Book D, 1766–1777; Wharton and Humphreys Ship Yard Accounts, 1773–1795; "Plan of Philadelphia," *PMHB* 80 (1956): 183; Charles H. Browning, *Welsh Settlement of Pennsylvania* (1912; rpt. Baltimore: Genealogical Publishing Co., 1967), 152–53, 155–160, 286–87; Scharf and Westcott, *Philadelphia*, 1: 254–57; "Cousin" in Henry H. Humphreys, "Who Built the First United States Navy?" *PMHB* 40 (1916): 385–411 (386).

14. Joshua Humphreys (Sr.), account for timber, 1767–68 with Thomas Wharton, Thomas Wharton to J. Humphreys, 22 August, 26 November 1767, 3 May 1768, Wharton-Willing Papers, box 2; Deed, Peregrine Hogg to J. Humphreys (Sr.), 30 October 1760, Deed Book I 11: 18–21; Joshua Humphreys Ledger Book "D" (1766–1777) "Some Accounts of the Humphreys Family" (c. 1839); Memo, 20 June 1739, Humphreys Correspondence, box 1; Dock Ward Tax List (1774), 27; Charles Barker, "Family Charts 18 Welsh Families," Genealogical Records, Gen. Box 16, HSP; Hampton L. Carson, "The Humphreys Family," *PGSP* 8 (March 1922): 121–38.

15. Southwark Tax List (1774), particularly 436, 438, 440, 447; Wharton and Humphreys Ship Yard Accounts (1773–1795), 1; Humphreys, "Who Built the First Navy?" 386.

16. John Wharton's mother's family, the Dobbinses, and the Dennis families were connected in business and probably related in marriage; see Southwark Tax List (1774), 449; "Francis Grice MS No. 2," Joseph Grice Papers. Robert Morris provided financial backing for the Wharton and Humphreys shipyard partnership; see Deed, Wharton to Morris, 5 July 1774, Deed Book I 16: 357–58; Wharton and Humphreys Ship Yard Accounts (1773–1795), 1–2, entry 31 January 1774, "Some Accounts of the Humphreys Family," both in Humphreys Papers; Browning, *Welsh Settlement*, 155, 160, 245; Scharf and Westcott, *Philadelphia*, 1: 273; John W. Jordan, *Colonial and Revolutionary Families of Pennsylvania* (Baltimore: Genealogical Publishing Co., 1978), 1: 532–33; Anne H. Wharton, "The Wharton Family," *PMHB* 1 (1876): 324–29.

17. "James and John Wharton: Non Importation Agreement," Gratz Collection, box 2. See Richard Alan Ryerson, *The Revolution Is Now Begun: The Radical Committees of Philadelphia, 1763–1776* (Philadelphia: University of Pennsylvania Press, 1978).

18. Adams to Elbridge Gerry, 7 June 1775, *Delegates Letters*, 1: 450–51.

19. Jefferson to Eppes, 4 July 1775, John Adams Notes of Debates, 24 September and 7 October 1775, *Delegates Letters*, 1: 580–81; 2: 50, 130–31.

20. Franklin to Deane, 27 August 1775, *Franklin Papers*, 20: 184–85n.

21. Committee Minutes, 30 June 1775, *Pennsylvania Colonial Records*, 10: 279.

22. Franklin to Deane, 27 August 1775, *Delegates Letters*, 1: 709–10; Committee Minutes, 30 June 1775, *Pennsylvania Colonial Records*, 10: 279–80.

23. Committee Minutes, 4, 6, 7, and 24 July and 5 August 1775, *Pennsylvania Colonial Records*, 10: 290; Erskine to Franklin, 16 August (1775) *Franklin Papers*, 22: 563–65.

24. Committee Minutes, 4, 6, and 7 July 1775, *Pennsylvania Colonial Records*, 10: 283–84; see also Humphreys Ledger Book "D" (1766–1777); Chandler, "Early Shipbuilding," 14–15; John W. Jackson, *The Pennsylvania Navy, 1775–1781: The Defense of the Delaware* (New Brunswick, N.J.: Rutgers University Press, 1974); Jeffery M. Dorwart, *Fort Mifflin of Philadelphia* (Philadelphia: University of Pennsylvania Press, 1998).

25. Committee Minutes, 15 July and 8 October 1775, *Pennsylvania Colonial Records*, 10: 287, also in *NDAR*, 2: 363–66; Goldenberg, *Shipbuilding*, 69.

26. John Rice's shipyard lay between those of Emanuel Eyre and Benjamin Eyre, Northern Liberties Tax List (1774), 361; Wharton and Humphreys Ship Yard Accounts (l773–37), 29–37, and Humphreys Ledger Book "D" (1766–77), *Experiment* cost "£241.0.7 for labor, £6 for 30 gallons of rum, and £1.6 for "113 brush brooms"; *Washington* cost £355.4.2 ; List of the Pennsylvania Row Galleys, 12 October 1775, *NDAR*, 1: 1072–74; 2: 428.

27. Entry, 5 August 1775, Wharton and Humphreys Ship Yard Accounts, 31–32, comment on row galley *Experiment*, Humphreys to Secretary of the Navy, 16 August 1798, Humphreys Letter Books; Committee Minutes, 8, 10, 19, and 26 July and 29 September 1775, B. Franklin, "An Estimate of Moneys already expended, 29 September 1775," all in *Pennsylvania Colonial Records*, 10: 284–85, 289, 291, 350–51; Macpherson to the Pa. Committee of Safety, 29 September 1775, *NDAR*, 2: 240. Macpherson was trying to interest the Continental Congress in his own secret naval weapon to defend the river, a floating mine system; see John Adams, Diary entry, 18 September 1775, *Delegates Letters*, 2: 28. See also Charles Oscar Paullin, *The Navy of the American Revolution* (1906; rpt. New York: Haskell House, 1971), 388–89; Scharf and Westcott, *Philadelphia*, 1: 303.

28. J. Adams to Josiah Quincy, 6 October 1775, and J. Adams to James Warren, 8 October 1775, *Delegates Letters*, 2: 128, 143; "Editorial Notes on the Pennsylvania Committee of Safety," *Franklin Papers*, 22: 72–74, 224–25; Committee Minutes, 2 and 3 August 1775, 14 and 16 March 1776, *Pennsylvania Colonial Records*, 10: 297–98, 514, 519.

29. Committee Minutes, 17 July 1775 and 9 and 26 February 1776, *Pennsylvania Colonial Records*, 10: 288, 480, 495; Committee Minutes, 30 October 1775, *NDAR*, 2: 652.

30. Committee Minutes, 1 and 6 September 1775, 16 February and 19 and 30 March 1776, *Pennsylvania Colonial Records*, 10: 329–31, 487, 489, 520, 530.

31. John Adams, Notes of Debates, 7 and 12 October 1775, Silas Deane's Proposals for Establishing a Navy, 16 October 1775, Deane to Mumford, 16 October 1775, *Delegates Letters*, 2: 131, 166, 182–84, 190–92; Paullin, *American Revolution*, 35–41, 80–83.

32. John Adams, Notes of Debates, 12 October 1775, Silas Deane's Proposals for Establishing a Navy, 16 Octo-

ber 1775, Deane to Mumford, 16 October 1775, all in *Delegates Letters*, 2: 166, 182–84, 190–92; also Paullin, *American Revolution*, 35–41, 80–83.

33. John Adams, Notes of Debates, 21 October 1775, Naval Committee to Deane, 7 November 1775, Deane to Elizabeth Deane, 26 November 1775, S. Ward to Henry Ward, 16 November 1775, all in *Delegates Letters*, 2: 220–23, 315, 391, 355; Paullin, *American Revolution*, 83.

34. Connecticut Delegation to Gov. J. Trumbull, 5 December 1775; Silas Deane for Conn. Delegates to Trumbull, 5 December 1775, Robert Smith Diary entry, 20 December 1775, all in *Delegates Letters*, 2: 440, 441–43, 501.

35. Entries, 4 May and 30 December 1775, Wharton and Humphreys Ship Yard Accounts, 29, 42, 44; Cost of Outfitting the Continental Ship *Alfred* (and *Columbus*), Philadelphia, 30 December 1775, *NDAR*, 3: 607–10.

36. Cost of Outfitting the Continental Brig *Andrea Doria*, Partial Cost of Outfitting the Continental Brig *Cabot*, Cost of Outfitting the Continental Sloop the *Providence*, 30 December l775, all in *NDAR*, 3: 610–11; "Ship Sally" in Transcript of Listing of Ships in Humphreys Notebook, c. 1800, Humphreys Shipyard Papers, 4, ISML provides evidence that Wharton and Humphreys built *Columbus*. For argument that the builder was "unknown," see Howard Irving Chapelle, *The History of the American Sailing Navy: The Ships and Their Development* (New York: Bonanza Books for W.W. Norton, 1949), 53, 55; John J. McCusker, *Alfred: The First Continental Flagship, 1775–1778* (Washington, D.C. Smithsonian Institution Press, 1973); *DANFS*, 1-A: 291–94; 2: 4, 150; 5: 395–97; William Bell Clark, *Gallant John Barry, 1745–1803: The Story of a Naval Hero of Two Wars* (New York: Macmillan, 1938), 67–69, 87–88; Hope S. Rider, *Valour Fore & Aft: Being the Adventures of the Continental Sloop Providence, 1775-1779, formerly, Flagship Katy of Rhode Island's Navy* (Annapolis, Md.: Naval Institute Press, 1977).

37. Deane to Elizabeth Deane, 15 December 1775, Hewes to Robert Smith, 8 January 1776, Hancock to Commander Pennsylvania Battalion, 2 December 1775, *Delegates Letters*, 2: 421, 488; 3: 57; *DANFS*, 5: 395–97.

CHAPTER 2. CONTINENTAL SHIPYARD

1. Richard Smith's diary entry, 13 December 1775, *Delegates Letters*, 2: 438–84; Vote of the Continental Marine Committee, 16 December 1775, and Vote of Committee for Building Philadelphia's Frigates, 9 January 1776, *NDAR*, 3: 130, 694; William M. Fowler, Jr., *Rebels Under Sail: The American Navy During the Revolution* (New York: Scribner, 1976), 246–47; Howard I. Chapelle, *The History of the American Sailing Navy: The Ships and Their Develop-* *ment* (New York: Bonanza Books, 1949), 55–57; Marion V. Brewington, "The Designs of Our First Frigates," *American Neptune*, 8 (January 1948): 11–25.

2. Samuel Ward to Henry Ward (brother), 14 December 1775, S. Adams to J. Adams, 22 December 1775, J. Bartlett to New Hampshire Committee of Safety, 21 December 1775, *Delegates Letters*, 2: 487, 506–7, 501.

3. Frigate Commissioners to Master Builders in Philadelphia for Building of the Frigates, enc. Humphreys to Gentlemen, 9 January 1776, Joshua Humphreys Correspondence, 2: 1, Humphreys Papers.

4. Adams to Horatio Gates, 27 April 1776, Morris to Committee of Secret Correspondence, 16 December 1776, *Delegates Letters*, 3: 587; 5:609; 2: 470; see also Clarence L. Ver Steeg, *Robert Morris: Revolutionary Financier with an Analysis of His Earlier Career* (1954; rpt. New York: Octagon, 1972).

5. Committee of Secret Correspondence Statements, 1 and 18 October 1776, Committee of Secret Correspondence to William Bingham, 3 June 1776, *Delegates Letters*, 10: 398; 5: 273–74n.5, 351–52; 4: 126–29.

6. Southwark Tax List (1774), 436; Frigate *Delaware* built by Coat[e]s and Marsh, Samuel Humphreys Dimensions of Ships Book, Humphreys Shipyard Records, folder 66.84.1, MS 16, p. 3a; Commissioners for Building the Continental Frigates in Philadelphia to Wharton & Humphreys, 9 January 1776, Wharton and Humphreys Agreement, c. 1 January 1776, *NDAR*, 3: 696–97n.1, 1039–40; Willing, Morris & Co. contract, Committee of Secrecy, minute, 14 December 1775, *Delegates Letters*, 2: 483–86.

7. See Northern Liberties Tax List (1774), 361. Rice was undoubtedly carpenter on *Effingham*, rather than "Grice & Co." as suggested in Brewington, "Design of the First Frigates," 12–13. Rice and the Eyres had an arrangement similar to Warwick Coates and Joshua Humphreys, who built *Randolph* and *Delaware* on adjacent stocks. This was a familiar way to build frigates. "I suppose it will be no objection if both these Vessels are put upon the Stocks at the same place," Richard Henry Lee wrote about the two frigates laid down at Gosport, Virginia in 1776; see Lee to James Maxwell, 1 December 1776, *Delegates Letters*, 5: 562–63.

8. Morris, "Organization for Building the Four Philadelphia Frigates," 1 January 1776; Commissioners for Building the Continental Frigates in Philadelphia to Wharton and Humphreys, 9 January 1776; Timber Commissioners advertisement, *Pennsylvania Gazette*, 4 January 1776, all in *NDAR*, 3: 561–62, 693–97, 607n.1; for Nixon and Nesbitt as naval mobilizers, see Committee Minutes, *Pennsylvania Colonial Records*, 10.

9. Frigate Commissioners to Master Builders, Humphreys to Gentlemen, 9 January 1776; Contract Frigate 2, Humphreys Ledger Book (1776–1780); "Several draughts of

the Several Ships of War laid before us & approved one for each Contractor to be Made," Bartlett to Langdon, 13 January 1776, *Delegates Letters*, 3: 88; see also *NDAR*, 3: 696–97.

10. See Wharton and Humphreys Ship Yard Accounts (1773–1795); Charles Lyon Chandler, "Early Shipbuilding in Pennsylvania, 1683–1812," in Charles Lyon Chandler, Marion V. Brewington, and Edgar P. Richardson, *Philadelphia Port of History, 1609–1837* (Philadelphia: Maritime Museum, 1976), 14–15; Henry H. Humphreys, "Who Built the First United States Navy?" *Journal of American History* 10 (1916): 48–89; Chapelle, *Sailing Navy*, 58.

11. Jonathan Grice, in entry 1 April 1776, Wharton and Humphreys Ship Yard Accounts (1773–1795), Francis Grice, MS 2, Grice Papers; J. Adams to C. Adams, 30 March 1777, *Delegates Letters*, 6: 506–7. Chapelle, *Sailing Navy*, 56–58, 98–99 doubts that Joshua Humphreys had the skill to design the frigates, but Fowler, *Rebels Under Sail*, 220–21 argues that it is more important to understand that the plans for the frigates came from the politically well-connected and influential Wharton and Humphreys firm, than who actually drew them. See also Tyrone G. Martin and John C. Roach, "Humphreys's Real Innovation," *Naval History* 8, 2 (March–April 1994): 32–37.

12. Richard Smith diary, entry 30 January 1776, Morris to Secret Committee, 19 February 1777, *Delegates Letters*, 5: 172; 7: 322; diary entries, 5, 6, and 27 April 1776, William Duane, ed., *Extracts from the Diary of Christopher Marshall kept in Philadelphia and Lancaster during the American Revolution, 1774–1781* (Albany, N.Y.: Joel Munsell, 1877; rpt. New York: Arno Press, 1969), 65, 67; see Hampton L. Carson, "The Humphreys Family," *PGSP* 8 (March 1922): 121–38 (129).

13. Middle Department Navy Board to Messrs Humphreys, Wharton, Master Builders at the Continental Ship Yard, 1 July 1779, Humphreys Correspondence, 1; C. Biddle to G. Taylor, 12 May 1776, Marine Committee of the Continental Congress to Capt. John Ashmead, 12 July 1776, *NDAR*, 5: 71–72, 1045–45; George Read to Gertrude Read, 14 July 1776, *Delegates Letters*, 4: 454–55; William Bell Clark, *Gallant John Barry, 1745–1803: The Story of a Naval Hero of Two Wars* (New York: Macmillan, 1938), 73; *DANFS*, 6: 27–28.

14. Bill of Repairs to the Continental brig *Lexington* n.d.; Repairs to sloops *Fly* and *Sachem*; Joshua Humphreys Ship Yard Account Against the Continental Sloop *Hornet*, 12 April 1776; Joshua Humphreys Bill for Repairs to Continental schooner *Wasp*, April 1776; *NDAR*, 5: 172, 923–24; 4: 789–90, 700; Committee Minutes, 25 and 28 March 1776, *Pennsylvania Colonial Records*, 10: 524, 527–28; Clark, *Barry*, 72–86.

15. Commissioners of the Continental Navy in Account with the Brigantine *Lexington*, 28 March 1776, John Barry,

Commander to the Commissioners of the Navy, 28 March 1776, Commissioners of the Continental Navy in Account Sloop *Hornet*, 21 May 1776; *NDAR*, 4: 548–51; 5: 193–95.

16. Marine Committee to Falconer, 13 October 1776, Marine Committee to Rhode Island Frigate Committee, 9 October 1776, Secret Committee Minutes, 5 September, 3 October 1776, Morris to Mumford, 22 August 1776, Richard Smith diary, entry 6 February 1776, Bartlett to Langdon, 7 October 1776, *Delegates Letters*, 5: 338, 299, 327–28, 108, 299, 49–50, 312–13; Charles Oscar Paullin, *The Navy of the American Revolution: Its Administration, Its Policy and Its Achievements* (Cleveland: Burrows, 1906; rpt. New York: Haskell, 1970), 91.

17. Morris (Secret Committee) to Deane, 1 October 1776, Matthew Thornton (New Hampshire) to Meshech Weare, 23 January 1777, Morris to Charles Lee, 17 February 1776, Morris to Bradford, 8 October 1776, *Delegates Letters*, 5: 280; 6: 139; 5: 450; 3: 267; 5: 321: 847.

18. William Ellery to William Lemon, 7 November 1776, Executive Committee (Morris) to Hancock, 4 February 1777, J. Adams to C. Adams, 30 March 1777, *Delegates Letters*, 5: 450; 6: 212, 506; Paullin, *American Revolution*, 122, 96–100; Jackson, *Pennsylvania Navy*, 289, 313.

19. Richard Smith diary, entries, 13, 16, and 18 March 1776, Adams to Abigail Adams, 12 August 1776, Morris to Bingham, 4 December 1776, 25 April 1777, Morris to Deane, 31 January 1777, Morris to Hancock, 23 January 1777, *Delegates Letters*, 3: 375, 387, 397, 5: 572–73; 6: 651, 177, 137; Franklin to Deane, 1 October 1776, Franklin to Samuel Cooper, 25 October 1776, *Franklin Papers*, 22: 664, 670. See William Bell Clark, *Ben Franklin's Privateers: A Naval Epic of the American Revolution* (Baton Rouge: Louisiana State University Press, 1956); Paullin, *American Revolution*, 94–95.

20. Hooper to Jos. Hewes, 16 November 1776, S. Adams to James Warren, 25 December 1776, *Delegates Letters*, 5: 499, 660; Journal of Continental Congress, 12 December 1776, in *NDAR*, 7: 188.

21. Morris to Bradford, 24 December 1776, Morris to Hancock, 13, 14, 16, and 23 December 1776, Hancock to Morris, 14 January 1777, *Delegates Letters*, 5: 658, 604, 611, 608–9, 642–43; 6: 98–100; Journal of Continental Congress, 12 December 1776, *NDAR*, 7: 188.

22. Morris to Hancock, 23 December 1776, Morris to Washington, 23 December 1776, *Delegates Letters*, 6: 642–43, 649.

23. Morris to Pennsylvania Council of Safety, 20 and 21 December 1776, Morris to Richard Henry Lee, 29 December 1776, Morris to Deane, 20 December 1776, *NDAR*, 7: 534, 545, 622, 531.

24. Hooper to Morris, 28 December 1776, Harrison to Morris, 29 December 1776, R. H. Lee to Morris, 24 December 1776, Morris to R. H. Lee, 27 December 1776, *Dele-*

gates Letters, 5: 687–90n.5, 691, 656–57, 682–83.

25. Hancock to the Massachusetts Council, 28 December 1776, Morris to Hancock, 30 December 1776, Roger Sherman to Jonathan Trumbull, Sr., 23 April 1777, *Delegates Letters,* 5: 687n.2, 697–700; 6: 642.

26. Hancock to Morris, 24 January 1777, *NDAR,* 7: 1032–33; Marine Committee to Morris, 24 January and 5 February 1777, Morris to Hancock, 26 January 1777, Whipple to Langdon, 7 November 1776, *Delegates Letters,* 6: 141, 223, 149–50; 5: 451; Minutes Council of Safety, 15 February 1777, Supreme Executive Council Minutes, 27 March and 6 June 1777, *Pennsylvania Colonial Records,* 11: 125, 191–93, 216; see also Jackson, *Pennsylvania Navy.*

27. Hamond to Captain Henry Bellew (HMS *Liverpool*), 7 December 1777, Journal of HMS *Liverpool,* 8 December 1777, *NDAR,* 10: 681–82, 686; Supreme Executive Council Minutes, 15 June and 30 October 1778, *Pennsylvania Colonial Records,* 11: 513, 610, 745; "A Plan of the City and Environs of Philadelphia with the Works and Encampments of His Majesty's Forces," J. Thomas Scharf and Thompson Westcott, *A History of Philadelphia, 1609–1884* (Philadelphia: L.H. Everts, 1884), 1: betw. 360–61.

28. John Eyre petition, 22 April 1779, Supreme Executive Council Minutes, 22 April 1779, *Pennsylvania Colonial Records,* 11: 756; Francis Grice MS 2, Joseph Grice Papers; Peter D. Kayser, ed., "Memorials of Col. Jehu Eyre," *PMHB* 3, part 2 (1879): 412–25 (424); Thomas McKean (Delaware) to wife Sarah McKean, 16 May 1778, *Delegates Letters,* 9: 668–89; Clark, *John Barry,* 122–25, 127–33, 155; Jackson, *Pennsylvania Navy,* 296–97; Steven Rosswurm, *Arms, Country, and Class: The Philadelphia Militia and "Lower Sort" During the American Revolution, 1775–1783* (New Brunswick, N.J.: Rutgers University Press, 1978), 152.

29. Ellery to William Vernon, 11 March 1778, Ellery to Whipple, 25 April 1778, *Delegates Letters,* 9: 300–301, 488–89.

30. Marine Committee to the Eastern Navy Board, 30 May 1778, Marine Committee to Loyall and Maxwell, 17 October 1778, 12 March 1779, Marine Committee to John Wereat (naval agent, Georgia), 29 July 1778, John Fell Diary, entries, 1 February and 8 April 1779, *Delegates Letters,* 9: 775–78; 11: 69–70; 10: 369–70; 12: 3, 308; Paullin, *American Revolution,* 113–15.

31. Supreme Executive Council Minutes, 29 January, 7, 9, 14, and 18 December, 25 September 1778, *Pennsylvania Colonial Records,* 11: 408, 585, 638, 645, 640, 643.

32. Supreme Executive Council Minutes, 31 July, 14 August, 30 October, 28 November, 23 and 24 December 1778, *Pennsylvania Colonial Records,* 11: 541, 552, 610, 629, 650–51, 664–65; Scharf and Westcott, *Philadelphia,* 1: 403.

33. Middle Department Navy Board to Humphreys and Wharton, 1 July 1779, Humphreys Correspondence Book I(A): 6; Jay to Washington, 26 April 1779, *Delegates Letters,* 12: 386–87; Morris to President of Congress (McKean), 8 September 1781, E. James Ferguson ed., *The Papers of Robert Morris, 1781–1784* (Pittsburgh: University of Pittsburgh Press, 1975), 2: 214–15; Paullin, *American Revolution,* 102ff; launch day 10 April 1780, *DANFS,* 6: 334–36; William Bell Clark, *The First Saratoga: Being the Saga of John Young and His Sloop-of-War* (Baton Rouge: Louisiana State University Press, 1953), 24–25.

34. Entries, June 1781, 4 September and 20 December 1782, Wharton & Humphreys Ship Yard Accounts (1773–1795): 60, 62, 64, 86, 96, 112; John A. McManemin, *Captains of the Privateers During the Revolutionary War* (Spring Lake, N.J.: Ho-Ho-Kus Publishing, 1985), 308, 316–17, 353–54; Scharf and Westcott, *Philadelphia,* 1: 409.

35. Humphreys Account Book 2 (1784–1813), 1; Eugene S. Ferguson, *The Life of Commodore Thomas Truxtun, U.S. Navy, 1755–1822* (Baltimore: Johns Hopkins University Press, 1956); Philip Chadwick Foster Smith, *The Empress of China* (Philadelphia: Philadelphia Maritime Museum, 1984), 20–23.

36. Entries, 6 February 1785, March 1785, Humphreys Account Book, II: 5–6, 11; John Nixon and Elizabeth Hutton (executrix for John Hutton, ship carpenter) to Joshua Humphreys, 8 August 1798 (to pay debts of John Hutton), Deed Book D 76: 4–8; Southwark Town, Philadelphia County, *First Census of the United States* (1790): 212–13; Ferguson, *Truxtun,* 48–87.

37. Entries 9 December 1787, 23 August 1788, Humphreys Account Book 2: 57–59; Humphreys to Harrison 10 September 1801, Humphreys Letter Book 3: 62; Ferguson, *Truxtun,* 51.

38. "Federalist Nos. 24, 11, 23," *Hamilton,* 4: 422, 343–44, 412–22.

CHAPTER 3. UNITED STATES SHIPYARD

1. J. Thomas Scharf and Thompson Westcott, *History of Philadelphia, 1609–1884* (Philadelphia: L.H. Everts, 1884), 1: 447–49; Charles Lyon Chandler, "Early Shipbuilding in Pennsylvania, 1683–1827," *Philadelphia Port of History, 1809–1837* (Philadelphia: Philadelphia Maritime Museum, 1976), 23.

2. Humphreys Account Book 1, Section 2 (1792–1806): 3–5, 18, 25, 36–37, 72,134, 110, Articles of Agreement between Joshua Humphreys, Junior of Southwark, Shipwright, and Andrew Clow and David Cay of the City of Philadelphia, Merchant, 11 July 1789, Humphreys Correspondence 1; Abraham Ritter, *Philadelphia and Her Merchants* (Philadelphia: published by the author, 1860), 38; William Bell Clark, "James Josiah, Master Mariner," *PMHB*

79 (October 1955): 468; William Bell Clark, *Gallant John Barry, 1745–1803: The Story of a Naval Hero of Two Wars* (New York: Macmillan, 1938), 330, 335; Eugene S Ferguson, *The Life of Commodore Thomas Truxtun, U.S. Navy, 1759–1822* (Baltimore: Johns Hopkins University Press, 1956), 52, 62, 87, 27n8.

3. Deed, Samuel Church to Joshua Humphreys, Jr., Deed Book D26: 283–85; Whipple to Hamilton, 19 December 1789, Lincoln to Hamilton, 22 December 1789, *Hamilton Papers*, 6: 21, 27–28; Irving H. King, *George Washington's Coast Guard Cutters: Origins of the U.S. Revenue Cutter Service 1789–1801* (Annapolis, Md.: Naval Institute Press, 1978), 41.

4. "Contract with United States of America," Humphreys Account Book 1: 134; Hamilton to Washington, 10 September 1790, Hamilton to Sharp Delaney, 22 January 1791, *Hamilton Papers*, 7: 31–32, 447.

5. See Craig L. Symonds, *Navalists and Antinavalists: The Naval Policy Debate in the United States, 1785–1827* (Newark: University of Delaware Press, 1980); Marshall Smelser, *The Congress Founds the Navy, 1787–1798* (Notre Dame, Ind.: University of Notre Dame Press, 1959).

6. Humphreys to Morris, 6 January [(1794)], Humphreys Letter Book 1: 1 (1793–97): Transcript of Listing of Ships in Humphreys's notebook, c. 1800, MS. 17, Humphreys Shipyard Papers. Humphreys wrote Morris in response to the capture of *President*, 23 October 1793, and Knox invited Humphreys to War Office on 3 February 1794, not 1793, as suggested in Howard I. Chapelle, *The History of the American Sailing Navy: The Ships and Their Development* (New York: Bonanza Books for W.W. Norton, 1949), 126–27, Clark, *Barry*, 366; Ferguson, *Truxtun*, 87; Smelser, *Congress Founds the Navy*, 23ff. Seizure of *President*, in Gardner W. Allen, *Our Navy and the Barbary Corsairs* (1905; rpt. Hamden, Conn.: Archon Books, 1965), 16; *President* misidentified as *Prudent*, "List of American Vessels Captured by Algerines in October & November, 1793," *Barbary Wars Naval Documents*, 1: 56; Scharf and Westcott, *Philadelphia*, 1: 477–78, used the original Humphreys list and identified *President* correctly.

7. Fitzsimmons Report to House of Representatives, 20 January 1794, "An Act to Provide a Naval Armament, 27 March 1794," both in *Barbary Wars Naval Documents*, 1: 60–61, 69–70; Smelser, *Congress Founds the Navy*, 52–53; also Symonds, *Navalists and Antinavalists*.

8. Humphreys to Morris, 6 January (1794), Humphreys Letter Book 1: 1; Knox to Humphreys, 3 February 1794, Humphreys Correspondence, 1, box B; Knox to Hamilton, 21 April 1794, *Hamilton Papers*, 16, 308n.; Chapelle, *American Sailing Navy*, 118–19; Clark, *Barry*, 366; Ferguson, *Truxtun*, 109.

9. Knox to Humphreys, 3 February 1794, Humphreys

Correspondence 1, box A; J. Humphreys to son Samuel Humphreys, 20 August 1827, Humphreys Letter Book 3; Marion V. Brewington, "The Design of Our First Frigates," *American Neptune* 8 (January 1948): 16–25; Merle T. Westlake, Jr., "Josiah Fox, Gentleman, Quaker, Shipbuilder," *PMHB* 88 (July 1964): 316–27, 323; Tyrone G. Martin and John C. Roach, "Humphreys's Real Innovation," *Naval History* 8, 2 (March/April 1994): 32–37.

10. Humphreys to Fox, 18 May 1795, Humphreys to Secretary of War [June 1795], "The Strengthening of Ships," Humphreys Memo, 18 February 1796, Humphreys Letter Book 1: 158, 165–66, 197; Knox to Humphreys, 12 April, 24 July 1794, Humphreys Correspondence IA: 7, 107; Knox to Hamilton, 21 April 1794, *Hamilton Papers*, 16: 306; Knox to Humphreys, 28 June and 24 July 1794, *Barbary Wars Naval Documents*, 1: 75, 78, 76; Hampton L. Carson, "The Humphreys Family," *PGSP* 8 (March 1922): 129.

11. Knox to Humphreys, 24 July 1794, Tench Coxe to Humphreys, 1 August 1794, Truxtun to Humphreys, 3 October 1794, 30 April 1795, Humphreys Correspondence 1: 10–12, 21; Secretary of War to Secretary of Treasury, 21 April 1794, Secretary of War to Barry, 7 August 1794, *Barbary Wars Naval Documents*, 1: 71–74, 79; Hamilton to Coxe, 8 May 1792, Coxe to Knox, 14 April 1794, Knox to Hamilton, 21 April 1794, 25 June, and 9 July 1794, Coxe to Hamilton, 22 and 25–27 December 1794, *Hamilton Papers*, 16: 241n.1, 304–7, 526, 582–83; 17: 452–59, 468; Chapelle, *American Sailing Navy*, 118–20.

12. John T. Morgan to Humphreys, 30 August and 21 October 1794, Humphreys Letter Book 1: 28–29; Knox to Hamilton, 21 April, 12 May, 25 June 1794, Coxe to Hamilton, 22 and 25–27 December 1794, Coxe to Knox, 7 June 1794, *Hamilton Papers*, 16: 304–7, 406–7, 526; 17: 468, 452–59, 468; 16: 455n.1; Virginia Steele Wood, *Live Oaking: Southern Timber for Tall Ships* (Boston: Northeastern University Press, 1981), 25–32.

13. Knox to Humphreys, 28 June 1794, 2 May 1795, Coxe to Humphreys, 22 August 1794, Humphreys Correspondence IA; Humphreys to Knox, 20 December 1794, Humphreys to Morgan 18 November and 18 December 1794, Humphreys to Joseph Henderson, 13 February and 31 March 1795, Humphreys to Francis, 25 January 1797, Humphreys Letter Book 1: 6–9, 29–31, 97, 256; entry, 23 October 1794, Humphreys Ledger Book (1794–1801); Deeds, Richard Wells, Miers Fisher, and Daniel Smith to Humphreys the younger, 6 October 1794, Daniel Martin and wife Agnes to Humphreys, 2 January 1795, Deed Book D 49, 500–503, 505–8; Knox to Hamilton, 12 May 1794, Coxe to Hamilton, 25–27 December 1794, Knox to Coxe, 14 December 1794, *Hamilton Papers*, 16: 406–7; 17: 472–75, 473n.31; Clark, *Barry*, 371.

14. Humphreys to Morgan, 18 December 1794, 18 April 1795, Humphreys to William Pennock (Norfolk), 8 Septem-

ber and 26 November 1794, Humphreys to John Blagge (New York), 17 September and 20 November 1794, Humphreys to Henry Jackson (Boston) 20 September and 3 November 1794, Humphreys to Jacob Sheafe (Portsmouth), 4 December 1794, Humphreys to Secretary of War, 19 May 1795, Humphreys Circular to the Different Mast Builders and Capt. Truxtun, 17 April 1795, Humphreys to Col. James Hackett, 5 November 1795, Humphreys Letter Book 1: 31–45, 48–55, 152, 155, 159, 188–91; Morgan to Humphreys, 30 August 1794, Blagge to Humphreys, 10 December 1794, Humphreys Correspondence I; Memo, 6 December 1797, Miscellaneous Correspondence, Box 1, Humphreys Papers.

15. Pickering to Humphreys, 20 May 1795, Humphreys Correspondence I; Humphreys to James Owner, 4 November 1795, Humphreys to [Claghorn], 11 January 1796, Humphreys to Secretary of War, 26 June 1795, Humphreys Letter Book 1: 193, 187–88, 173–74; Coxe to Hamilton, 18 and 23 September, 22 December 1794, *Hamilton Papers*, 17: 247, 256–60, 452–59; Washington to Secretary of War, 20 November 1795, *Washington's Writings*, 34: 366.

16. Humphreys to T. Francis, 22 July 1795, Humphreys to Truxtun, 29 July 1796, Humphreys Letter Book 1: 177, 236.

17. Washington to Secretary of War, 20 November 1795, *Washington's Writings*, 35: 366; see also Smelser, *Congress Founds the Navy*; Symonds, *Navalists and Antinavalists*.

18. Humphreys to James Hackett, 26 May 1796, Humphreys Letter Book 1: 210; Washington to the Senate and House of Representatives, 15 March 1796, *Washington's Writings*, 34: 499–500; Smelser, *Congress Founds the Navy*, 77–80.

19. Hamilton to Wolcott, 16 June 1796, Wolcott to Hamilton, 28 June 1796, "Draft of George Washington's Eighth Annual Address to Congress," 10 November 1796, Washington to Hamilton, 12 November 1796, Hamilton to Washington, 19 November 1796, Hamilton to Sedgwick, 20 January 1797, *Hamilton Papers*, 20: 228, 242–43, 385, 394–95n.2, 408, 474n.1.

20. Humphreys to Hackett, 5 November 1795, "Report on the State of the Frigate UNITED STATES now Building in the Port of Philadelphia," 28 December 1796;, Humphreys to Secretary of War, 1 February 1797, Humphreys Letter Book 1: 188–91, 246, 257–58; 2: 1; Humphreys to Sharp Delany, 10 February 1797, Oliver Wolcott to Humphreys, 9 February 1797, Humphreys Correspondence IA: 33; Secretary of War to Parker, 12 January and 2 February 1797, "Statement of the Progress made in Building a Frigate At Philadelphia to Carry Forty Four Guns, enc. In An Estimate of the Probable Cost of a Site for a Navy Yard, and the Buildings necessary thereto," *Barbary Wars Naval Documents*, 1: 187–91, 194–95.

21. Humphreys to Swanwick, Parker, and Bingham, 1 February 1797, Humphreys to Swanwick, 6 February 1797, Humphreys Letter Book 1: 259–64; McHenry to Hamilton, [December 1796/January 1797], Hamilton to Sedgwick, 20 January 1797, *Hamilton Papers*, 20: 455–56n.1, 473–74; Roland M. Baumann, "John W. Swanwick: Spokesman for 'Merchant Republicanism' in Philadelphia, 1790–1798," *PMHB* 97 (April 1973): 131–82; Smelser, *Congress Founds the Navy*, 77–80, 90–97, 101n.37; Symonds, *Navalists and Antinavalists*, 39–49.

22. Humphreys to Secretary of War [May/June 1795], 6 and 11 May 1797, Humphreys to G. Washington, 24 April 1797, Humphreys Letter Book 1: 162; 2: 20, 25–27; Doc. No. 6—Three Frigates Directed to Be Completed, 17 March 1796, Doc. No. 7—Progress Made in Building Frigate, The Establishment of a Navy Yard, and the Purchase of Live Oak Plantation, 25 January 1797, *American State Papers: Naval Affairs*, 1: 25–26.

23. Humphreys to Truxtun, 25 October 1796, Humphreys to Secretary of War, 11 May 1797, Humphreys Letter Book 1: 245, 2: 26–27; Pickering to Humphreys, 20 May 1795, Humphreys Correspondence I; Memorandum of Secretary of War, 10 May 1797, Secretary of War to Humphreys, 23 March 1798, *Quasi-War Naval Documents*, 1–5, 45; Clark, *Barry*, 384–88; Chapelle, *American Sailing Navy*, 129.

24. Smelser, *Congress Founds the Navy*, 108–13, 117–18.

25. Humphreys to Secretary of War (June 1795), Humphreys to Fox (clerk in the Marine Department War Office), 25 July 1797, Humphreys Letter Book 1: 165–66; 2: 33; S. Humphreys to Editor *National Gazette*, December 1837, Newspaper folder, box 1, Humphreys Papers; Secretary of War [Fox] to Humphreys, 25 July 1797, *Quasi-War Naval Documents*, 1: 9; Edmund H. Quincy (Portsmouth) to Fox, 5 November 1797, *Barbary Wars Naval Documents*, 1: 221–22.

26. McHenry to Humphreys, 25 July 1797, Humphreys Correspondence IIB; Humphreys to Truxtun, 11 June 28 and 31 July 1797, Humphreys, David Stodder and Truxtun to Secretary of War, 28 July 1797, Humphreys to David Stodder, 31 July 1797, Humphreys to Secretary of War, 4 October 1797, Humphreys to Wolcott, 12 October 1797, Humphreys Letter Book 2: 28, 35, 36–37: Secretary of War to Humphreys, 25 July 1797, Truxtun to Humphreys, 7 September 1797, *Quasi-War Naval Documents*, 1: 9, 17; Ferguson, *Truxtun*, 129–35.

27. A Jeffersonian Republican representative opposed to the navy "came down to the Yard & in conversation with him, he declared to me he would sooner see her in flames than see her built," Humphreys notes, c. 1827 on Naval Affairs, Humphreys Correspondence IB: 128; Humphreys to Fox, 16 September 1796, Humphreys to Secretary of War, 7 February 1795, 26 March 1797, 6 May 1797, Humphreys

Letter Book 1: 222, 77–78; 2: 17–18, 25; Ferguson, *Truxtun*, 136; Smelser, *Congress Founds the Navy*, 97, 101n. 53.

28. Humphreys to Francis, 28 March 1797, Humphreys to Captain John Rice, 13 February 1797, Humphreys to Secretary of War, 26 March and 20 February 1797, Humphreys Letter Book 2: 18, 2, 17, 10, Fort Mifflin timber yard, in Humphreys Ledger [Waste Book] Accounts, 1794–1801, Philadelphia Navy Yard Records; Jeffery M. Dorwart, *Fort Mifflin of Philadelphia* (Philadelphia: University of Pennsylvania Press, 1998), 78.

29. Pickering to Humphreys, 14 July 1797, Humphreys Correspondence IA; Humphreys to Wolcott, 11 August 1797, Humphreys to Pickering, 19 July and 22 August 1797, Humphreys to O'Brien, 2 September, 6 November, and 16 December 1797, Humphreys to Truxtun, 6 September 1797, Humphreys to Builders of Philadelphia, 19 July 1797, Humphreys Letter Book 2: 32, 39–40, 43, 41, 62, 67, 44; Secretary of War to Thomas Thompson (Portsmouth, N.H.), 7 July 1797, Secretary of War to Fox, 1 September 1796, Estimate by Naval Constructors Joshua Humphreys and Josiah Fox of Cost of Building and Equipping a Frigate of 36 guns, and navigating same to Algiers, 29 November 1796, O'Brien to Dey of Algiers, 4 December 1797, *Barbary Wars Naval Documents*, 1: 204, 172, 181, 223; Allen, *Barbary Corsairs*.

30. Humphreys to Wolcott, 11 August 1797, Humphreys to Pickering, 19 July and 25 September 1797, Humphreys to Francis, 2, 22, and 25 September, 20 October 1797, Humphreys to McHenry, 22 September 1797, Humphreys Letter Book 2: 39–40, 32, 58, 42, 51, 48, 47; Chapelle, *American Sailing Navy*, 93, 135–38.

31. McHenry to Hamilton, 12 February 1798, Adams Message to Congress, 19 March 1798, *Hamilton Papers*, 21: 352n.3, 368–69n.1; Smelser, *Congress Founds the Navy*, 128–31.

32. Pickering to Hamilton, 25 March, 9 April 1798, *Hamilton Papers*, 21: 376–77, 409; Smelser, *Congress Founds the Navy*, 128–33, 136–37; Clark, *Barry*, 383, 375; Scharf and Westcott, *Philadelphia*, 1: 480, 493; Hampton L. Carson, "The Humphreys Family," *PGSP* 8 (March 1922): 132–33.

33. McHenry to Hamilton, 12 May 1798, Hamilton to McHenry, 17 May 1798, Wolcott to Hamilton, 18 May 1798, *Hamilton Papers*, 21: 459–61, 426, 465; Smelser, *Congress Founds the Navy*, 147–56.

34. McHenry to Humphreys, 20 March 1798, Humphreys Correspondence IIB; Wolcott to Hamilton, 18 May 1798, McHenry to Hamilton, 12 May 1798, *Hamilton Papers*, 21: 459–61, 465; Secretary of War to Humphreys, 14 June 1798, Deed of transfer for Purchase of Ship *Ganges*, 3 May 1798, *Quasi-War Naval Documents*, 1: 116, 63–64; *DANFS*, 3: 17.

35. List of ships converted at Humphreys Yard during Quasi-War, "Vessels in Humphreys's Ms.," Humphreys Shipyard Papers, MS.18, ISML; Secretary of War to Humphreys, 14 June 1798, Robert Gill to Secretary of Navy, 8 August 1798, *Quasi-War Naval Documents*, 2: 116, 282; Michael A. Palmer, *Stoddert's War: Naval Operations During the Quasi-War with France, 1798–1801* (Columbia: University of South Carolina Press, 1987), 20; Chapelle, *American Sailing Navy*, 142–43, 533; Smelser, *Congress Founds the Navy*, 182–83.

36. Humphreys to [unknown], 15 February 1797; Wolcott to Humphreys, 25 May 1798, Humphreys Correspondence, IA:38; Humphreys to Wolcott, 29 March and 24 May 1798, Humphreys to Secretary of the Treasury [March, 1798], Calculation of Number of Tons, n.d., Humphreys Letter Book 2: 6–9, 95–97, 114–17.

37. Secretary of Navy to Yellot, 6 July 1798, *Quasi-War Naval Documents*, 1: 173; Palmer, *Stoddert's War*, 22–23; John J. Carrigg, "Benjamin Stoddert, 18 June 1798–31 March 1801," in Paolo E. Coletta, ed., *American Secretaries of the Navy*, 2 vols. (Annapolis, Md.: Naval Institute Press, 1980), 1: 62, 65; Chapelle, *American Sailing Navy*, 158; Charles Oscar Paullin, *Paullin's History of Naval Administration, 1775–1911: A Collection of Articles from the U.S. Naval Institute Proceedings* (Annapolis, Md.: Naval Institute Press, 1968), 103.

38. Humphreys to Secretary of Navy, 19 and 22 June 1798, Humphreys Letter Book 2: 119–35; Palmer, *Stoddert's War*, 12–17; Scharf and Westcott, *Philadelphia*, 1: 494; Richard Gibbs Miller, "Philadelphia: The Federalist City" (Ph.D. diss., University of Nebraska, 1970), 212, 218.

39. Stoddert to Humphreys, 17 August 1798, Humphreys Correspondence IA; Humphreys to Secretary of the Navy, 26 August 1798, Humphreys Letter Book 2: 147; Scharf and Westcott, *Philadelphia*, 1: 438.

40. Humphreys to Francis, 3 June 1797, Mast Yard Ledger Book, 1797–1806, Humphreys Papers; Humphreys to Francis, 2 October 1797, Humphreys to Hugh Campbell, 2 September 1798, Humphreys to James Watson, 20 September 1798, Humphreys to Secretary of the Navy, 16 and 29 August and 10 September 1798, Humphreys to Secretary of State, 11 October 1798, Humphreys Letter Book 2: 149; Stoddert [in Trenton] to Humphreys [care of post master at Chester near Marcus Hook], 30 August 1798, Humphreys Correspondence IIB.

41. Humphreys to Stoddert, 16 and 19 September and 1 October 1798, Humphreys to James and Edward Watson [New York], 26 October and 16 November 1798, Humphreys to Secretary of State, 2 and 8 November 1798, Humphreys Letter Book 2: 154, 156, 164, 174–75, 172, 177.

42. James Dodds and James Moore, *Building the Wooden Fighting Ship* (New York and Bicester, England: Facts on File, 1984), 7–12; Secretary of Navy to Secretary of Treasury, 26 December 1798, Tingey to Secretary of Navy,

20 November 1798, *Quasi-War Naval Documents*, 2: 126.

43. Humphreys to Secretary of Navy, 16 November 1798, Humphreys to Cottinger, 27 December 1798, Humphreys Letter Book 2: 178; Stoddert to Secretary of Treasury, 26 December 1798, *Quasi-War Naval Documents*, 2: 126; Palmer, *Stoddert's War*, 126–27; Paullin, *Naval Administration*, 112; Chapelle, *American Sailing Navy*, 172.

44 Humphreys to Secretary of Navy, 26 December 1798, Humphreys Letter Book 2: 180–82.

CHAPTER 4. "SECOND AND NEXT BEST" NAVY YARD

1. Humphreys to Stoddert, 4 and 5 June, 9 and 18 July, 3 August, 3 September 1800, Humphreys to William Pennock (naval agent, Norfolk), 27 July 1799, 26 May 1799, Humphreys to James and Ebenezer Watson (naval agents, New York), 31 July 1799, Humphreys to Steven Hyginson & Co. (Boston) and Jacob Sheafe (Portsmouth), 2 September 1799, Humphreys to Marbury (Washington, D.C.) and to Ebenezer Jackson (Charleston), 24 August 1799, all in Joshua Humphreys Letter Book 2: 232–68.

2. Stoddert to Humphreys, 7 June 1800, Humphreys Correspondence, IA; Humphreys to Stoddert, 9 and 18 July, 3 August, and 3 September 1800, Humphreys to Stoddert, 9 and 18 July, 3 August, and 3 September 1800, Humphreys Letter Book 2 (no page numbers after May 1800); Humphreys to Stoddert, 2 December 1800, 3 May 1801, Humphreys to the Secretary of War, 22 April 1801, William Bainbridge to Humphreys, 22 April 1801, Humphreys to George Harrison, 3 May 1801, J. Humphreys to S. Humphreys, 1801, Humphreys Letter Book 3; G. Terry Sharrer, "The Search for a Naval Policy, 1783–1812," in Kenneth J. Hagan, ed., *In Peace and War: Interpretations of American Naval History, 1775–1984* (Westport, Conn.: Greenwood Press, 1984), 34–38.

Ships laid up or overhauled at Philadelphia in 1800/1801 included frigates *United States*, *Constellation*, and *Philadelphia*, brig *Patapsco*, armed ship *Experiment*, revenue cutter *Scammel*, former French privateer *Delaware*, converted ship *George Washington*.

3. Humphreys to Marbury, 13 and 16 November 1799, Humphreys to Stoddert, 6 and 30 October 1799, Humphreys Letter Book 2; K. Jack Bauer and Leslie Kingseed, "District of Columbia, Washington Navy Yard, 1800–," in Paola E. Coletta, ed., *United States Navy and Marine Corps Bases, Domestic* (Westport, Conn.: Greenwood Press, 1985), 181–82; Charles Oscar Paullin, *Paullin's History of Naval Administration, 1775–1911: A Collection of Articles from the U.S. Naval Institute Proceedings* (Annapolis, Md.: Naval Institute Press, 1968), 114–15.

4. Stoddert to Humphreys, 29 January 1800, Humphreys Correspondence IA; *Hamilton Papers*, 25: 225–26; Michael A. Palmer, *Stoddert's War: Naval Operations During the Quasi-War with France, 1798–1801* (Columbia: University of South Carolina Press, 1987), 169, 229–30; Paullin, *Naval Administration*, 114–15.

5. Humphreys Report on Potential Navy Yard Sites, April 1800, Humphreys to Stoddert, 16 May 1800, Humphreys Letter Book 2: 308–27.

6. "Soundings of the River Delaware (with map)," January 1800, Humphreys Letter Book 2: 292–94; Humphreys to Peter Muhlenberg (port warden in Philadelphia), 10 July 1804, Humphreys Letter Book 3.

7. Humphreys to Alexander J. Dallas, December 1815, Humphreys Letter Book 3.

8. Humphreys to Stoddert, 9 July 1800, Humphreys to Bonsall and Shoemaker (Allen's lawyers), 11 December 1800, Humphreys Letter Book 3.

9. Humphreys to Stoddert, 4 June, 9 and 18 July 1800, Humphreys Letter Book 2.

10. Stoddert to J. Humphreys, 8 June and 30 August 1800, Michael Halyday (Kensington land agent) to Humphreys, 12 and 13 September 1800, Humphreys Correspondence IA; Humphreys to Stoddert, 4 June, 9 and 15 September 1800, Humphreys Letter Book 2.

11. Stoddert to Humphreys, 4 and 17 November 1800, Humphreys Correspondence IA; Paullin, *Naval Administration*, 116.

12. Humphreys to Stoddert, 2 December 1800, and 6, 16, and 26 February 1801, Humphreys to Anthony Morris, 8 December 1800, Humphreys to Webb, 6 December 1800, Humphreys Letter Book 3; Stoddert to Humphreys, 16 December 1800, 14 February, and 2 March 1801, Humphreys Correspondence IA; Deeds, William Allen and Margaret his wife, and John Allen to U.S., 20 January 1801, Luke Morris and Wife to U.S.A., 20 February 1801, Anthony Morris and Wife to U.S.A., 20 February 1801, Deed Book EF 7: 1–5. The official date of government acquisition was 3 March 1801; see Nelson M. Blake, acting chief archivist, to Commander Philadelphia Naval Shipyard, 27 July 1950, file A3-1/NY4, PNSR, Central Subject Files, 1927–1954, box 4.

The three lots, totaling 11 acres, cost $37,000. The cost for all six original navy yards stood at $135,848, see Paullin, *Naval Administration*, 114; John J. Carrigg, "Benjamin Stoddert," in Coletta, ed., *Secretaries of the Navy*, 1: 71; J. Thomas Scharf and Thompson Westcott, *History of Philadelphia, 1609–1884*, 3 vols. (Philadelphia: L.H. Everts, 1884), 3: 2340.

13. Humphreys to Stoddert, 8 December 1800, Humphreys to S. Humphreys, 4 May 1801, Humphreys Letter Book 3:; Carrigg, "Stoddert," 75n.56 suggested that Stoddert misused government funds through a "dubious interpretation" of the naval law of 1799 to build 74s. Even

navy yard partisan Assistant Secretary of the Navy Gustavus V. Fox questioned the legality of the transaction. "These navy yards were finally bought without express authority from Congress, and without any appropriation, but were necessarily established to carry out the law of congress directing the building of six frigates"; see A New England Man, *Advantages of League Island for a Naval Station, Dockyard, and Fresh-Water Basin for Iron Ships, and other Vessels of War, as Recommended by Public Authorities* (Philadelphia: Sherman & Co., Printers, 1866), 29.

14. Entry, 17 March 1801, Humphreys Ledger Accounts [Waste Book] (1794–1801): 138–45, PNSR; Humphreys to Stoddert, 26 February and 16 March 1801, Humphreys to the Secretary of the Navy, 6 May 1801, Humphreys Letter Book 3.

15. Stoddert to Humphreys, 17 November 1800, 14 February and 2 March 1801, Humphreys Correspondence IA; Humphreys to Stoddert, 2 December 1800, Humphreys to Secretary of the Navy, 8 May, 9 June, and 27 July 1801, Humphreys to Secretary of War, 22 April 1801, Humphreys Letter Book 3; Paullin, *Naval Administration*, 127.

16. Humphreys to Turner, 1 April, 12 and 19 November 1801, Bainbridge and Humphreys to Secretary of War, 22 April 1801, Humphreys to Secretary of the Navy, 8 May, 9 June, and 27 July 1801, Humphreys Letter Book 3; S. Smith, acting secretary of the Navy to Humphreys, 11 June 1801, Humphreys Correspondence IA; Abishai Thomas (acting for S. Smith) to George Harrison, 14 April 1801, S. Smith to Truxtun, 1 May 1801, *Barbary Wars Naval Documents*, 1: 433, 441.

17. R. Smith to Humphreys, 12 and 13 August, 21 September 1801, S. Smith to Humphreys, 11 June 1801, Humphreys Correspondence IA; Humphreys to R. Smith, 10 and 19 August 1801, Humphreys to Jacob Sheafe (Portsmouth, NH), 20 December 1800, Humphreys Letter Book 3; Frank L. Owsley, Jr., "Robert Smith, 27 July 1801–7 March 1809," in Coletta, ed., *Secretaries of the Navy*, 1: 77; Paullin, *Naval Administration*, 121–22.

18. Humphreys to R. Smith, 10 and 19 August 1801, Humphreys Letter Book 3. The base closure committee of 1801 included Republican U.S. Representative William Jones of Philadelphia, naval captains Stephen Decatur (Senior) and John Barry, Kensington Shipbuilder Samuel Bowers, merchant Thomas Fitzsimmons, Penrose, and Humphreys, see Harrison to Secretary of the Navy, 17 March 1802, *Barbary Wars Naval Documents*, 2; Paullin, *Naval Administration*, 126.

19. Smith to Humphreys, 26 October 1801, Humphreys Correspondence IA; Humphreys to Secretary of the Navy, 29 October 1801, Humphreys to John Templeman (naval agent, Georgetown), 11 December 1801, Humphreys Letter Book 3; Paullin, *Naval Administration*, 125–27; Owsley,

"Smith," 79.

20. Humphreys Draft of a Report for Mr. Harrison, n.d. [1802], Humphreys to A. J. Dallas, n.d. [December 1815], Humphreys Letter Book 3; Secretary of the Navy Circular to Navy Agents, 22 February 1802, Harrison to Secretary of the Navy, 17 March 1802, *Barbary Wars Naval Documents*, 2: 64, 89–90.

21. Report for Mr. Harrison, n.d. [February/March 1802], Humphreys to Dallas, December 1815, Humphreys Letter Book 3; Harrison to Secretary of the Navy, 17 March 1802, Extract from Midshipman F. Cornelius DeKrafft's Journal, 6 August 1803, *Barbary Wars Naval Documents*, 2: 89–90, 506. Howard Irving Chapelle, *The History of the American Sailing Navy: The Ships and Their Development* (New York: Bonanza Books for W.W. Norton, 1949), claimed that the Navy Yard built twelve gunboats, but no evidence could be found to substantiate this.

22. Doc. No. 52—Jefferson to House of Representatives, 18 February 1806, *American State Papers: Naval Affairs*, 1: 84–87; Statement of Expenditures of the U.S. Navy, 1798–1805," *Barbary Wars Naval Documents*, 6: 329; Hamilton to Samuel Miles Hopkins, 17–24 April 1801, *Hamilton Papers*, 25: 378–79; Paullin, *Naval Administration*, 119–35; Craig L. Symonds, *Navalists and Antinavalists: The Naval Policy Debate in the United States, 1785–1827* (Newark: University of Delaware Press, 1980), 157–68.

23. Doc. No. 27—Expenditures of the Navy and Navy Yards, 1798–1805, *American State Papers: Naval Affairs*, 1: 85–86.

24. Smith to President, 20 January 1802, Mitchill to President, 10 March 1802, in Doc. No. 27— Expenditures of the Navy and Navy Yards, 84–87.

25. Humphreys to Smith, 21 January 1802, Humphreys to Fitzsimmons, 30 March 1802, Humphreys to Harrison, n.d., Humphreys Letter Book 3.

26. Humphreys to John Rutledge, 15 January 1803, Humphreys Letter Book 3; Paullin, *Naval Administration*, 129–30.

27. Gunboat letters in, *American State Papers: Naval Affairs*, 1: 194–200; *Gunboats Nos.* 116, 120, 121, 125, 129, 132, and 135 operated from the Philadelphia Navy Yard in 1812, see William S. Dudley and Michael J. Crawford, eds., *The Naval War of 1812: A Documentary History* (Washington, D.C.: Naval Historical Center: U.S. Government Printing Office, 1985–), 2: 200n.2; Paullin, *Naval Administration*, 133–35.

28. Doc. No. 74—Condition and Disposition of the Naval Force, Secretary of the Navy Paul Hamilton to the Senate, 25 May 1809; Doc. No. 118—Statement of the Expenditures at the Navy Yard at Philadelphia, as also the cost of repairs on vessels, &c. made for account of the United States in the years 1811 and 1812, by George Harrison, Navy Agent

at Philadelphia, 1814, *American State Papers: Naval Affairs*, 1: 193, 337–38; Dudley, *Naval War of 1812*, 1: 454; Paullin, *Naval Administration*, 142, for bill to close Portsmouth, Philadelphia, and Washington Navy Yards in 1810.

29. Meeting 3 January 1815, Defense Committee Minute Book 2: 138, HSP; Jones to Harrison, 5 March 1813, Jones to Murray, 8 July 1813, in Dudley and Crawford, *Naval War of 1812*, 2: 46, 181.

30. Thomas Leiper to Jones, 2 September 1814, Jones to Leiper, 16 September 1813, Minutes and General Orders, 24, 26, and 31 August 1814, Defense Committee Minute Book 1: 16, 11, 40, 45; Dr. Edward Cutbush to Charles W. Goldsborough, 4 January 1813, Jones to Biddle, 31 March 1813, Biddle to Jones, 28 April 1813, Eyre to Jones, 9 May 1813, Jones to Eyre, 12 May 1813, Jones to Madison, 6 June 1813, Murray to Secretary of the Navy, 3, 12, July, 25 August 1813, Master Commandant Samuel Angus to Murray, 29 July 1813, Sailing Master William W. Sheed to Angus, 6 August 1813, Dudley and Crawford, *Naval War of 1812*, 2: 9–10, 83–85n.1, 115, 117, 145–46, 180, 182–83, 199–201, 204; Doc. No. 111—Jones Report on condition of the Navy, 18 March 1814, *American State Papers: Naval Affairs*, 1: 305; Paullin, *Naval Administration*, 142.

31. Murray to Jones, 3 July 1813, Cutbush to Goldsborough, 4 January 1813, Dudley and Crawford, *Naval War of 1812*, 2: 179–81n.1, 9–10; Leonard F. Guttridge and Jay D. Smith, *The Commodores* (New York: Harper and Row, 1969), 67; Paullin, *Naval Administration*, 145, 168.

32. Jones to Murray, 8 July 1813, Dudley and Crawford, *Naval War of 1812*, 2: 181; Doc. No. 313—On the Tenure by Which the Site of the Navy Yard at Philadelphia is Held, Amount of Taxes Paid Thereon to the City and County, Jurisdiction Over the same not being Granted to the United States by the Legislature of Pennsylvania, *American State Papers: Naval Affairs*, 2: 707–21.

33. Minutes and General Orders, 24 August 1814, Minutes of Meetings 26 and 31 August, 1 September 1814, Defense Committee Minute Book, 1: 16, 11, 40, 45; Secretary of the Navy to Harrison, 9 April 1813, Southwark Commissioners to Harrison, 2 September 1813, *American State Papers: Naval Affairs*, 2: 713–14.

34. Humphreys to Jonathan Williams, 30 August 1814, Humphreys to Charles Biddle, 6 September 1814, Humphreys Letter Book 3: 130–32; Meetings 1 and 7 September 1814, Defense Committee Minute Book 1: 55, 82; entry 20 June 1814, Breck Diary, Nicholas B. Wainwright, ed., "Diary of Samuel Breck," *PMHB* 102 (October 1978): 475; Gail E. Farr and Brett F. Bostwick, *John Lenthall, Naval Architect: A Guide to Plans and Drawings of American Naval and Merchant Vessels, 1790–1874* (Philadelphia: Maritime Museum, 1991), 15.

35. Seybert to Humphreys, 14 February 1814, 13 and 27 January 1815, Humphreys Correspondence IA: 116, 119, 120; Humphreys to Seybert, 8 and 24 January 1815, Humphreys Letter Book 3.

36. Doc. No. 409—Rodgers to Crowninshield, 2 May 1815, Secretary of the Navy John Branch to Andrew Stevenson (Speaker of House), On Number, Extent, and Arrangements of Navy Yards and Dry Docks, 1 February 1830, *American State Papers: Naval Affairs*, 3: 493–94; Edwin M. Hall, "Benjamin W. Crowinshield," in Coletta, ed., *Secretaries of the Navy*, 1: 112–20, 117–18; Charles O. Paullin, *Commodore John Rodgers: Captain, Commodore and Senior Officer of the American Navy, 1773–1838* (1910; rpt. Annapolis Md.: Naval Institute Press, 1967); Paullin, *Naval Administration*, 167–73; Chapelle, *American Sailing Navy*, 305–8.

37. Humphreys to Dallas, December 1815, Humphreys Letter Book 3: 136–41.

38. Entry, 21 August 1815, "Breck Diary," 480.

CHAPTER 5.
BUILDING THE WOODEN SAIL AND STEAM NAVY

1. John Barry, William Jones, and Joshua Humphreys first developed the idea of a navy board with three officers and two civilians; see Charles Oscar Paullin, *Paullin's History of Naval Administration, 1775–1911: A Collection of Articles from the U.S. Naval Institute Proceedings* (Annapolis, Md.: Naval Institute Press, 1968), 164–68, 176–77, 172–73; Edwin M. Hall, "Benjamin W. Crowinshield, 16 January 1815–30 September 1818," in Paolo E. Coletta, ed., *American Secretaries of the Navy*, 2 vols, (Annapolis, Md.: Naval Institute Press, 1980), 1: 114.

2. Entries, 15 April, 25 June, 6, 12, and 18 May, 16 July, 10, 11, and 14 December 1818, 4 January, 2, 3, and 24 April 1819, Samuel Humphreys Journal, 1; log entries, 1, 3, and 22 March and 27 December 1819, Station Log, Navy Yard, Philadelphia, PNSR. Howard Irving Chapelle, *The History of the American Sailing Navy: The Ships and Their Development* (New York: Bonanza Books for W.W. Norton, 1949), 313–15 claimed that Doughty not Samuel Humphreys designed *North Carolina*.

3. Entries 2, 3, and 24 April, 11 May, and 13 July 1818, 10 May, 13, 16, and 29 November 1819, 5 and 10 February, 29 May 1820, Samuel Humphreys Journal; Chauncey, in log entry, 10 February 1819, Station Log.

4. Woolwich Dockyard influence, in entries, 27 March, 20 May 1819 and 7 September 1820, Samuel Humphreys Journal; Bainbridge to Southard, 6 February 1827, Doc. No. 336—Cost of, and Expenditures at, the Several Navy Yards,

from 1819 to 1826, *American State Papers: Naval Affairs*, 3: 41; James Dodds and James Moore, *Building the Wooden Fighting Ship* (New York: Facts on File, 1984), 43, 63–75.

5. Log entries, ll April 1820 and 28 November 1821, Station Log; Gail E. Farr and Brett F. Bostwick, *John Lenthall, Naval Architect* (Philadelphia: Philadelphia Maritime Museum, 1991), 18; David F. Long, *Ready to Hazard: A Biography of Commodore William Bainbridge, 1774–1833* (Hanover, N.H.: University Press of New England, 1981), 72–77, 262, 311–12; Craig Symonds, "William S. Bainbridge, Bad Luck or Fatal Flaw?" in James C. Bradford, ed., *Command Under Sail: Makers of the American Naval Tradition, 1775–1850* (Annapolis, Md.: Naval Institute Press, 1985), 97–125.

6. Log entries, 15 and 16 August and 12 November 1821, 25 October 1827, Station Log; Ship House Carpenter Philip Justice, in Thomas Wilson, *Philadelphia Directory and Strangers Guide* (Philadelphia: Wilson and Vaubaun, 1825), 77; J. Thomas Scharf and Thompson Westcott, *History of Philadelphia, 1609–1884*, 3 vols. (Philadelphia: L.H. Everts, 1884), 3: 2340; "Philadelphia Naval Station to Mark Its 150th Year Tomorrow," *Philadelphia Evening Bulletin*, 21 April 1972, [Philadelphia Navy Yard] History 1960– File, box 173A, Mounted Clippings.

7. Bainbridge to Secretary of the Navy, 28 June 1827, Navy Commissioners to Commandant Navy Yard, Philadelphia, 3 March and 28 June 1827, Rodgers to Bainbridge, 10 October 1827, 22 October and 4 November 1828, Commissioners Letters; Stewart to Paulding, 8 and 9 June 1840, Commandants Letters 2; log entry, 25 October 1827, Station Log; B. Armitage, architect, Drawing of Buildings for Navy Yard, 1827, Photo Collection Bb 615, A734, HSP; Paullin, *Naval Administration*, 182.

8. George C. Read to William A. Graham, 24 October 1850, Commandants Letters, 3; Doc. No. 196—Condition of the Navy and Its Expenses, 25 January 1821; Doc. No. 217—Exhibit of Labor performed at the Navy Yard in Philadelphia during the years 1820, 1821, and 1822; Doc. No. 217—Expense of Building Each Vessel Authorized by Act of 2 January 1813, in James Monroe to Senate, 3 January 1823; Doc. No. 219—Register of the Navy of the United States for 1823, *American State Papers: Naval Affairs*, 1: 715–16, 847, 832, 854; Long, *Bainbridge*, 299–300.

9. Bainbridge to Southard, 16 November 1836, Miscellaneous Papers, box 1, Humphreys Papers; Board of Navy Commissioners to Bainbridge, 3 July, 13 September and 1 November 1827, Circular Letter, 14 September 1827, Commissioners Letters; for *Cyane*, log entry, 5 November 1835, Station Log; Doc. No. 370—Annual Report of the Secretary of the Navy for 1828, Doc. No. 409—Branch to House of Representatives, 28 January 1830, *American State Papers: Naval Affairs*, 3: 234, 493.

10. Rodgers to Bainbridge, 11 October 1828, Commissioners Letters; Stewart to Paulding, 9 June 1840, John Grimm to Paulding, 14 January 1840, Elliott to Warrington, 10 May 1845, Commandants Letters, 2; *Philadelphia Public Ledger and Daily Transcript*, 1 December 1841; Paul Barron Watson, *The Tragic Career of Commodore James Barron, U.S. Navy, 1769–1851* (New York: Coward-McCann, 1942); Leonard F. Guttridge and Jay D. Smith, *The Commodores* (Annapolis, Md.: Naval Institute Press, 1984), 298–99; Nicholas B. Wainwright, "Commodore James Biddle and his Sketch Book," *PMHB* 90 (January 1966): 24; Edwin M. Hall, "Samuel Lewis Southard," in Coletta, ed, *Secretaries of the Navy*, 1: 132; Paullin, *Naval Administration*, 183–84.

11. Entries, 4 and 7 November and 5 December 1818, Samuel Humphreys Journal; Doc. No. 313—Thompson to Harrison, 23 August 1823, Harrison to Thompson, 25 August 1823, Southard to Harrison, 5 March 1824, Southard to Joseph Hemphill, Samuel Breck, and D. H. Miller, 28 January 1825; Doc. No. 313—On the Tenure by Which the Site of the Navy Yard at Philadelphia is Held, Amount of Taxes Paid Thereon to the City and County, in Southard to Speaker of House, 28 April 1826, *American State Papers: Naval Affairs*, 2: 707–21; Hall, "Southard," 131–40; Scharf and Westcott, *Philadelphia*, 1: 590.

12. Bainbridge to Secretary of the Navy, 23 June 1828, Southard to Bainbridge, 1 November 1828, John Rodgers to Bainbridge, 29 May, 10 July, and 6 October 1828, Barron to Woodbury, 12 December 1832, Commissioners Letters 1; Doc. No. 313—Vaughan to Southard, 20 April 1826, *American State Papers: Naval Affairs*, 2: 721; Hall, "Southard," 131–40.

13. Barron to Secretary of the Navy, 22 September and 27 November 1832, Hunter to Secretary of the Navy, 9 September 1833, Stewart to Dickerson, Frederick Engle to Abel P. Upshur, 31 August 1842, Commandants Letters 2; Agnes Addison Gilchrist, *William Strickland: Architect and Engineer, 1788–1854* (Philadelphia: University of Pennsylvania Press, 1950), 73–76; Park Benjamin, *The United States Naval Academy* (New York and London: G.P. Putnam's, 1900), 119–27; Wainwright, "Biddle and His Sketch Book," 32–33.

14. Bainbridge to Southard, 15 November 1828, Bainbridge to Branch, 25 February 1831, Commandants Letters; Doc. No. 394—Navy Board to Branch, 19 October 1829, *American State Papers: Naval Affairs*, 3: 354; Patrick Straus, "John Branch, 9 March 1829–12 May 1831," in Coletta, ed., *Secretaries of the Navy*, 1: 145–46; Long, *Bainbridge*, 301ff; Symonds, "Bainbridge," 97–125; Scharf and Westcott, *Philadelphia*, 1: 606, 608, 616; Paullin, *Naval Administration*, 173.

15. Bainbridge to Branch, 25 February and 5 March 1831, Commandants Letters 1; Doc. No. 394—Navy Board to Branch, 19 October 1829, *American State Papers: Naval Affairs*, 3: 354; Patrick Straus, "John Branch, 9 March 1829 to 12 May 1831," in Coletta, ed., *Secretaries of the Navy*, 1: 145–46; Long, *Bainbridge*, 301–8.

16. Barron to Woodbury, 27 December 1831, Commandants Letters; Woodbury to Rodgers, 7 January 1833, Toucey to Barron, 14 May 1835, Woodbury to Barron, 28 March, 1 August 1833, Philadelphia Navy Yard Collection; log entries, 26 May 1830, l0 December 1835 and 14 September 1836, Station Log; Doc. No. 564—Annual Report of the Secretary of the Navy for 1835, *American State Papers: Naval Affairs*, 4: 747, 612; Hall, "Southard," 138; Scharf and Westcott, *Philadelphia*, 1: 636.

17. Barron to Secretary of the Navy, 24 February, 31 December 1832 and 5 February 1833, Commandants Letters 1; Harold D. Langley, "The Grass Roots Harvest of 1828," *USNIP* 90 (October 1964): 51–59; Hall, "Southard," 137.

18. Hunter [for Barron] to Secretary of the Navy, 16 January 1835, Commandants Letters, 1; Dickerson to Barron, 13 January 1837, Philadelphia Navy Yard Collection; Stewart to Secretary of the Navy, 27 January 1841, Commandants Letters 2; Daniel Bowen, *A History of Philadelphia* (Philadelphia: Daniel Bowen, 1839); Frederick M. Binder, "Pennsylvania Coal and the Beginnings of American Steam Navigation," *PMHB* 83 (October 1959): 420–43; Nicholas B. Wainwright, "The Age of Nicholas Biddle, 1825–1841," in Russell F. Weigley, *Philadelphia: A 300-Year History* (New York: W.W. Norton, 1982), 270–74.

19. Master of Yard H. S. Connor (for Barron) to Secretary of the Navy, 7 July 1835, Barron to Secretary of the Navy, 13 November 1834, Barron to Woodbury, 5 December 1831, Barron to Secretary of the Navy, 20 August 1833, Barron to Dickerson, 21 July 1835, Commandants Letters 1; Edwin M. Hall, "Smith Thompson," in Coletta, *Secretaries of the Navy*, 1: 124; Weigley, *Philadelphia*, 270–74; Scharf and Westcott, *Philadelphia*, 1: 582–83.

20. Stewart to Paulding, 29 March 1839, 18 December 1840, Commandants Letters 2; Binder, "Pennsylvania Coal," 440; Paullin, *Naval Administration*, 178–80.

21. Ray W. Irwin, "Charles Stewart," *Dictionary of American Biography*, 18: 6–7; Hall, "Southard," 136; Guttridge and Smith, *Commodores*, 266–67; Scharf and Westcott, *Philadelphia*, 1: 748.

22. Dickerson to Stewart, 13 October 1837, Board of Navy Commissioners to Stewart, 31 October 1838, Philadelphia Navy Yard Collection; Stewart to Dickerson, 4 September, 2 and 9 October 1837, Commandants Letters 2; *Public Ledger*, 19 July 1837, 10 October 1839; Farr and Bostwick, *Lenthall*, 7–8, 22; Chapelle, *American Sailing Navy*, 354, 362–64, 402.

23. Morris (for Board of Navy Commissioners) to Stewart, 19 February and 14 May 1840, Philadelphia Navy Yard Collection; *Public Ledger*, 10 October 1839; Farr and Bostwick, *Lenthall*, 7–8; Frank M. Bennett, *The Steam Navy of the United States: A History of the Growth of the Steam Vessel of War in the U.S. Navy, and of the Naval Engineer Corps* (1896, rpt. Westport, Conn.: Greenwood Press, 1974), 35–36; Donald L. Canney, *The Old Steam Navy*, vol. 1, *Frigates, Sloops, and Gunboats, 1815-1885* (Annapolis, Md.: Naval Institute Press, 1990), 11–16.

24. Morris to Stewart, 18 April, 13 June, and 1 and 12 December 1840, Philadelphia Navy Yard Collection.

25. Warrington to Stewart, 21 September 1841, Philadelphia Navy Yard Collection; log entries, 5 May and 13 June l843, Station Log; *Public Ledger*, 17 July 1837, 4 May 1841; Farr and Bostwick, *Lenthall*, 24, 27; Bennett, *Steam Navy*, 69–98; Binder, "Pennsylvania Coal," 441.

26. Warrington to Read, 15 July 1842, McLane (for Board of Navy Commissioners) to Read, 22 February 1842, Philadelphia Navy Yard Collection; Farr and Bostwick, *Lenthall*, 25.

27. Log entry, 13 June 1843, Station Log; *Public Ledger*, 17 July 1837, 5 and 6 May 1841, 13 and 14 June 1843.

28. *Public Ledger*, 19 July 1837, 14 June 1843.

29. Log entries, 2 August 1842, 1 December 1843, 1 and 9 February, 2 March, and 16 April 1844, Station Log; *Public Ledger*, 5, 6, and 8 September 1843; Bennett, *Steam Navy*, 67–69; Canney, *Old Steam Navy*, 1: 21–23; Chapelle, *American Sailing Navy*, 428–30; Scharf and Westcott, *Philadelphia*, 1: 662.

30. Gauntt to Warrington, 4 and 9 September 1845, Read to Warrington (Chief Bureau of Yards & Docks [Warrington]), 1 April 1845, Elliott to Warrington, 10 May 1845, Yards and Docks Letters; Mason to Elliott, 1 April 1845, Regulations for Commanders Steam Vessels, 26 February 1845, Philadelphia Navy Yard Collection; *Public Ledger*, 25 June and 24 August 1846; Charles Oscar Paullin, "Jesse Duncan Elliott," *Dictionary of American Biography*, 6: 96–97; Guttridge and Smith, *Commodores*, 229ff; Scharf and Westcott, *Philadelphia*, 1: 598.

31. Engle to Mason, 10 May 1844, Commandants Letters 2; Morris to Stewart, 13 February and 19 March 1841, Philadelphia Navy Yard Collection; Scharf and Westcott, *Philadelphia*, 1: 638–39, 654–55, 667–70; Michael Feldberg, *The Philadelphia Riots of 1844* (Westport, Conn.: Greenwood Press, 1975), 143–61.

32. Minutes of Meeting of General Committee, 27 and 28 January 1846, Memorial to Senate and House of Representatives from Citizens of Philadelphia City and Country, 17 February 1846, W. Potter to Cadwalader, 24 February 1846 and 23 January 1846, Cadwalader Military Papers; U.S. Congress, Floating Docks, Basin, and Railways, January

19, 1848, 30th. Cong., 1st sess., House Report 106, Doc. No. 307—On the Construction of Docks for the Preservation and Repair of United States Vessels, in Southard to House of Representatives, 30 March 1826; Doc. No. 318—On the Expediency of Constructing Dry Docks for Repairing Ships-of-War, 8 May 1826, *American State Papers: Naval Affairs*, 2: 700–701, 726; *Great Meeting of the Citizens of the City and County of Philadelphia in Favor of a Dry Dock, 24 July 1837* (Philadelphia, 1837), pamphlet, Library Company of Philadelphia; Charles B. Stuart, *The Naval Dry Docks of the United States* (New York: Charles B. Norton, Irving House, 1852), 7–11.

33. Woodburne Potter to Gen. George Cadwalader, 16 March 1846, Cadwalader Military Papers; *Speech of Mr. Levin of Philadelphia, Pa. on an Amendment to the Naval Appropriation Bill, directing the Construction of a Sectional Floating Dry-Dock, Basin, and Railways, at the Philadelphia Navy-Yard. Delivered in the House of Representatives, Saturday, June 13, 1846*, pamphlet, Library Company of Philadelphia; *Public Ledger*, 16 June 1846; Levin, in Elizabeth M. Geffen, "Industrial Development and Social Crisis, 1841–1854," in Weigley, *Philadelphia*, 358.

34. Mason to Stewart, 29 October and 8 December 1846, Acting Secretary of the Navy John Appleton to Stewart, 7 June 1847, Mason to Stewart, 31 July 1847, Philadelphia Navy Yard Collection; Floating Docks, 3–16; *Public Ledger*, 8 April 1850; Paullin, *Naval Administration*, 224–25; Canney, *Old Steam Navy*, 27; Bennett, *Steam Navy*, 52.

35. *Public Ledger*, 8 April 1850; Canney, *Old Steam Navy*, 32–33; Farr and Bostwick, *Lenthall*, 8, 30.

36. Read to Smith, 3 May, 20 June, and 19 September 1851, Stewart to Smith, 19 December 1848, Yards and Docks Letters; entries, 13 January 1853, 25 December, 30 November 1850, 25 March 1851, Station Log; *Public Ledger*, 8 April 1850, 12 June and 30 June 1851.

37. Read to Smith, 1 October 1851, Stewart to Smith, 18, 28 and 29 November and 19 December 1848, Read to Graham, Ritchie to Smith, 18 June 1850, Yards and Docks Letters; Read to Graham, 8 July 1851, Commandants Letters 3; entries, 8 and 14 October 1853, Station Log; *Public Ledger*, 21 July, 1 August 1851.

38. Lenthall to Stewart, 23 June and 23 September 1854, Philadelphia Navy Yard Collection; entry, 2 October 1852, Station Log; *Public Ledger*, 8 April 1850, 1 and 2 May 1855; Stuart, *Dry Docks*, 12; Bennett, *Steam Navy*, 141–42; Canney, *Old Steam Navy*, 45–51.

39. Entry, 15 October 1858 recorded 1,476 civilian laborers, entries, 20 October 1852, 30 October 1858, Station Log; *Public Ledger*, 21 October 1858, 20 January and 10 October 1859; Bennett, *Steam Navy*, 154, 158, 166–70, 173–74; Canney, *Old Steam Navy*, 61, 66–67, 84–86; Scharf and

Westcott, *Philadelphia*, 1: 728–29.

40 Carl M. Cochran, "James Queen, Philadelphia Lithographer," *PMHB* 82 (April 1958): 138–75 (150).

41. *Public Ledger*, 1 May 1855.

CHAPTER 6. CIVIL WAR AND TWO NAVY YARDS

1. Du Pont to William Whetten (New York businessman), *Du Pont Civil War Letters*, 1: 15–16; see Russell F. Weigley, "The Border City in Civil War, 1854–1865," Weigley, ed., *Philadelphia: A 300-Year History* (New York: W.W. Norton, 1982), 363–416; J. Matthew Gallman, *Mastering Wartime: A Social History of Philadelphia During the Civil War* (New York: Cambridge University Press, 1990; rpt. Philadelphia: University of Pennsylvania Press, 2000).

2. Du Pont to Henry Winter Davis, 30 December 1860, Du Pont to George Smith Blake (commandant, Naval Academy), 27 September 1861, Du Pont to Samuel Mercer (commander, *Powhatan*), 24 December 1860, Du Pont to Whetten, 5 January and 23 June 1861, Fox to Du Pont, 22 May 1861, Fox to Bache, 30 May 1861, Bache to Du Pont, 25 November 1862, *Du Pont Civil War Letters*, 1: 14–15, 155, 13n.5, 15–16, 78–79n.2, 71–73; 2: 370n.23.

3. Du Pont to Welles, 23 April 1861, Welles to Secretary of War Simon Cameron, 27 April 1861, Du Pont to Kautz, 8 May 1861, *Official Records*, ser. I, vol. 4: 3, 318–19, 341, 382.

4. Du Pont to Mrs. S. F. Du Pont, 19 April [1861], Welles to Du Pont, 21 April 1861, *Du Pont Civil War Letters*, 1: 53, 56; Du Pont to Welles, 20, 23, 24, and 25 April 1861, Welles to T. T. Craven, 15 August 1861, *Official Records*, ser. I, vol. 4: 287, 319, 321, 333–34, 612–14; for *Keystone State*, see ser. II, vol. 1: 120.

5. List of vessels purchased by Philadelphia Navy Yard Commandants Du Pont, Garrett Pendergrast, and Cornelius K. Stribling, *Official Records*, ser. II, vol. 1: 27–246; "List of Vessels Repaired at the United States Navy Yard Philadelphia since March 1, 1861," *Reports of the Secretary of the Navy and the Commission by Him Appointed on the Proposed New Navy Yard at League Island* (Philadelphia: Collins, 1863), Appendix D: 53–54.

6. Log entries, 24 August, 1 October 1864, Station Log; 14 and 23 June and 26 August 1861; Donald L. Canney, *The Old Steam Navy*, vol. 1, *Frigates, Sloops, and Gunboats, 1815–1885* (Annapolis, Md.: Naval Institute Press, 1990), 71–74; Frank M. Bennett, *The Steam Navy of the United States: A History of the Growth of the Steam Vessel of War in the U.S. Navy, and of the Naval Engineer Corps* (1896; rpt. Westport, Conn.: Greenwood Press, 1974), 218; J. Thomas Scharf and Thompson Westcott, *History of Philadelphia, 1609–1884*, 3 vols. (Philadelphia: L.H.

Everts, 1884), 1: 777; Charles Oscar Paullin, *Paullin's History of Naval Administration, 1775–1911* (Annapolis, Md.: Naval Institute Press, 1968), 260–63.

7. Log entry, 24 August 1861, Station Log; *Public Ledger*, 26 August 1861; Scharf and Westcott, *Philadelphia*, 1; 777.

8. Log entries, 24 August and 16 November 1861, 20 March, 10 July, 8 December 1862, 7 May and 29 September 1863, 19 March 1864, Station Log; *Public Ledger*, 26 August 1861, 21 March, 14 June, 11 July, and 9 December 1862; 7 May, 30 September 1863, 21 March 1864; Scharf and Westcott, *Philadelphia*, 1: 799; 3: 2339; Canney, *Old Steam Navy*,1: 71–74, 94–111, 122–25.

9. Lenthall to Du Pont, 3 February 1861, Fox to Ericsson, 27 February 1864, *Du Pont Civil War Letters*, 1: 30n.s, 314n.5; Du Pont to Fox, 12 November 1861, Lenthall to Du Pont, 11 May 1862, *Official Records*, ser. I, vol. 12: 341–42, 813–14; Edward William Sloan, III, *Benjamin Franklin Isherwood, Naval Engineer: The Years as Engineer in Chief, 1861–1869* (Annapolis, Md., Naval Institute Press, 1965), 61–74; Paullin, *Naval Administration*, 287.

10. *Public Ledger*, 10 and 11 May 1863; Donald L. Canney, *The Old Steam Navy*, vol. 2; *The Ironclads, 1842–1855* (Annapolis, Md.: Naval Institute Press, 1993), 68–69 on *Tonawanda*; Bennett, *Steam Navy*, 347; Scharf and Westcott, *Philadelphia*, 1: 797.

11. Log entries, 22 and 31 March 1864, 14 January, 23 February, 25 July 1865, Station Log; *Public Ledger*, 22, 26, and 29 March, 8 May 1864; Scharf and Westcott, *Philadelphia*, 1: 814; Bennett, *Steam Navy*, 347, 628.

12. "List of Buildings & Rooms, U.S. Navy Yard Philadelphia, 8 Nov. 1867," Civil Engineers Letters, Yards and Docks Letters 3; Joseph C. Wilson, "Old Plans of Historic Ships," Society of Naval Architects and Marine Engineers, *Historical Transactions* (1938): 27, figure 26 for *Shackamaxon* blueprint.

13. "List of Buildings & Rooms, Yards & Docks Letters 3; A New England Man, *Advantages of League Island for a Naval Station, Dockyard, and Fresh-Water Basin for Iron Ships* (Philadelphia: Sherman & Co., printers, for the Board of Trade of Philadelphia, 1866), vi.

14. Log entries, 13, 14 and 30 September 1863 Station Log; *Public Ledger*, 14 September 1863; Scharf and Westcott, *Philadelphia*, 1: 810; Weigley, "Border City," 398.

15. *Public Ledger*, 11 June 1862; Scharf and Westcott, *Philadelphia*, 1: 758–86, 807; George H. Burgess and Miles C. Kennedy, *Centennial History of the Pennsylvania Railroad Company, 1846–1946* (Philadelphia: Pennsylvania Railroad Company, 1946), 270–71, 395.

16. Special Committee Minutes, Meeting, 16 January 1862, *Select Council Journal*, 24; Du Pont to Welles, 25 April 1861, S. M. Felton (president, Philadelphia, Wilmington, and Baltimore Railroad) to Du Pont, 22 April 1861,

Official Records, ser. I, vol. 4: 314, 334, 394–95; Du Pont to H. W. Davis, 8 October 1861, *Du Pont Civil War Letters*, 1, 163; *Public Ledger*, 18 June 1862; Scharf and Westcott, *Philadelphia*, 1: 775, 797, 799, 811; 3: 2190; Burgess and Kennedy, *Pennsylvania Railroad*, 344–45, 787–89; "Men and Things," *Evening Bulletin*, 7 February 1927, [Philadelphia Navy Yard] History Prior to 1960 file, Mounted Clippings.

17. Meetings, 16 September, 1 October, and 21 November 1861, Meetings, 11, 16, 17 June 1862, Special Committee Minutes, *Select Council Journal*, 319–25, 333–36; meeting, 17 June 1862, *Common Council Journal*, 333; Scharf and Westcott, *Philadelphia*, 1: 775; Historic Preservation Studio, *Source Document*, phase I: 12; the Pennsylvania Company originally loaned Charles Wharton $60,000 in 1835 to purchase League Island, but foreclosed on the mortgage in 1842 after Wharton's death.

18. Meeting, 17 June 1862, Special Committee Minutes, *Select Council Journal*, 319–22; meeting, 17 June 1862, *Common Council Journal*, 333–35; entry, 3 December 1862 *Welles Diary*, 1: 185; *Public Ledger*, 17 and 18 June 1862; Arthur Menzies Johnson, "The Genesis of a Navy Yard," *USNIP* 81 (September 1955): 993–95; Scharf and Westcott, *Philadelphia*; 3: 775; Paullin, *Naval Administration*, 287, 295–96.

19. Special Committee Minutes, Meeting, 11 June 1862, Appendix No. 131, *Select Council Journal*, 319–25, 675; entry, 3 December 1862, December 1863, *Welles Diary*, 1: 185–86, 483; Johnson, "Genesis of a Navy Yard," 995–97.

20. Du Pont to Mrs. S. F. Du Pont, 17 September 1861, Bache to Du Pont, 25 November 1862, *Du Pont Civil War Letters*, 1: 149n.3; 2: 370n.23; entries, 3 December 1862, 22 December 1861, *Welles Diary*, 1: 185–86; 3: 489; *Commission on Proposed New Iron Navy Yard*, 13; *Advantages of League Island*, 21; Johnson, "Genesis," 995–97.

21. Entries, 3 December 1862, 13 January and 12 June 1863, *Welles Diary*, 1: 186, 222–23, 327; *Commission on Proposed New Iron Navy Yard*, 6; Johnson, "Genesis," 996–98.

22. Special Committee Minutes, Meeting, 2 April 1863, *Select Council Journal*, 162–63, 236; meeting, 25 June 1863, *Common Council Journal*, 379; *Public Ledger*, 21 March 1864; Scharf and Westcott, *Philadelphia*, 1: 813; Johnson, "Genesis," 998–99.

23. Special Committee Minutes, Meetings, 11 and 18 February, 9 June 1864, Report, *Select Council Journal*, 88, 112, 455, 659; Kelley, in *Welles Diary*, 3: 16.

24. Log entries, 25 and 26 November, 3 December 1864 and 9 January 1865, Station Log; Olcutt to Welles, 18 and 21 April 1863, *Official Records*, ser. I, vol. 14: 156–57; entries, 7 February, 25 and 26 November, 3 December 1864, 9 and 14 January, 7 February, 20 and 24 December 1865, *Welles Diary*, 1: 483, 511, 539–41, 547; 2: 224, 238, 400;

Public Ledger, 22 March, 5, 6, and 14 December 1864; Paullin, *Naval Administration*, 305–6.

25. Entries, 14 and 28 January 1865, 22 December 1865, 10 and 27 January 1866, *Welles Diary*, 2: 224, 231, 401–2, 413, 418–19; *Advantages of League Island*, 20.

26. Log entries, 15 April, 25 July, 9 and 11 September 1865, Station Log; Scharf and Westcott, *Philadelphia*, 1: 826.

27. Log entries, 23 May and 5 October 1865, Station Log; *Public Ledger*, 24 May 1865, 17 December 1866; Entries, 3 October 1865, 10 January and 26 September 1866, *Welles Diary*, 2: 376–77, 413, 603; Canney, *Old Steam Navy*, 133–42; Bennett, *Steam Navy*, 629.

28. Entries, 7 March, 2 and 26 July 1866, *Welles Diaries*, 2: 445–46; 547, 563.

29. Entry, 7 March 1866, *Welles Diaries*, 3: 445–46; *Advantages of League Island*, 23–24.

30. Du Pont to King, 20 May 1862, *Du Pont Civil War Letters*, 2: 132–33n.21; *Advantages of League Island*, 24; *Source Document*, Phase I: 15.

31. Prindle to Emmons, 27 March 1871, Commandants Letters, Yards and Docks Letters, 3; *Public Ledger*, 17 December 1866; *A Concise Statement of the Action, in Relation to a Navy Yard for Iron Clad Vessels: with Extracts from Reports Made on the Subject, and the Comparative Expense of a Navy Yard at League Island and New London* (New London, Conn., 1866), Miscellaneous Pamphlet Collection, Library of Congress, Washington, D.C.

32. "An Act Ceding to the United States of America, the Right of Exclusive Legislation over League Island, in the Delaware River, in the County of Philadelphia, arrived February 10, 1863, supplemented by a further Act approved April 4, 1866, the City of Philadelphia, by its Indenture date December 12, 1868," in "Report Relative to Navy Yard Proper," Charles W. Parks (civil engineer) to Commandant W. S. Benson, 20 February 1914, file 598, Public Works Department Records, PNSR; City of Philadelphia to U.S.A., 12 December 1868, Deed Book I.T.O. 42: 16; entry, 22 December 1868, *Welles Diary*, 3: 489; Grant attended meeting, in *Public Ledger*, 23 December 1868; Paullin, *Naval Administration*, 317–18; Burgess and Kennedy, *Pennsylvania Railroad*, 787–88.

33. Entries, 27 July, 27 August 1868, 24 March 1868, *Welles Diary*, 3: 416–17, 422–23, 560.

34. Entry, 24 March 1869, *Welles Diary*, 3: 560; Robert G. Albion, "Adolf E. Borie, 9 March 1869–25 June 1869," in Paolo E. Coletta, ed., *American Secretaries of the Navy*, 2 vols. (Annapolis, Md.,: Naval Institute Press, 1980), 363–66; Nathaniel Burt and Wallace E. Davies, "The Iron Age, 1876–1905," in Weigley, ed., *Philadelphia*, 477.

35. Entry, 24 March 1869, *Welles Diary*, 3: 560; Albion, "Borie," 363–65.

36. Entry 30 May 1869, *Welles Diary*, 3: 588; Albion,

"Borie," 363–65; *Public Ledger*, 23 December 1868, 11 June 1869; *DANFS*, 2: 164; Bennett, *Steam Navy*, 477; Canney, *Old Steam Navy*, 1: 125–27.

37. Lance C. Buhl, "Maintaining 'An American Navy,' 1865–1889," in Kenneth J. Hagan, ed., *In Peace and War: Interpretations of American Naval History, 1775–1984* (Westport, Conn.: Greenwood Press, 1984), 145–48, Paullin, *Naval Administration*, 321–26, 341; Albion, "Robeson," 369–78.

38. "Investigation by the Committee on Naval Affairs," *House Miscellaneous Document No. 170* (Washington, D.C.: U.S. Government Printing Office, 1876); Leonard Alexander Swann, Jr., *John Roach, Maritime Entrepreneur: The Years as Naval Contractor, 1862–1886* (Annapolis, Md.: Naval Institute Press, 1965), 138–42, 239–41; Charles Merriam Knapp, *New Jersey Politics During the Period of the Civil War and Reconstruction* (Geneva, N.Y.: W.F. Humphrey, 1924), 145n.9, 149n.23, 176; Albion, "Robeson," 369–78.

39. Prindle to Chief Bureau of Yards & Docks, 27 October 1869, Simmons to Prindle 15 March 1871, Civil Engineers Letters, Yards and Docks Letters; *Official Records*, ser. I, vol. 1: 27–246; Thomas R. Heinrich, *Ships for the Seven Seas: Philadelphia Shipbuilding in the Age of Industrial Capitalism* (Baltimore: Johns Hopkins University Press, 1997); Swann, *John Roach*, 138–42, 239–41.

40. Civil Engineer (Prindle) to Emmons, 27 March 1871, Civil Engineers Letters; A. R. Ritter, "A Brief History of the Philadelphia Navy Yard from Its Inception to December 31, 1920," typescript, Bureau of Yards and Docks Library, Philadelphia Navy Yard, PNSR.

41. *Source Document*, Phase I: 15; Paullin, *Naval Administration*, 352.

42. *Source Document*, Phase I: 15–17; Ritter, "Philadelphia Navy Yard."

43. Prindle to Emmons, 27 March 1871, 25 January 1872, Mordecai S. Endicott (Assistant Civil Engineer) to Emmons, 13 and 18 December 1872, Civil Engineers Letters; log entries, 26 September, 29 November, 1 December 1871, 5 January 1874, 3 September and 22 October 1875, Station Log; Scharf and Westcott, *Philadelphia*, 1: 714, 813, 3: 2340; Paullin, *Naval Administration*, 318.

44. The number of employees on the payroll at Southwark during the early 1870s included 875 personnel in Construction and Repair, 132 in Equipment, 95 in Steam Engineering, and 92 in Yards and Docks to lay up the screw sloop *Chattanooga* in 1871 at League Island, where it sank in the ice, log entries, 25 August, 21 and 22 September, 2 October 1875, Station Log; Swann, *Roach*, 152.

45. Log entries, 17 May and 28 September 1875, Station Log; *Camden Democrat*, 4 December 1875, 15 January 1876; *Public Ledger*, 29 September 1875; Henry M. Vallette, "History and Reminiscences of the Philadelphia Navy

Yard," tenth paper, *Potter's American Monthly* 7, 58 (October 1876): 264; Canney, *Old Steam Navy*, 151; Bennett, *Steam Navy*, 628; Paullin, *Naval Administration*, 344–46.

46. Log entries, 24 September, 13 and 27 November 1875, Station Log; *Public Ledger*, 13 and 15 November, 1875.

47. *Public Ledger*, 2 and 3 December 1875; Vallette, "History and Reminiscences," 266.

48. Log entries, 1, 5, and 7 January 1876, 30 December 1875, and 7 January, 11 July 1876, Journal of Transaction, U.S. Navy Yard League Island, Station Log; *Public Ledger*, 13 and 15 November 1875; Vallette, "History and Reminiscences," 267–68; Paullin, *Naval Administration*, 353.

CHAPTER 7. LEAGUE ISLAND NAVY YARD

1. Log entries, 8 January, 17 February, 12 May, and 2 July 1876, Journal of Transactions, U.S. Navy Yard League Island, Pa., 1876, Station Log; Ray Ginger, *Age of Excess: The United States from 1877 to 1914* (New York: Macmillan, 1970).

2. Log entries, 28 June 1879 and 6 December 1883, Naval, Marine and Civil Officers, and Attaches of the U.S. Navy Yard, League Island, Pa., 1 January 1882, Station Log; Philadelphia Census Schedule 1, 1st District, 1st Ward, League Island, U.S. Bureau of the Census, *Tenth Census of the United States* (Washington, D.C.: U.S. Government Printing Office, 1880); "Manning retires," *Philadelphia Evening Bulletin*, 20 July 1933, History Prior to 1960 File, box 173A, Mounted Clippings.

3. A[mandus] R. Ritter, "A Brief History of the Philadelphia Navy Yard from Its Inception to December 31, 1920," (Philadelphia: Bureau of Yards and Docks Library, 1921), 2–3, 19–20; Douglas C. McVarish, Thomas M. Johnson, John P. McCarthy, and Richard Meyer, *A Cultural Resources Survey of the Naval Complex Philadelphia: Philadelphia, Pennsylvania*, 2 vols. (Westchester, Pa: John Milner Associates for TAMS Consultants, New York, and Department of the Navy Northern Division Naval Facilities Engineering Command, 1994), 2: App. 3, Quarters A.

4. Crosby [Commandant, 1877–80] to Chief, Bureau of Yards and Docks, 14 August 1879, Commandants Letters; entries, 15 January and 22 July 1877, Station Log; Ritter, "Philadelphia Navy Yard," 74–77.

5. League Island assumed a war readiness status during the Railroad Strike of 1877 that included arming the monitor *Nahant* with ammunition from Fort Mifflin Ammunition Depot, mounting a field gun on tugboat *Glance*, drilling a landing force from the training ship *Constitution*, and dispatching an armed guard to the grain elevator at Girard Point at the mouth of the Schuylkill River, log entries, 23 and 25 July 1877 and 1 January 1882, Station Log; Census

of League Island, *Tenth Census*; *Evening Bulletin*, 20 July 1933, History Prior to 1960 File, Mounted Clippings; McVarish et al., *Cultural Resources Survey*, 2, App. 3: Building 1.—Yards and Docks Storehouse (1875), Building 2.—Boiler House (1875); Ritter, "Philadelphia Navy Yard History," 3, 16–18.

6. Entries, 28 June 1879, 1 January 1882, Station Log; McVarish et al., *Cultural Resources Survey*, 2, App. 3: Building 4 (1875–77).

7. Entry, 1 January 1882, Station Log; Philip Hichborn, *Report on European Dock-Yards* (Washington, D.C.: U.S. Government Printing Office, 1886); McVarish et al., *Cultural Resources Survey*, 2, App. 3: Building 3.—C & R Iron Plating Shop (1877); Ritter, "Philadelphia Navy Yard," 17–18.

8. Stratton to Crosby, 23 October 1878, Crosby to Secretary of Navy, 23 October 1878, Commandants Letters; log entry, 18 September 1876, Station Log; Manning's son recollects great flood and photograph of Ship House No. 5, *Beacon*, 31 March 1978.

9. Crosby to Chief, Bureau of Yards and Docks, 28 June, 2 May 1879, and 8 January 1880, Commandants Letters; Leonard Alexander Swann, Jr., *John Roach, Maritime Entrepreneur: The Years as Naval Contractor, 1862–1886* (Annapolis, Md.: Naval Institute Press, 1965), 240.

10. Entry, 18 April 1876, Station Log; Charles Oscar Paullin, *Paullin's History*, 346–47, 405; John Davis Long, *The New American Navy*, 2 vols. (New York: Outlook Company, 1903).

11. Simpson to Secretary of the Navy, 27 December 1881, 9 August, 14 October 1882, Commandants Letters; entry, 1 March 1883, Station Log; Arthur Menzies Johnson, "Genesis of a Navy Yard," *USNIP* 81 (September 1955): 1002.

12. Simpson to Secretary of Navy, 27 December 1881, 9 August, 14 and 20 October, and 23 December 1882, Commandants Letters; entries, 27 April and 21 May 1883, Station Log; "The Shipyard Borders on Closure in Late 1800's," *Beacon*, 27 August 1993.

13. Entries, 21 and 24 May 1883, Station Log; Henry M. Vallette, "History and Reminiscences of the Philadelphia Navy Yard," *Potter's American Monthly* 6–7 (January–October 1876).

14. Entries, 27 April 1883, 20 February, 25 and 30 July 1884, Station Log; Commandant League Island to Secretary of Navy, 18 January 1883, Simpson to Chandler, 8 January 1884, Commandants Letters; newspaper clipping opposing navy yard closing, 19 January 1884, attached to Vallette, "Reminiscences"; *Beacon*, 27 August 1993.

15. Swann, *Roach*, 241, 479–80; Paullin, *Naval Administration*, 387–95; Thomas R. Heinrich, *Ships for the Seven Seas: Philadelphia Shipbuilding in the Age of Industrial Capitalism* (Baltimore: Johns Hopkins University Press, 1997), 99–100; H. E. Rossell, "Types of Naval

Ships," in Society of Naval Architects and Marine Engineers, *Historical Transactions, 1893–1943* (New York: The Society, 1945), 270–71; Nathaniel Burt and Wallace E. Davies, "The Iron Age, 1876–1905," in Russell F. Weigley, *Philadelphia: A 300-Year History* (New York and London: W.W. Norton, 1982), 479–81.

16. Potter to Whitney, 17 May and 14 July 1887, Seely to Secretary of Navy, 8 September 1888, 31 May and 3 June 1890, Commandants Letters; entry, 14 September 1891, Station Log.

17. "The First Dry Dock," *Evening Bulletin*, 29 September 1890, [Philadelphia Navy Yard] Dry Dock File, box 173A, Mounted Clippings; Paullin, *Naval Administration*, 387–95, 409–11; Johnson, "Genesis of a Navy Yard," 998, 1002.

18. Peary to Commandant, 18 February and 4 April 1889, 7 April 1890, Peary to Bradford, 3 March 1891, Press Copies of Letters Sent Relating to the Construction of a New Timber Dry Dock, 1887–91, Bureau of Yards and Docks Correspondence, box 34, PNSR; entries, 28 February, 4 March and 5 June 1891, Inspection Board for Dry Dock No. 1 Captain R. F. Bradford, senior member, Chief Engineer Henry B. Jones, Naval Constructor John F. Hanscom, Civil Engineers Francis C. Prindle and W. McCollom, Station Log.

19. H. Gerrish Smith, "Shipyard Statistics," in H. G. Gassett, ed., *The Shipbuilding Business in the U.S.A.* (1948; rpt. New York: Library Edition, 1970), 179, table 88; Paullin, *Naval Administration*, 409–11.

20. Chief, Bureau of Yards and Docks to Commandant, Navy Yard, League Island, 27 March 1900, Civil Engineer Wolcott endorsement, 30 March 1900, to Commandant, 30 March 1900, Letters Received from Bureau of Yards and Docks, Correspondence with Secretary of Navy and the Navy Department Bureaus, Bureau of Yards and Docks Correspondence, box 19, PNSR; for list of buildings, McVarish et al., *Cultural Resources Survey*, 2, App. 3: Building 139 (1891), Piers 1 and 2 (1891), Officers Quarters B and C (1892), and Reserve Basin (1897-98) and Building 8 (1899); Ritter, "Philadelphia Navy Yard," 4.

21. List of Trades, in Kirkland to Secretary of Navy, 18 April 1892, List of Trades for Registration, Navy Yard, League Island, 24 May 1892, Commandants Letters; Frederick R. Black, *Charlestown Navy Yard, 1890–1973* (Boston: National Historical Park, National Park Service, U.S. Department of the Interior, 1988), 1: 128–38.

22. Kirkland to Secretary of Navy, 17 July and 30 September 1891, 8 March 1892, Commandants Letters; Arnold S. Lott, *A Long Line of Ships: Mare Island's Century of Naval Activity in California* (Annapolis, Md.: Naval Institute Press, 1954); Peter Karsten, *The Naval Aristocracy: The Golden Age of Annapolis and the Emergence of Modern American Navalism* (New York: Free Press,

1972); Ritter, "Philadelphia Navy Yard," 12–13.

23. Entries, 24 February, 1, 4, and 31 March, 16 April and 1 June 1898, Station Log. Civilian employment during the war with Spain peaked on 16 April 1898 at 2,020.

24. Entries, 28 April and 1 June 1898, Station Log; Jack F. Ayers and Henry A. Vadnais, Jr., *Fifty Years of Naval District Development, 1903–1953* (Washington, D.C.: Naval History Division, Office of the Chief of Naval Operations, 1956), 3; *DANFS*, 1: 353–54.

25. Entries, 28 July 1898, 6 June, 6 November, and 13 December 1899, Station Log; Frank Bart Freidel, *The Splendid Little War* (Boston: Little Brown, 1958).

26. Endicott to Commandant, 3 July 1900, box 19, Commandants Letters; *Source Document*, 17–18, McVarish et al., *Cultural Resources Survey*, 2, App. 3: Buildings No. 6, 99, 100, M2-4, 148, 14 (1903), 17, 19, 10, 11, 12, 15, 18 and Dry Dock #2.

27. Minutes of Public Works Conference, 10 July 1900, Civil Engineer Miscellaneous Correspondence, box 2, PNSR; Order 468, 18 April 1900 on numbering of streets, Commandants Office Building File, 12 June 1900, Commandants Letters; Paullin, *Naval Administration*, 478–79; *Beacon*, 31 March 1944. The Philadelphia Navy Yard streets Porter, Rowan, Davis, Philip, and Preble probably were named after Commandant Samuel F. Du Pont's Civil War staff, blockade planners, and members of a Naval Advisory Board for League Island Stephen H. Rowan, Charles H. Davis, John Woodward Philip, David Dixon Porter, and George Preble (the last commandant at Southwark), rather than Commodores David Porter and Edward Preble of War of 1812 fame as popularly supposed.

28. Oliver Crosby, American Hoist & Derrick Co., to Wolcott, 31 January 1901, Catt to Endicott, 28 April 1900, Endicott to Commandant, 16 May 1900, 31 January 1901, Supplementary Report, Board on Increased Cost of Dry Dock (Converse Board) at the U.S. Navy Yard , League Island, Pa., December 17, 1900, Captain George A. Converse, senior member, Joseph H. Linnard, constructor, Christopher C. Wolcott, Harry H. Rousseau, and Leonard M. Cox to Secretary of Navy, 17 December 1900, Civil Engineer Miscellaneous Correspondence, box 2; "Expenditures, Yards and Docks, League Island, PA., 1902–1907," Press Copies of Letters Sent to the Bureau of Yards and Docks, 1902–1907, PNSR; "Philadelphia Navy Yard Pictorial," *USNIP* 81 (September 1955): 1005; Ritter, "Philadelphia Navy Yard," 43–46; Paullin, *Naval Administration*, 480.

29. Census of Military and Naval Population, League Island, Schedule No.1, U.S. Bureau of the Census, *Twelfth Census of the United States* (1900) shows no Italian Americans working on League Island, but six appeared in the 1909 List of Men in Public Works, Memo for Captain of the Yard, 7 December 1909, File 598, Civil Engineer Miscellaneous Correspondence; *New York Times*, 26 October

1902, 31 December 1904, 9 August 1906, and 17 March 1909; *Public Ledger*, 26 October 1902 and 17 March 1909; *Philadelphia Times*, 26 October 1903; Lloyd M. Abernethy, "Progressivism 1905–1919," in Weigley, ed., *Philadelphia*, 527–32, 548–50.

30. Surgeon Biddle to Commanding Officer United States Receiving Ship *Puritan*, 16 September 1903, Goodrich to Commandant, 13 May 1903, File 525, box 16, Civil Engineer Miscellaneous Correspondence; *New York Times*, 11 May 1903; *Public Ledger*, 12 and 15 May 1903.

31. Goodrich to Sigsbee, 8 June 1903, Sigsbee to Bureau of Navigation, 29 July 1903, Secretary of Navy to Commanding Officer U.S. receiving ship *Puritan*, 29 July 1903, Civil Engineer R. E. Bakenaus to Commandant, 4 December 1903, File 525, box 16, Civil Engineer Miscellaneous Correspondence.

32. Log entries 8 and 9 April, 19 May and 28 June 1904, Station Log; *New York Times*, 13 January 1904; "Monitors in Ordinary League Island, circa 1900," *USNIP* 81 (September 1955): 1005; *DANFS*, 5:3.

33. Chief Bureau of Construction & Repair Washington Capps to Commandant League Island, 23 October 1907, Letters Received from Secretary of Navy: Correspondence with Bureau of Construction & Repair, 1906–1908, PNSR; *New York Times*, 6 and 7 December 1907; Ritter, "Philadelphia Navy Yard," 42–46; "Opening of the Large Dry Dock at League Island," *Scientific American* 97 (24 August 1907): 132; Franklin Matthews, *With the Battle Fleet: Cruise of the Sixteen Battleships of the United States Atlantic Fleet from Hampton Roads to the Golden Gate, December 1907–May 1908* (New York: B.W. Heubsch, 1908).

34. Log entry 21 November 1902, Station Log; *Public Ledger*, 3, 4, 6, and 9 December 1909 and 7 January 1910; *New York Times*, 2 and 10 December 1909; "Scientific Management on Sea and Shore," *Scientific American* 105 (3 June 1911): 542. Battleships *Alabama, Maine, Kansas, Mississippi, Idaho, New Hampshire, South Carolina,* and *Michigan*; armored cruisers *Colorado, Pennsylvania, Washington, Denver, Tennessee,* cruiser *St. Louis*; torpedo boat destroyers *Bainbridge, Barry, Chauncey, Hopkins, Hull, Smith, Preston, McCall, Burrow, Warrington, Mayrant,* and *Ammen*.

35. *Public Ledger*, 12 November 1909; *New York Times*, 19 April, 3 and 18 May 1909, 25 June and 10 December 1911, and 7 November 1910, the headings on the Station Log changed in mid-1908 from League Island to "Navy Yard, Philadelphia," Station Log; S. E. *Zubrow*, "The History of the Philadelphia Navy Yard," *United States Naval Administration in World War II, Commandant, Fourth Naval District*, 6 vols. (Washington, D.C.: Naval History Division, 1946), 3, Part IV, Sect. 1: 36; George von Lengerke Meyer, "The Business Management of the Navy," *Scientific American* 105 (9 December 1911), 513; Black, *Charlestown Navy Yard*, 1: 187–96.

36. U.S. Congress, House Naval Affairs Committee, *Hearings on the Proposed Reorganization of the Navy Department before the Committee on Naval Affairs of the House of Representatives*, Appendix 2, 802–4; *New York Times*, 19 April and 3 and 18 May 1909.

37. Stanford to Meyer, 20 June 1910, File 598, Civil Engineer Miscellaneous Correspondence; log entries, 21 November 1902 and 17 May 1909. Station Log; Josiah S. McKean, "War and Policy," *USNIP* 40 (January–February 1914): 3–15; *New York Times*, 19 April, 3 and 18 May 1909, 9 January 1910, 2 December 1911; Paolo E. Coletta, "George Von Lengerke Meyer, March 1909–4 March 1913," in Coletta, ed., *American Secretaries of the Navy*, 1: 495–522.

38. Stanford to Meyer, 20 June 1910, Stanford to Pay Inspector, 11 November 1909, Meyer to Commandant, Navy Yard Philadelphia, 30 December 1909, File 598, Civil Engineer Miscellaneous Correspondence; diary entry, 1 May 1909, Meyer to Theodore Roosevelt, 9 July 1909, in M. A. DeWolfe Howe, *George von Lengerke Meyer: His Life and Public Service* (New York: Dodd, Mead, 1920), 432–33, 438; "The Business Management of the Navy"; Coletta, "Meyer," 497–98.

39. Commandant to Engineer, Construction, Accounting Officer, Inspection Office, General Storekeeper, Public Works Officer, 28 February 1913, "Report of Board on Location of Wireless Telegraph Station, December 22, 1908 of Which Commander T. S. Rodgers was Senior Member," File 598, Civil Engineer Miscellaneous Correspondence; log entries, 17 and 25 June, 9, 10, and 13 August, 11 October, 5, 15, 16, and 26 November 1909, 12 June and 17 July 1911, Station Log; T. D. Parker, "Inspection Duty of Navy Yards," *USNIP* 38 (June 1912), 397–424; W. B. Tardy, Discussion of "Minimum Navy Yard Manufacturing Costs," *USNIP* 148 (1913): 1645; Tardy obit., *New York Times*, 1 December 1932; Ritter, "Philadelphia Navy Yard," 27, 107; Johnson, "Genesis of a Navy Yard," *USNIP* 81 (September 1955), 1006–8.

40. Stanford E. Moses, Discussion of "Navy Yard Problems," *USNIP* 143 (September 1912): 1101–2; *Public Ledger*, 12, 23, and 24 November 1909; *New York Times*, 22 December 1911.

41. Stanford to Commandant, 14 June 1910, File 598, Civil Engineer Miscellaneous Correspondence; *Public Ledger*, 13 November 1909, 12 November 1911; *New York Times*, 19 July 1910.

42. Joint Special Committee of Philadelphia City Council and Mayor's Citizens Committee on the Greater League Island Navy Yard, *The Philadelphia Navy Yard Its Advantages as a Naval Station and Shelter for the Reserve Fleet of the United States Navy, Prepared for Congressional Committees on the Occasion of their Visit, May 10–11, 1912* (Philadelphia: Dunlap, 1912); *Public Ledger*,

13, 18, and 20 November and 4 December 1911.

43. John Hood, "Naval Policy as it Relates to the Shore Establishment, and the Maintenance of the Fleet," *USNIP* 150 (1914): 319–44; Meyer, "Business Management of the Navy," 513; Coletta, "Meyer," 499–512; Black, *Charlestown Navy Yard*, 1: 249–99.

CHAPTER 8. NEUTRALITY AND WORLD WAR

1. Parks to Thomas Reilly, 7 July 1913, Stanford to A. W. Grant, 14 July 1913, File 1–2, vol. 108, Civil Engineer Miscellaneous Correspondence, box 79; E. David Cronon, ed., *The Cabinet Diaries of Josephus Daniels, 1913–1921* (Lincoln: University of Nebraska Press, 1963), 59–73; Josephus Daniels, *The Wilson Era: Years of Peace, 1910–1917* (Chapel Hill: University of North Carolina Press, 1944), 261ff, 273–74, 386–403.

2. Machinery Division Instructions, Navy Yard, Philadelphia, May 1917, "Notes on How to Get Work Done at Navy Yard," File NYF (1916–22), Central Subject Files; entries, 27 March, 3 and 5 April, and 28 May 1938, *Daniels Diaries*, 15, 21, 23, 74; Daniels, *Wilson Era*, 345; Paolo E. Coletta, "Josephus Daniels, 5 March 1913 – 5 March 1921," in Coletta, ed., *American Secretaries of the Navy*, 2 vols. (Annapolis, Md.: Naval Institute Press, 1980), 2: 525–81.

3. Entry, 6 March 1913, *Daniels Diaries*, 4; "Report on the Atlantic Navy Yards, *Army and Navy Journal* (6 September 1913), 18; Frank Freidel, *Franklin D. Roosevelt: The Apprenticeship* (Boston: Little Brown, 1952); Daniels, *Wilson Era*, 126–33.

4. George W. Norris, *Ended Episodes* (Philadelphia: John C. Winston, 1937), 98; Lloyd M. Abernethy, "Progressivism, 1905–1919," in Russell F. Weigley, ed., *Philadelphia: A 300-Year History* (New York: W.W. Norton, 1982), 552; Freidel, *Apprenticeship*, 160, 194–95; John Hood, "Naval Policy as It Relates to the Shore Establishment, and the Maintenance of the Fleet," *USNIP* 150 (1914): 319–44; *Philadelphia Evening Bulletin*, 6 June 1914, Philadelphia Navy Yard Dry Dock File, box 173A, Mounted Clippings.

5. *Evening Bulletin*, 17, 19, and 20 December 1913, 5 and 18 February, 6 June 1914, Dry Dock File, Mounted Clippings.

6. Entry, 4 March 1913, Station Log; *Public Ledger*, 22 September 1914; A. R. Ritter, "A Brief History of the Philadelphia Navy Yard from Its Inception to December 31, 1920" (Philadelphia: Bureau of Yards and Docks Library, 1921), 81, 22, 29; "Philadelphia Navy Yard (pictorial)," *USNIP* 81 (September 1955): 1009.

7. *Public Ledger*, 22 September 1914; Mary Klachko, *Ad-*

miral William Shepherd Benson, First Chief of Naval Operations (Annapolis, Md.: Naval Institute Press, 1987), 24, 26, 30; Daniels, *Wilson Years*, 244–45. Approximately one-quarter of the work force in shipyards along the Delaware on the eve of World War I were unskilled recent Italian, Polish, or Russian-American immigrant; see Thomas R. Heinrich, *Ships for the Seven Seas: Philadelphia Shipbuilding in the Age of Industrial Capitalism* (Baltimore: Johns Hopkins University Press, 1997), 190.

8. Charles W. Parks to R. E. Bakenhus [Civil Engineer, Boston], 15 June 1914, Frederick Harris to L. M. Cox [New York], 27 May 1915, File 48B, Civil Engineer Miscellaneous Correspondence; "Professional Notes on Transport No. 1," *USNIP* 148 (December 1913): 1762; Ritter, "Philadelphia Navy Yard," 81, 94–96; Daniels, *Wilson Era*, 324; Allan R. Millett, *Semper Fidelis: The History of the United States Marine Corps,* rev. ed. (New York: Free Press, 1991), 276–92.

9. Daniels, *Wilson Era*, 439–40; draft of *Sussex* Pledge, 12 April 1916, enclosed, Wilson to Lansing, 17 April 1916, in Arthur S. Link, ed., *The Papers of Woodrow Wilson* (Princeton, N.J.; Princeton University Press, 1981), 36: 490–96.

10. Richard M. Watt [Bureau of Construction and Repair] to Navy Department, 8 October 1914, Frank F. Fletcher to Secretary of Navy, 5 October 1914, File 641-1 [*Minnesota*], box 38, Ship Files; Stanford to William S. Vare, 27 June 1916, File 17–32; Harris to Archibald L. Parson, 30 June 1915, Harris to Clinton D. Thurber, 4 August 1915, File 48; FDR to Benson, 3 May 1915, File 713, Civil Engineer Miscellaneous Correspondence.

11. Entries, 10 May, 13 June 1915, 28 January, 12 February, 1 May, 1 August, and 1 September 1916, Station Log.

12. Bureau of Construction and Repair Memorandum, 17 April 1916, File 862-22(1), [*South Carolina*], box 54, Ship Files; entries, 23 February 1915, 20 April, 1 May, 24 June 1916, Station Log; Court of Inquiry included DeWitt Coffman, Thomas Snowden, Hugh Rodman, Frederick L. Oliver. Battleships berthed at League Island, 1 May 1916 included *Iowa, Massachusetts, Indiana* (out of commission); *Alabama, Wisconsin, Missouri, Ohio, North Dakota* (Atlantic Reserve Force); and *Illinois, Connecticut, Minnesota, Kansas, Michigan, South Carolina* (in commission).

13. Entries, 29 January, 29 February 1916, Station Log; entry, 21 April 1917, *Daniels Diaries*, 138-39; Daniels, *Wilson Era*, 240-41; Robert Lee Russell obituary, *New York Times*, 9 May 1934.

14. FDR to Commandant, Navy Yard, Philadelphia, 29 June 1916, File 713, Chief, Bureau of Supplies and Accounts McGowan to Commandant, Philadelphia Navy Yard, 11 November and 13 December 1916, File 26-30/40, box 129, Civil Engineer Miscellaneous Correspondence; entry, 29 June 1916, Station Log; *New York Times*, 13 January,

10 and 19 February 1916; *Evening Bulletin*, 25 April, 24 May 1916, Dry Dock File, Mounted Clippings (Dry Dock No. 3 was completed in 1919).

15. Operations [Washington] to Commandant, Navy Yard, Phila., 27 June 1916, File SP 829 [*Santa Catalina*], box 78, Daniels to Captain of Yard C. B. Price, 15 October 1915, File 424 [*Dixie*], box 11, Ship Files; entries, 1, 15, and 23 May, 23 and 24 June, 1 September 1916, Station Log; *New York Times*, 27 April, 5 July 1916; Jeffery M. Dorwart, *Cape May County, New Jersey: The Making of an American Resort Community* (New Brunswick, N.J.: Rutgers University Press, 1992), 186–87.

16. Harris to Snow, 3 April 1916, File 48-S, Civil Engineer Miscellaneous Correspondence; entries, 1, 15, and 23 May 1916, Station Log; Russell to Secretary of Navy, 27 December 1916, File 624-1 [*Maine*], box 36, Ship Files; Frederick R. Black, *Charlestown Navy Yard, 1880–1978*, 2 vols. (Boston: National Park Service, 1988) 1: 306ff.

17. Ritter, "Philadelphia Navy Yard," 5, 46–47; "Report on the Atlantic Navy Yards," *Army and Navy Journal* (6 September 1913): 18; J. S. McKean, "On Storekeeping at the Navy Yards," *USNIP* 122 (1907): 815.

18. Entry, 1 October 1916, Station Log; *Public Ledger*, 2 October 1916; "Professional Notes," *USNIP* 157 (1915): 986–87; Ritter, "Philadelphia Navy Yard," 539; *Evening Bulletin*, 26 September 1916; Philadelphia Navy Yard German Ships and Internees, 1916–17 File, box 173A, Mounted Clippings; *DANFS*, 3: 254; 7: 560–63.

19. Grant to Bureau of Engineering, 17 April 1917, File 381 [*Chicago*], box 8, Ship Files; entries, 2 February, 16 March 1917, Station Log; *Public Ledger*, 2 October 1916; *New York Times*, 23 November 1915, 30 January and 4 February 1917; *DANFS*, 6: 258–59.

20. Entries, 10 and 27 March 1917, *Daniels Diaries*, 3: 123; *Public Ledger*, 11 and 13 March 1917; *New York Times*, 11 and 13 March, 14 April 1917, 23 October 1916; *Evening Bulletin*, 4 April 1917, German Ships and Internees File, box 173A, and 8 February, 3 and 9 March 1917, World War I File, box 173B, Mounted Clippings.

21. Commanding Officer to Commandant, 7 and 20 February 1917, File 695-1 [*North Dakota*], box 42, Ship Files; *New York Times*, 1, 4, and 10 February 1917; *Evening Bulletin*, 3 March 1917, World War I File, Mounted Clippings.

22. *Public Ledger*, 4 and 6 February 1917; *New York Times*, 1, 4, and 10 February 1917; *Evening Bulletin*, 9, 17, 23 February 1917, World War I File, Mounted Clippings.

23. Stanford to Public Works Department, Phila., 20 June 1917, File 19-40, box 125, Civil Engineer Miscellaneous Correspondence; Black, *Charlestown*, 1: 342–43.

24. Stanford to Grant, 6 February 1917, File 48G, box 16, Civil Engineer Miscellaneous Correspondence; entries, 10, 15, and 27 March, 21 April 1917, *Daniels Diaries*, 11, 114,

122–23, 138–39; *Public Ledger*, 13 March 1917; *New York Times*, 10 February, 11 and 13 March 1917; *Evening Bulletin*, 8 February, 9 March 1917; German Ships and Internees File, Mounted Clippings.

25. Daniels to Russell, 15 January 1917, File 624-1 [*Maine*], box 36, Ship Files; entries, 26 and 28 March 1917, Station Log; entries, 3 and 18 April 1917, *Daniels Diaries*, 128, 137; *Public Ledger*, 7 April 1917; *New York Times*, 13, 17, 18, and 20 March 1917.

26. Entries, 3 and 18 April, 7 July 1917, *Daniels Diaries*, 128, 137, 174.

27. Entries, 25, 26, and 29 May 1917, Station Log; entries, 5 and 7 May, 31 August 1917, *Daniels Diaries*, 147–48, 200; *DANFS*, 7: 560–63; 3: 254.

28. "On arrival these vessels will be immediately available for repairs and equipment for distant service and will have the right of way over other work at the yards excepting for other destroyers of six and seven divisions previously ordered to the yards," OPNAV Radio Telegram to Navy Yard, Phila., 30 April 1917, CNO (McKean) to Commandant, Navy Yard, Phila., 2 May 1917, Daniels to Tappan, 10 May 1917, File 315-1 [*Ammen*], box 1, Ship Files; entries, 6 and 24 April, 1 and 26 May 1917, Station Log; entries, 24 and 27 April, 6 December 1917, *Daniels Diaries*, 140–42, 247; Elting E. Morison, *Admiral Sims and the Modern American Navy* (Boston: Houghton Mifflin, 1942), 343–55; *DANFS*, 1: 179–80.

29. Russell to Construction Officer, 9 April 1917, Snow to Tappan, 24 April 1917, File 381 [*Chicago*], box 8, Ship Files; Captain A. R. Niblack, "Naval Stations and Bases Needed by Our Fleet," *New York Times Magazine Section*, 4 February 1917, part V: 3–4; *Public Ledger*, 8 April 1917; Morrison, *Admiral Sims*, 355; *DANFS*, 3: 493.

30. "Mobilization Plan," 15 March 1917, File 999 [*Woodbine*], box 79, Ship Files; *Public Ledger*, 8 April 1917; *New York Times*, 10 June 1916, 17 March 1917; Norman Friedman, *U.S. Small Combatants Including PT-Boats, Subchasers, and the Brown-Water Navy* (Annapolis, Md.: Naval Institute Press, 1987), 21–22.

31. Webb to Commandant, 13 October 1917, File 1010, box 125, Civil Engineer Miscellaneous Correspondence; entries, 21 March, 22 August 1917, *Daniels Diaries*, 119n.61, 194n.28; *Public Ledger*, 11 and 22 March 1917; *New York Times*, 5 May, 10 June 1916, 17, 21, and 22 March 1917; *Evening Bulletin*, 23 April, 19 June 1917, World War I File, 18 November 1918, Improvements and Expansion Prior to 1940 File, Mounted Clippings; Abernethy, "Progressivism 1905–1919," in Weigley, *Philadelphia*, 550–52; James J. Walsh, *America's Forgotten Heroine* (private printing, 1992), photocopy, Navy Department Library, Washington Navy Yard; Jean Ebbert and Marie-Beth Hall, *Crossed Currents: Navy Women from WWI to Tailhook* (Washington, D.C.: Brassey's, 1993), 3–5. Philadelphia's

Naval Coast Defense Reserve Force included James Henry Roberts Cromwell, Stotesbury's stepson, speed-boat racer, and member of the Corinthian Yacht Club; John R. Fell, son-in-law of Navy League leader J. Alexander Van Rensselaer of New York; Anton Ahlers of the Keystone Yacht Club of Philadelphia; Samuel Doyle Riddle, Philadelphia textile baron, horse racing, and yachting enthusiast; Cyrus H. K. Curtis, publishing magnate; and Samuel H. Collom.

32. Entries, 11, 17, and 25 April, 1 May 1917, Station Log; Fifteen ALNAV, Washington, to Navy Radio Phila., 2 April 1917, and Sixteen ALNAV, 6 April 1917, Russell to All Heads of Departments, Divisions, Officers and Retired Officers in the Fourth Naval District, 3 and 7 April 1917, File 994, box 125, Civil Engineer Miscellaneous Correspondence; *Public Ledger*, 11 March 1917; General Order 314 of 28 July 1917 designated privately owned vessels as Scout Patrol (SP) Boats, *DANFS*, 4: 710, 124, 5: 37.

33. Du Pont to Henry Hall Porter, 28 May 1917, File SP 58 [*CTBs*], box 68, Record of Telephone Conversation Grady and Goss, 13 July 1917, File SP 384 [*Rehoboth*], box 68, Foley, Operations, to Goss, 10 August 1917, File SP 383 [*City of Lewes*], box 74, Arthur G. Kavanagh Memo. to Commandant, 23 February 1918, File SP 1126 [*Conestoga*], box 78, Ship Files; McKeever's trawlers became *Mine Sweepers* 3, 4, and 7, *DANFS*, 2: 329; 4: 304; 6: 617; also 2: 121; 4: 256; 6: 62.

34. Construction Officer to Commandant, 10 May 1917, Snow to Commandant, 14 May 1917, File 497 [*Guthrie*], box 79, Ship Files; log entries, 2, 9, and 10 May, Station Log.

35. Board of Investigation Lloyd Bankson, Alva B. Courts, J. M. Luby, Benjamin F. Hutchinson, Commandant to Bankson, 10 May 1917, File 497 [*Guthrie*], Ship Files; Medical Inspector Robert A. Bachmann headed Board of Inquest into death of Hadlock, entries, 9 and 10 May 1917, Station Log; *Evening Bulletin*, 13 and 16 July 1917, World War I File, Mounted Clippings; Jack F. Ayers, and Henry A. Vadnais, Jr., *Fifty Years of Naval District Development, 1903–1953* (Washington, D.C.: Naval History Division, Office of the Chief of Naval Operations, 1956), 4–5.

36. Special Order No. 13, Commandants Order, 14 December 1917, File 7-20, Civil Engineer Miscellaneous Correspondence; diary entry, 2 April 1918, *Daniels Diaries*, 296n2; *Evening Bulletin*, 25 and 26 June, 16 July 1917, World War I File, Mounted Clippings.

37. Diary entry, 3 August 1917, *Daniels Diaries*, 186; Ritter, "Philadelphia Navy Yard," 98–99; William F. Trimble, *Wings for the Navy; A History of the Naval Aircraft Factory, 1917–1956* (Annapolis, Md.: Naval Institute Press, 1990).

38. Parson to Tappan, 12 May 1917, Tappan to Public Work Officer, 12 May 1917, File 14-39, box 111, Civil Engineer Miscellaneous Correspondence; Ritter, "Philadelphia

Navy Yard," 98–102; Trimble, *Wings for the Navy*, 31.

39. Stanford to Field, 15 June 1917, File 14-31, box 111, Webb to Fraser, Brace and Company, 17 July 1917, File 1001-1, Civil Engineer Miscellaneous Correspondence; Ritter, "Philadelphia Navy Yard," 6, 38–39, 88–92.

40. McKay to Stanford, 7 May 1917, File 48M, box 16, Chief, Bureau of Supplies & Accounts to Commandant, Philadelphia Navy Yard, 28 November and 27 December 1917, Public Works Officer to Commandant, 26 November 1917, File 26, box 129, President Roche Construction to C. W. Parks, 19 February 1918, File 27-6, box 141, Civil Engineer Correspondence.

41. Ritter, "Philadelphia Navy Yard," 40–41, 76–79, 85–86.

42. "The World's Largest Crane," *Scientific American* (21 August 1920); R. D. Gatewood, "Uncle Sam's Greatest Navy Yard," *American Machinist* 52 (11 March 1920); *Evening Bulletin*, 1 December 1919 and 22 February 1920, [Philadelphia Navy Yard] Cranes File, box 173A, Mounted Clippings; Ritter, "Philadelphia Navy Yard," 47–48, 29–31, 107–8.

43. Record of Telephone Conversation Pinney and Kavanagh, 8 July 1918, File 862-1 [*South Carolina*], box 54, Record of Telephone Conversation between Benson and Goss, 9 January 1918, File 624-1 [*Maine*], box 36, Ship Files; *Beacon*, 6 March 1942; Heinrich, *Ships for the Seven Seas*.

44. Gatewood, "Uncle Sam's Greatest Navy Yard," 539.

CHAPTER 9. BETWEEN WORLD WARS

1. Hughes to Engineer and Construction Officer, 22 November 1918, Hughes to Commanding Officers *Vermont*, *Ohio*, *Minnesota*, File 641-1 [*Minnesota*], box 38, Ship Files; entry 6 December 1918, *Daniels Diaries*, 353.

2. See entries, 1919–20, Station Log; Frederick R. Black, *Charlestown Navy Yard, 1890–1973*, 2 vols. (Boston: National Historical Park, National Park Service, U.S. Department of the Interior, 1988), 2: 336; Paolo E. Coletta, *The American Naval Heritage in Brief*, 2nd ed. (Washington, D.C.: University Press of America, 1980), 277–82; Mary Klachko, *Admiral William Shepherd Benson, First Chief of Naval Operations* (Annapolis, Md.: Naval Institute Press, 1986), 156–57.

3. Record of telephone conversation between Frank Lucius Pinney, Operations, and Albert G. Kavanagh, Aide to Commandant, Navy Yard, Phila., 12 December 1918; Hughes to Commanding Officer *Minnesota*, 11 December 1918, Record of telephone conversation Reuben Rockwell Clarke (CC), Bureau of Construction and Repair, and William

Pierre Robert (CC), Head Hull Division, Navy Yard Phila., 31 January 1919; Daniels to Hughes, 4 December 1918, File 641-1 [*Minnesota*], box 38, Ship Files.

4. Telephone conversation, W. P. Robert and Robert Ferrell (CC), 8 February 1922, File 357-1 [*Constitution*], box 5, Ship Files; Acting Public Works Officer to Construction Officer, 19 December 1919, File 850, box 26, Public Works Office Miscellaneous Correspondence; entries, 28 April 1919 (*Sandpiper*), 26 May 1919 (*Vireo*), 30 July 1919 (*Warbler*), 11 September 1919 (*Willet*), 23 December 1919, Station Log; diary entry, 13 March 1913, *Daniels Diaries*, 8; A. R. Ritter, "Brief History of the Philadelphia Navy Yard from Its Inception to December 31, 1920," (Philadelphia: Yard and Docks Library, 1921), 67, 81.

5. "Appropriation 9124, Improving and Equipping Navy Yards for Construction of Ships, 1 July 1918," for Navy Yard, Philadelphia $3.1 million; New York, $3 million; Hampton Roads, Virginia, $2 million; Norfolk, $800,000; Boston, $500,000; Mare Island, $250,000; Puget Sound, $250,000; Base No. 25, $80,000; in McGowan (Chief Bureau of Supplies and Accounts) to Commandant, Navy Yard, Philadelphia, 11 January 1919, File 2001, box 81, Ship Files; entries, 24 March and 18 April 1920 (Dry Dock No. 3 tested with Cruiser *Quincy*), 29 April, 3 September (*Dickerson*), 20 October (*Jacob Jones*), 10 December (*Smith Thompson*), 6 April 1920 (*J. D. Edwards*), 23 April (*Whipple*), 17 May (*Fox*), 18 June (*Brooks*), 16 August (*Bulmer*), 15 December (*Pillsbury*), 30 December (*Ford*), 1920, 19 January 1921 (*Groff*), Station Log; Ritter, "Philadelphia Navy Yard," 29–31, 47–48.

6. Record of telephone conversation Charles A. Blakely and Albert G. Kavanagh, 19 January 1919, Daniels to Bureaus, 28 October 1919, CTB 1 (ex-*Foote*), CTB 2 (ex-*Rodgers*), CTB 3 (ex- *Dupont*), CTB 4 (ex-*Dahlgren*), CTB 8 (ex-*Bailey*), CTB 11 (ex-*Barney*), CTB 12 (ex-*Biddle*), CTB 15 (ex-*Shubrick*), CTB 17 (ex-*Tingey*), File CTB [CTB 11] box 68, Ship Files.

7. Construction Officer Gatewood to Commandant Hughes, 2 August 1919, Record of Telephone Conversation, Harry L. Brinser to George W. Simpson, 10 February 1920, File 624-1 [*Maine*], box 36, Ship Files; Christopher Morley, *Travels in Philadelphia* (Philadelphia: David McKay, 1920), 130–31.

8. Coontz to Commandant, Navy Yard, Phila., 20 November 1919, File 424 [*Dixie*], box 11, Ship Files; entries, 8 January, 14 June, l September, and 1 December 1920, Station Log; Yates Stirling, *Sea Duty: The Memories of a Fighting Admiral* (New York: Putnam's, 1939), 200–201.

9. Earle to Commandant, Navy Yard, Phila., Commander Flotilla B, Destroyer Force, and Commander, Naval Ammunition Depot, Fort Mifflin, 19 July 1919, File 315-1 [*Ammen*], box 1, Ship Files; Acting Commander, Fort Mifflin to Commandant, Navy Yard, Phila., 24 December 1920,

File 220-1, box 19, Public Works Office Miscellaneous Correspondence; entries, 7 December 1920, 11 June 1923, Station Log; Jeffery M. Dorwart, *Fort Mifflin of Philadelphia: An Illustrated History* (Philadelphia: University of Pennsylvania Press, 1998), 145–46.

10. Comerford to Bureau of Ordnance, 24 July 1919, Earle to Naval Inspector of Powder, East Coast, 1 August 1919, File SP 724 [*Fearless*], box 77, Ship Files.

11. H. L. Tissot to Commandant, Navy Yard, Phila., 25 August 1919, File SP 724 [*Fearless*], Ship Files; entry, 8 September 1936, Station Log.

12. Order No.4, 13 November 1920, Commandants Orders, box 6, Public Works Office Miscellaneous Correspondence; Ritter, "Philadelphia Navy Yard," 112.

13. Log entries, 8 January and 14 May 1919, Station Log.

14. Jackson to Acting Secretary of Navy Theodore Roosevelt, Jr., 31 March 1921, Secretary of the Navy to Commandant, Navy Yard, Phila., 4 April 1921, File 17, Box 10, Public Works Office Miscellaneous Correspondence.

15. Chief Bureau Yards & Docks to Navy Yard, Phila., 14 January 1921, Secretary of Navy to Commandant, Navy Yard, Phila., 23 March 1921, T. Roosevelt, Jr., to Commandant, Navy Yard, Phila., 6 June 1921, File 47, box 15; Record of Permanent Employees, 15 August 1920, Statement of Time, 31 July 1921, File 825, box 26, Public Works Office Miscellaneous Correspondence; Personnel attached to Board of Survey, Appraisal and Sale, 31 October 1919, File A3–1, box 13, Central Subject Files; Enumeration District No. 1, Philadelphia, *Fourteenth Census of the United States* (Washington, D.C.: U.S. Government Printing Office, 1920).

16. Denby to Chiefs of Bureaus, 29 March 1922, File 424 [*Dixie*], box 11; Harding to T. Roosevelt, Jr. to Chief of Bureaus, 1 July 1921, File 624–1 [*Maine*], box 36, Ship Files.

17. Harold Sprout and Margaret Sprout, *Toward a New Order of Sea Power*, 2nd ed. (Princeton, N.J.: Princeton University Press, 1943); Yamato Ichihashi, *The Washington Conference and After* (1928; rpt. New York: AMS Press, 1969), 147; Philip T. Rosen, "The Treaty Navy, 1919–1937," in Kenneth J. Hagan, ed., *In Peace and War: Interpretations of American Naval History, 1775–1984* (Westport, Conn.: Greenwood Press, 1984), 221–36; Dudley W. Knox, *The Eclipse of American Sea Power* (New York: Army and Navy Journal, 1922).

18. Log entries, 1 November and 1 December 1919, Station Log showed eighteen battleships berthed at Philadelphia Navy Yard: *Alabama, Kearsarge, Kentucky, Illinois, Iowa, Kansas, Louisiana, Massachusetts, Michigan, Minnesota, Maine, New Hampshire, Nevada, Ohio, Wisconsin, Connecticut, Indiana, Missouri*; "Replacement and Scrapping of Capital Ships," U.S. Senate, *Conference on the Limitation of Armament*, Doc. 9, 67th Cong., 2d sess., sect. II: 98–99; *Philadelphia Evening Bulletin*, 10

and 20 February, 4 October 1922, [Philadelphia Naval Base] Closing Rumors prior to 1964 File, box 173A, Mounted Clippings.

19. Industrial Manager W. P. Robert to Chief Bureau of Engineering O. G. Murfin, 3 February 1922, box 13, Nulton to T. Roosevelt, Jr., 17 August 1921 and Nulton to Capt. H. T. Wright, 21 September 1921, box 14, Captain Henry Williams (CC), "The Organization of an Industrial Navy Yard," 6 January 1937, box 16, File A3–1, Central Subject Files.

20. T. Roosevelt, Jr., to Chief Bureau Construction & Repair, 13 August 1921, T. Roosevelt, Jr., to Chiefs of Bureaus, 1 July 1921, Acting Chief Bureau of Ordnance to Commandant, 4th Naval District, 1 February 1922, Nulton to Captain of Yard, Manager, Supply Officer, and Inspection Officer, 7 February 1922, "Rules of Sale" attached to Hitner Salvage Corp, to Commandant, Navy Yard, Phila., 5 September 1923, Scales to Board of Inspection, 27 September 1923, Board of Inspection to Scales, 5 October 1923, Board of Inspection to Commandant, 17 December 1923, File 624-1 [*Maine*], box 36, Ship Files.

21. Taylor and Griffin Memo to Inspector of Hulls, Phila., 17 July 1921, File 357-1 [*Constitution*], box 5, Ship Files; Alan F. Pater, *United States Battleships: The History of America's Greatest Fighting Fleet* (Beverly Hills, Calif.: Monitor Book Co., 1968), 66, 192; John C. Reilly, Jr., and Robert L. Scheina, *American Battleships, 1886–1923: Predreadnought Design and Construction* (Annapolis, Md.: Naval Institute Press, 1980), 88–93; *DANFS*, 1: 210–11.

22. *Lexington* (Fore River), *Constellation* (Newport News), *Saratoga* (New York Ship), *Ranger* (Newport News), *DANFS*, 1: 210–11; Ernest Andrade, "The Battle Cruiser in the United States Navy," *Military Affairs* 44 (February, 1980): 18–23.

23. Industrial Manager, Philadelphia Navy Yard, to Commandant, 21 July 1921, Cleland Offley, Office of Inspector of Engineering Material Buffalo and Pittsburgh to Buffalo Steam Pump Company, 14 July 1921, Nulton to General Electric Company, 14 July 1921, File 357–1 [*Constitution*], box 5, Ship Files; Kaiser to Heads of all Departments, 2 July 1919, Kempff to All Officers, 12 February 1920, File 845, box 26, Public Works Office Miscellaneous Correspondence.

24. Manager to Commandant, 9 March 1922, Griffin to General Electric Company, 30 June 1921, File 357-1 [*Constitution*], box 5, Ship Files; Hull and Machinery Departments consolidated under Manager Order No. 8, Commandant's Office, Navy Yard, Philadelphia, June 27, 1921, William Pierre Robert (CC) [first Navy Yard Industrial Manager], File 7-1, box 6, Public Works Office Miscellaneous Correspondence; Julius Augustus Furer, *Administration of the Navy Department in World War II* (Washington, D.C.: Department of the Navy, 1959).

25. Nulton to Manager, 30 June 1921, Nulton Special Order No. 13, 9 February 1922, Nulton to George C. Schafer by direction Paymaster General to Commandant, Navy Yard, Phila., 20 May 1922, Contract between Edwin Denby, Secretary of Navy, in presence of J. L. Latimer, JAG, and the Steel Scrap Company, 8 November 1923, T. Roosevelt, Jr., to Commandants, Navy Yards New York, Philadelphia, Norfolk, and Mare Island, 28 June 1923, Robert Memorandum Regarding Scrapping Hulls of Battle Cruisers Nos. 5 and 6, 13 July 1923, File 357-1 [*Constitution*], box 5, Ship Files; Stirling, *Sea Duty*, 162.

26. J. D. Beuret to John Tompkins, 27 November 1923, in Secretary of Navy via Chief of Naval Operations, 27 November 1923, Wolcott Ellsworth Hall, Inspector of Ordnance, New York Ship, to Commandant, Navy Yard, Phila., 13 December 1923, Kintner to Commandant, 22 January 1924, Secretary of Navy to Commandant, Navy Yard, Phila., 22 December 1923, Beuret to Commandant, 5 March 1924, File 962-1 [*Washington*], box 65, Ship Files; log entry, 28 January 1924, Station Log.

27. Entries, 31 March, 26 April, and 31 May 1926, Station Log.

28. Entries, 1 and 4 April 1921, 22 December 1925, 20 March and 1 September 1920, 31 January 1926, Station Log; "Fire Guts U-140 in Back Channel," *Evening Bulletin*, 20 April 1921, [Philadelphia Navy Yard] Fires and Fire Department File, box 173A, Mounted Clippings; Morley, *Travels in Philadelphia*, 131; *DANFS*, 3: 580–83, 7: 291; Raymond G. O'Connor, *Perilous Equilibrium* (Lawrence: University of Kansas Press, 1962); Rosen, "Treaty Navy," 224.

29. Griffin to Commandant, Navy Yard, Phila, 13 January 1920, File 315-6 [*Ammen*], box 1, Ship Files; log entries for 1926–1929, Station Log; Rosen, "Treaty Navy," 223–24.

30. Entries, 4 May, 28 June, 14 July 1919, 5 August 1921, 6 February 1923, 17 September 1926, Station Log; Ritter, "Philadelphia Navy Yard," on NAF, 98–104; *Source Document*, phase I, 20; Rosen, "Treaty Navy," 225–26; William F. Trimble, *Admiral William A. Moffett: Architect of Naval Aviation* (Washington, D.C.: Smithsonian Institution Press, 1994), 10–11, 107; William F. Trimble, *Wings for the Navy: A History of the Naval Aircraft Factory, 1917–1956* (Annapolis, Md.: Naval Institute Press, 1990), 60–80, 110.

31. *Source Document*, phase I: 20; Roger K. Heller, "Curtis Dwight Wilbur, 19 March 1924–4 March 1929," in Coletta, *American Secretaries of the Navy*, 2: 605–30.

32. Log entries, 3, 11, 24 May, 2 and 28 June, 6, 25, and 31 August, 3 and 18 September, 5 and 12 October, 15 November, and 17 December 1926, Station Log; *New York Times*, 13 and 16 May 1926; Brazilian Naval Mission to Rio de Janeiro from Philadelphia Navy Yard included Furer, Production Officer L. M. Atkins, and Design Superintendent Edward E. Brady, *Beacon*, 16 January 1942.

33. *New York Times*, 25 September, 27 October, and 5 December 1926.

34. Magruder, "The Navy and Economy," *Saturday Evening Post*, 24 September 1927; *New York Times*, 24 September, 2 October, and 13 November 1927; Heller, "Wilbur," 607–9.

35. Magruder Correspondence, File A702(1), Secretary of the Navy Confidential Correspondence, 1926–1940, Records of the Secretary of the Navy, RG 80, National Archives and Record Administration, Washington, D.C.; *New York Times*, 28 and 29 September, 26 and 28 October, 16 December 1927.

36. *New York Times*, 26 and 28 October, 13 November, 16 and 22 December 1927.

37. *New York Times*, 16 and 22 December l927, 5 and 7 January 1928.

38. Entry, 1 July 1930, Station Log; "Closing Navy Yards," *New York Times*, 4 May 1932; Arthur P. Dudden, "The City Embraces 'Normalcy' 1919–1929," in Russell F. Weigley, ed., *Philadelphia: A 300-Year History* (New York: W.W. Norton, 1982), 585; Heller, "Wilbur," 612–13; William R. Braisted, "Charles Evans Hughes, 14 November 1927–17 September 1930," in Robert William Love, Jr., *The Chiefs of Naval Operations* (Annapolis, Md.: Naval Institute Press, 1980), 49–68; Gerald E. Wheeler, "Charles Francis Adams, 5 March 1929–4 March 1933," in Coletta, *American Secretaries of the Navy*, 2: 633–53 (644).

39. Entries, 24 June 1930, 27 June 1931, 14 November 1932, Station Log; S. E. Zubrow, "The History of the Philadelphia Navy Yard," *United States Naval Administration in World War II: Commandant, Fourth Naval District*, 6 vols. (Washington, D.C.: Naval History Division, 1946), 3, Part IV, Sect. I: 53–54; *Evening Bulletin*, 15 May 1931, [Philadelphia Navy Yard] Closing Rumors prior to 1964 File, *Evening Bulletin*, 24 May 1932, [Philadelphia Navy Yard] Accidents File, and *Evening Bulletin*, 9 February and 28 June 1931, [Philadelphia Navy Yard] Shipbuilding, 1930–39 File, in Mounted Clippings; *DANFS*, 6: 244; 1: 214; Wheeler, "Adams," 644; Joan Hoff Wilson, *Hoover, the Forgotten Progressive* (Boston: Little, Brown, 1975).

40. Log entries, 5 June 1933, 11 January, 12 February, and 5 March 1935, 30 March 1936, Station Log; Commandant to Chief of Naval Operations, 12 February 1935, File A14-7, in Zubrow, "Philadelphia Navy Yard History," 3, part IV, section III: 577–79; Rosen, "Treaty Navy," 232; decommissioned Coast Guard Rum Patrol destroyers *Cassin, McDougal, Davis, Tucker, Shaw, Porter*, and *Conyngham*.

41. FDR to Commandant 4th Naval District, 20 January 1920, File A3-1, box 16, Central Subject Files; *Philadelphia Navy Yard News* 1 (April 1936): 1–6; *Evening Bulletin*, 21 October 1934, [Philadelphia Navy Yard] Ships Scrapped and Sold File , box 173B, Mounted Clippings; Margaret B. Tinkcom, "Depression and War, 1929–1946," in

Weigley, *Philadelphia*, 617–19; David Palmer, *Organizing the Shipyards: Union Strategy in Three Northeast Ports, 1933–1945* (Ithaca, N.Y., and London: ILR Press, an imprint of Cornell University Press, 1998), 41.

42. *Philadelphia Navy Yard News* 1 (April 1936): 3–5; *Evening Bulletin*, 10 March 1935, 17 and 19 August 1937, Employees, Employment prior to 1941 File, Mounted Clippings.

43. Log entries, 4 April, 23 August, 18 October 1933, Station Log; *Philadelphia Navy Yard News* 1 (April 1936): 3–5; *Evening Bulletin*, 10 March 1935, 17 and 19 August 1937, Employees, Employment prior to 1941 File, Mounted Clippings.

44. W. C. Watts Order, 9 April 1935, File A3-2, box 21, Central Subject Files; log entries, 28 May, 29 October 1935, Station Log; A. Raymond Kirk, "Navy Yard Surroundings," *Philadelphia Navy Yard News* 3 (December 1936): 7; *Evening Bulletin*, 10 March 1937, Accidents File, box 173A, 8 August, 1 September 1933, 31 January, 22 August, 1 October 1934, 28 May 1935, Shipbuilding, 1930–39 File, box 173B, Mounted Clippings; Zubrow, "Philadelphia Navy Yard History," 3, part IV, sect. I: 55–56; Furer, *Administration*, 57.

45. *Evening Bulletin*, 8 August, 1 September 1933, Shipbuilding 1930–39 File, box 173B, Mounted Clippings; Frederick A. Hunnewell, "United States Coast Guard Cutters," *Transactions of the Society of Naval Architects and Marine Engineers* 45 (1937): 81–114 (111).

46. Log entries, 28 October, 18 November 1935, Station Log.

47. Log entry, 3 June 1936, Station Log; *Philadelphia Navy Yard News* 1, 2 (April and September 1936).

48. Log entries, 20 April, 10 and 25 July, 21 August 1935, 27 October and 17 November 1936, 18 October 1937, 2 February and 30 August 1938, Station Log.

49. W. T. Cluverius, Discussion, "Neutrality: Can It Be Maintained by a World Power?" *USNIP* 352 (June 1932): 901–2; "Discussions—The London Treaty and American Naval Policy," *USNIP* 345 (November 1931): 1557; Book Review, "*If War Comes* by Major R. Ernest Dupuy," *USNIP* 417 (November 1937): 1631.

50. Log entries, 20 May, 28 July, 13 September, 17 November 1937, 10 March, 5 April, 6, 14, and 22 June, 3 October 1938, Station Log; *Evening Bulletin*, 22 June 1937, Shipbuilding 1930–39 File, box 173-B, Mounted Clippings; Friedman, *Small Combatants*, 55.

51. Log entries, 28 January, 7 November 1932; Baum, Report of Participation in Observance of Navy Day, 27 October 1938, "Navy Day Program," 27 October 1938, Station Log; 17,600 visitors arrived by trolley, 4,690 by bus, 14,454 by automobile; 24,824 by foot, 3,352 by tugboat across the river and 89 by private boat for Navy Day 1938, total of 65,118; *Evening Bulletin*, 29 January, 22 June and 17 Oc-

tober 1937, Shipbuilding, 1930–39 File, Mounted Clippings.

52. *Source Document*, phase I: 20; Zubrow, "Philadelphia Navy Yard History," 3, Part IV, Sect. III: 570–72.

53. Log entry, 15 November 1938, Station Log; Cluverius, "Book Review," *USNIP* 417 (November 1937): 1631–32.

CHAPTER 10. LEAGUE ISLAND AT WAR

1. Local Shore Development Board to Mathewson, Enright, etc., 23 October 1939, File A1-1, box 4; Commandants Office, 1923–1953, "National Emergency . . . with Limits of Peacetime," ALNAV 42, 8 September 1939, in Schwartz to Commander, Fourth Naval District, 19 April 1948, File A16, box 86, Central Subject Files; list of fifty-six Priority I, forty-one Priority II, and eighteen Priority III destroyers in S. E. Zubrow, "The History of the Philadelphia Navy Yard," *U.S. Naval Administration in World War II, Commandant 4th Naval District*, 6 vols. (Washington, D.C.: Naval History Division, 1946), 3; sect. I: 67–68.

2. Robert to Chief Bureau of Engineering Murfin, 3 February 1922, File A-3, box 13, Central Subject Files; *Public Ledger*, 29 December 1939; Zubrow, "Philadelphia Navy Yard," 57–68, 571; Allan J. Chantry, Jr., "Launching of U.S.S. 'New Jersey' and U.S.S. 'Wisconsin'," *Transactions of the Society of Naval Architects and Marine Engineers* 52 (1944): 391–438.

3. Entries, 4 and 18 December 1939, June 1940, Station Log; Julius Augustus Furer, *Administration of the Navy Department in World War II* (Washington, D.C.: Naval History Division, 1959).

4. John D. Venable, *Out of the Shadow: The Story of Charles Edison* (East Orange, N.J.: Charles Edison Fund, 1978), 120–22; Allison W. Saville, "Charles Edison, 2 January 1940–24 June 1940," in Paolo E. Coletta, ed., *American Secretaries of the Navy*, 2 vols. (Annapolis, Md.: Naval Institute Press, 1980), 2: 672; Furer, *Administration in World War II*, 58, 219–21.

5. Patrick Abbazia, *Mr. Roosevelt's Navy: The Private War of the U.S. Atlantic Fleet, 1939–1942* (Annapolis, Md.: Naval Institute Press, 1975); James R. Leutze, *Bargaining for Supremacy: Anglo-American Naval Collaboration, 1937–1941*; John Major, "The Navy Plans for War, 1937–1941," in Kenneth J. Hagan, ed., *In Peace and War: Interpretations of American Naval History, 1775–1984* (Westport, Conn.: Greenwood Press, 1984), 235–62 (251–52).

6. Venable, *Charles Edison*, 129–31, 154–55; Furer, *Administration in World War II*, 55–59; George H. Lobdell,

"Frank Knox, 11 July 1940–28 April 1944," in Coletta, ed., *American Secretaries of the Navy*, 2: 677–727.

7. Entry, 16 September 1940, Station Log; *Evening Bulletin*, 16 September 1940, [Philadelphia Navy Yard] Shipbuilding, 1940–49, File, box 173B; Mounted Clippings; Paul Stillwell, *Battleship New Jersey: An Illustrated History* (Annapolis, Md.: Naval Institute Press, 1986), 11.

8. Log entry, 20 September 1940, Station Log; *Public Ledger*, 20 September 1940; Norman Friedman, *U.S. Small Combatants: Including PT-Boats, Subchasers, and the Brown-Water Navy* (Annapolis, Md.: Naval Institute Press, 1987), 492.

9. Entries, 15 May, 6, and 7 June 1941, Station Log; H. Schwartz to Commander, 19 April 1948, File A-16, box 86, Central Subject Files.

10. Entries, 15 September and 8 October 1941, Station Log; *Public Ledger*, 14 October 1941 on *Furious*; 17 February 1941, 9 and 12 May 1941; *Philadelphia Evening Bulletin*, 22 September, 9 October 1941, [Philadelphia Navy Yard] Foreign Ships, and Shipbuilding, 1940–49 File, Mounted Clippings; List of "Requests," in Zubrow, "Philadelphia Navy Yard History," appendix.

11. *Public Ledger*, 17 April 1940, 30 May and 3 October 1941; Watson quoted in *Beacon*, 28 August 1942; *Philadelphia Record*, 11 October 1941, [Philadelphia Navy Yard] Employees, Employment, 1941–45 File, box 173A, Mounted Clippings.

12. *Public Ledger*, 10 September 1940; *Evening Bulletin*, 6 May 1941, [Philadelphia Navy Yard] Housing 1940 File, 20 and 21 October 1941, Housing 1941–1959 File, Mounted Clippings; log entry 19 January 1942, Station Log; *DANFS*, 4: 76 for *League Island* [YF 20]; Margaret B. Tinkcom, "Depression and War, 1929–1945," in Russell F. Weigley, ed., *Philadelphia: A 300-Year History* (New York: W.W. Norton, 1982), 601–49.

13. *Public Ledger*, 10 and 21 April 1941; unidentified newspaper clipping, 2 March 1941, Philadelphia Naval Base, Miscellaneous, 1941–1949 File, box 173B, Mounted Clippings; Zubrow, "Philadelphia Navy Yard History," 79–80.

14. Readiness and Security in Present Emergency, 27 November 1941, Fourth Naval District Order No. 12-1941, File A3–2, box 21, Central Subject Files.

15. Gaylord Church (Civil Engineer Public Works Office) to Bureau of Yards and Docks, 21 March 1942, File A16-3, box 92, Mathewson to Heads of all Departments, 17 September 1942, Draemel to Chief of the Bureau of Yards & Docks , 12 October 1942, "Navy Yard Air Raid Battle Problem," File A16-3, box 93, Central Subject Files; log entries, 9 and 21 December 1941, 3 February 1942, 16 January 1942, Station Log for roof watcher stations—General Storehouse (5), Administration (6), Machine and Electrical Shop (16), Power House (23), Structural Shop (57), General

Storehouse (83), Boat Storage (121), Labor Board (501), Hangar (534), Structural Assembly Shop (541), Turret Shop (546), Aero Structural Lab (601), Material Assembly Shop (592), Station Log; for machine gun mount locations at Navy Yard, see Zubrow, "Philadelphia Navy Yard History," map 1; Wood reminiscences, *Beacon*, 22 November 1991.

16. Bureau of Yards and Docks to Commandant, Navy Yard, Philadelphia, 13 January 1942, Mathewson to Heads of Departments, 13 September 1944, File A16-1, box 88, Central Subject Files.

17. Michael Gannon, *Operation Drumbeat: The Dramatic True Story of Germany's First U-Boat Attack Along the American Coast in World War II* (New York: Harpers, 1991), xvii–xviii, 199, 271; Zubrow, "Philadelphia Navy Yard History," 112.

18. Jack F. Ayers and Henry A. Vadnais, Jr., *Fifty Years of Naval District Development, 1903–1953* (Washington, D.C.: History Division, Office of the Chief of Naval Operations, 1956), 27–28; Zubrow, "Philadelphia Navy Yard History," 95–102; sea frontiers in Furer, *Administration in World War II*, 21.

19. Log entries, 22 June 1938, 25 September 1939, 7, 15, 26, and 28 January 1942, February 1942, Station Log; Zubrow, "Philadelphia Navy Yard History," 97, 107–8; Destroyer Escort *Rudderow* built in 1943 at Philadelphia Navy Yard to honor *Cythera*'s commanding officer.

20. Farm Out Program, Zubrow, "Philadelphia Navy Yard History," 85, 88, 51ff, appendix; Robert H. Connery, *The Navy and the Industrial Mobilization in World War II* (Princeton, N.J.: Princeton University Press, 1951), 485.

21. ALNAV, 7 January 1942, and list of "farm out" companies in Philadelphia, Homer N. Wallin to Bureau of Ships, 2 September 1948, File A16-1, box 89, Central Subject Files; Jeffery M. Dorwart, *Eberstadt and Forrestal: National Security Partnership, 1909–1949* (College Station: Texas A&M University Press, 1991); Connery, *Industrial Mobilization*, 485.

22. *Survey of Industrial Department, Navy Yard, Philadelphia, Pa.*, Report No. 6 of Industrial Survey Division, 29 January 1945, Bureau of Ships to Commandant, Navy Yard, Philadelphia, 2 February 1945, File A3-1, box 20, Central Subject Files; Zubrow, "Philadelphia Navy Yard History," 95–96; *50 Years New York Shipbuilding Corporation, Camden, N.J.* (Camden, N.J.: New York Shipbuilding Corp., 1949), 51; "Sun Shipbuilding," *Transactions of the Society of Naval Architects and Marine Engineers* 53 (1945): 236–37; Cramp history, in Thomas R. Heinrich, *Ships for the Seven Seas: Philadelphia Shipbuilding in the Age of Industrial Capitalism* (Baltimore: Johns Hopkins University Press, 1997).

23. Hertzog to Draemel, 23 November 1943, File A3-1,

box 18, Central Subject Files; racial tensions in Philadelphia, in Peg Porter Oral History Transcript 9, Oral History Project.

24. *Beacon*, 2 July, 14 August, 2 and 30 October 1942. The Industrial Department, Shop Organization in World War II: Shop 01 (Supply Department), 02 (Transportation), 03 (Power Plant), 06 (Central Tool), 11 (Shipfitters), 17 (Sheet Metal), 23 (Forge), 26 (Welders), 31 (Machine [Inside]), 38 (Machine [Outside]), 41 (Boiler makers), 51 (Electrician), 53 (Copper and Pipe), 61 (Shipwrights), 63 (Joiners [Wood workers]), 68 (Boatbuilders), 70 (Building Trades), 71 (Paint Shop), 72 (Riggers and Laborers), 74 (Sail Loft), 81 (Foundry), 94 (Pattern makers).

25. *Industrial Department Survey*, 29 January 1945; Forrestal to All Bureaus, Boards, and Offices, Navy Department, Commandants All Naval Districts and Navy Yards, 29 July 1944, box 5, Commandants Office, 4th Naval District, Central Subject Files; Homer L. Ferguson, "The Past, the Present and the Future," *Transactions of the Society of Naval Architects and Marine Engineers* 51 (1963): 207.

26. Chantry to Commandant, Navy Yard, Philadelphia, 11 July 1944, Draemel to Commandant, 17 April 1945, Mathewson to Commandant, 17 July 1944, Albert E. Randall to Commandant, 6 July 1944, Draemel to Chief of Naval Operations, 19 September 1945, Carroll M. Hall, Receiving Station, Instructions for All Hands in Connection with Prisoners of War on the Receiving Station, 14 June 1945, Draemel to Paul Lewis, War Manpower Commission, 16 March 1945, File A16-2, box 92, Central Subject Files.

27. *Industrial Department Survey*, 29 January 1945; log entry, 21 July 1943, Station Log; *Beacon*, 19 January 1945, 19 December 1941; *New York Times*, 15 February 1943; *Evening Bulletin*, 26 March 1943, Women Workers File, box 173B, Mounted Clippings; Marie Hindsley and Charles J. Bradford Oral History Transcripts, Oral History Project.

28. Log entries, 30 June and 30 December 1942, Station Log; *Beacon*, 2 January 1942; Zubrow, "Philadelphia Navy Yard History," Table 6.

29. *Beacon*, 17 July, 7 August, 9 and 23 October 1942; 7 July 1944.

30. *Beacon*, 23 January and 9 October 1942, 12 February 1943; Tinkcom, "Depression and War," 617;

31. Log Entries, 19 March, 2 and 14 May, 15 July, 5 November and 7 December 1942, 14 January and 7 December 1943, Station Log; *Beacon*, 20 March and 15 May 1942, 10 September 1943.

32. Log Entries, 2 May 1942, 4 July 1944, Station Log; *Beacon*, 23 June and 7 July 1944, 14 November 1941, 26 January 1945; Zubrow, "Philadelphia Navy Yard History," 107.

33. *Beacon*, 12 and 19 June, 31 July, and 12 September

1942; Peg Porter Oral History Transcript, 9, Oral History Project.

34. *Beacon*, 20 March 1942, 21 and 28 May, and 9 July 1943; *Evening Bulletin*, 16 April and 17 July 1943, Recreation File, box 173B, Mounted Clippings.

35. *Industrial Department Survey* 3; Draemel to Michael J. Bradley, 26 July 1943, box 90, Central Subject Files; log entry, 31 August 1942, Station Log; *Beacon*, 16 July and 10 September 1943, 26 January 1945.

36. *Beacon*, 24 December 1941, 2 and 9 January 1942.

37. Zubrow, "Philadelphia Navy Yard History," 86–93; *Industrial Department Survey*; log entries, 9 September 1942 for commissioning of *Montpelier*, 15 October 1942 for *Denver*, 14 January 1943 for *Independence*, 28 May 1943 for *Cowpens*, 25 February 1943 for *Princeton*, 3 March 1943 for *Belleau Wood*, Station Log.

38. Log entries, 12 February, 15 July, 7 December 1942, and 7 December 1943, Station Log.; [Camden] *Courier-Post*, 31 July 1998; Alan J. Chantry, "Launching the 'New Jersey' and 'Wisconsin'," *Transactions of the Society of Naval Architects and Marine Engineers* 52 (1944): 431–35.

39. Log Entries, 29 May, 14 October, 21 July 1943, 12 February 1944, Station Log; *Beacon*, 28 May 1943; *DANFS*, 1: 336; Bruce Hampton Franklin, *The Buckley-Class Destroyer Escorts* (Annapolis, Md.: Naval Institute Press, 1999).

40. *Beacon*, 30 October 1942, 12 February 1943.

41. Log entries, 20 August and 1 September 1944, Station Log; *Beacon*, 5 and 12 January 1945.

42. *Beacon*, 29 June, 8 and 13 July 1945; Vincent Davis, *Postwar Defense Policy and the U.S. Navy, 1943-1946* (Chapel Hill: University of North Carolina Press, 1966), 115.

43. Log entries, 5, 17 and 18 January 1942, 19 February 1942, 11, 16, and 19 February 1942, 19 March, 23 April, 30 June, and 19 August 1942, Station Log; *Beacon*, 16 January 1942; *Evening Bulletin*, 2 September 1944, [Philadelphia Navy Yard] Accidents File, box 173A, Mounted Clippings.

44. E. R. Gayler, Joint Army Navy Ammunition Storage Board to Secretary of the Navy, 9 October 1944, Special Classified Files, 1927–1945, box 21, Central Subject Files; explosion killed chemical engineers Douglas P. Meigs and Peter R. Bragg, *Evening Bulletin*, 2 September 1944, Accidents File, box 173A, Karl Abraham, "Navy Yard Has Its Own Atom Secret," *Evening Bulletin*, 27 August 1967, Atom Bomb Activity File, box 173A, Mounted Clippings; Norman Polmar and Thomas B. Allen, *Rickover* (New York: Simon and Schuster, 1982), 199–200; James L. McVoy, Virgil W. Rinehart, and Prescott Palmer, " The Roosevelt Resurgence," in Randolph W. King, ed., *Naval Engineering and American Seapower* (Baltimore: Nautical & Aviation Publishing Company, 1989), 196–97; Richard Rhodes, *The Making of the Atomic Bomb* (New York: Touchstone, 1986).

45. Forrestal to Robert Patterson, 7 November 1945, Day File, box 89, James V. Forrestal Papers, Seeley Mudd Library, Princeton University; Polmar and Allen, *Rickover*, 121, 155–56; Abraham, "Atom Secret,"; Norman Polmar, *Atomic Submarines* (Princeton, N.J.: Van Nostrand, 1964).

46. Fourth Naval District Order No. 23-45, 21 November 1945, in Draemel to All Units, Activities and Offices of the 4th Naval District, 4 February 1946, File A3-2, box 21, Central Subject Files; *Evening Bulletin*, 11 August 1945, Shipbuilding, 1941–49 File, box 173B, Mounted Clippings; for rocket experiments William F. Trimble, *Wings for the Navy: A History of the Naval Aircraft Factory, 1917–1956* (Annapolis, Md.: Naval Institute Press, 1990) 276–87.

47. Forrestal to Patterson, 7 November 1945, Forrestal Day File, box 89, Forrestal Papers; entry, 18 April 1945, James Forrestal, *The Forrestal Diaries*, ed. Walter Millis (New York: Viking, 1951), 46 ; Davis, *Postwar Navy*, 103–12.

48. Dorwart, *Eberstadt and Forrestal*; Paolo E. Coletta, *The United States Navy and Defense Unification, 1947-1953*, (Newark: University of Delaware Press, 1981); Davis, *Postwar Navy*, 101–2.

CHAPTER 11. COLD WAR NAVY YARD

1. *Beacon*, 21 December 1945, 6 August 1948.

2. *Beacon*, 15 February 1946; Steven T. Ross, "Chester William Nimitz, 15 December 1945–15 December 1947," in Robert William Love, Jr., ed., *The Chiefs of Naval Operations* (Annapolis, Md.: Naval Institute Press, 1980), 181–91; Paolo E. Coletta, *The United States Navy and Defense Unification, 1947–1953* (Newark: University of Delaware Press, 1981).

3. "The Impact of Military Production During 1952," Address by Chairman J. D. Small, Munitions Board, Department of Defense Before Annual Forecaster's Conference Sponsored by the Chamber of Commerce of Greater Philadelphia at Philadelphia, Pennsylvania, January 15, 1952, file A16-1, box 91, Central Subject Files; *Beacon*, 10 May 1945; Robert Greenhalgh Albion, *Makers of Naval Policy, 1798–1947* (Annapolis, Md.: Naval Institute Press, 1980), 529–45; Coletta, *Defense Unification*.

4. McKee penciled note, on Mathewson to Commander

Philadelphia Naval Shipyard, 11 December 1945, File A3, box 4, Chantry to Commandant, 4 August 1943, Draemel to Secretary of the Navy via Chief of Naval Operations, 27 December 1945, attached to Draemel to All Units, Activities, and Offices of 4th Naval District, 4 February 1946, File A3, box 3, Central Subject Files; *Beacon*, 7 December 1945.

5. Forrestal to Commandant 4ND, 27 January 1946, Draemel to Secretary of the Navy via Chief of Naval Operations, 27 December 1945, File A3, box 4, Draemel to All Units, Activities, and Offices of 4ND, 4 February 1946, File A3, box 3, Central Subject Files.

6. "History of Philadelphia Naval Shipyard, 1 October 1945 to 30 September 1946," typescript, File NY4, box 191, Central Subject Files; *Beacon*, 15 February, 8 and 15 March, 7 June 1946; *California* and *Tennessee* bow-to-bow stored in Dry Dock No. 4, in Charles J. Bradford Oral History transcript 12, Philadelphia Naval Shipyard Oral History Project; Alan F. Pater, *United States Battleships* (Beverly Hills, Calif.: Monitor Books, 1968), 35–36, 169–73, 207–13, 215–21.

7. History of Philadelphia Naval Shipyard, Scrapping and Disposal, 1 October 1945–30 September 1946, File NY4, box 191, Central Subject Files; for atomic test on *New York* and *Prinz Eugen*, see *Beacon*, 8 and 15 March 1946.

8. "Surplus and Disposal—Real Estate," in Naval Shipyard History, part III, Contract Curtailment, part II, History of the Philadelphia Naval Shipyard, 1 October 1945–30 September 1966, typescript, File NY4, box 191, Central Subject Files; *Beacon*, 10 May and 9 August 1946, 25 April 1947; *Philadelphia Record*, 12 June 1946, Miscellaneous, 1940–1949 File, box 173B, Mounted Clippings.

9. Annual Reports, 1945–58, Navy Yard Command Histories, typescripts, box 1, PNSY Histories, PNSR; *Beacon*, 22 March, 19 April, 31 May, 5 July 1946.

10. *Beacon*, 31 May, 7 and 28 June, 5 July 1946.

11. *Beacon*, 15 and 22 March, 10 and 31 May, 7 June, 5 July 1946; recollections of female welder, in *Camden Courier-Post*, 25 June 1998, for layoffs at New York Ship "the day after Japanese surrender."

12. Kniskern to Chief Bureau of Ships, 9 July 1951, File A3/NY4, box 8, Central Subject Files.

13. Command History, 1 December 1945–1 December 1958, typescript, box 1, PNSY Histories; Dean C. Allard, "An Era of Transition, 1945–1953," in Kenneth J. Hagan, ed., *In Peace and War: Interpretations of American Naval History, 1775–1984* (Westport, Conn.: Greenwood Press, 2nd ed., 1984), 290–303.

14. Wallin to Heads of Departments, Divisions, and Offices, 25 March 1947, Wallin to Walter Roth, president Philadelphia Naval Shipyard Development Association, 25 March 1947, Files A1–2, A3–1/NY4, box 5, Central Subject Files.

15. Paolo E. Coletta, "Louis Emil Denfeld, 15 December 1947–1 November 1949," in Love, ed., *Chiefs of Naval Operations*, 193–206; Coletta, *Defense Unification*, 100–110; Ross, "Nimitz," 181–91; Jeffery M. Dorwart, *Eberstadt and Forrestal: A National Security Partnership, 1909–1949* (College Station: Texas A & M University Press, 1991).

16. James L. Kauffman, Statement on Merger of the 4th and 5th Naval Districts, 31 August 1948, File A3-1, box 5, Central Subject Files; *Beacon*, 11, 12, and 23 August 1949.

17. Kniskern to Chief Bureau of Ships, 9 July 1951, File A3/NY4, box 8, H. C. Zitzewitz by direction Commandant to Abbitt Machine Company of Philadelphia, 5 February 1951, File A16-1, box 91, Central Subject Files; *Beacon*, 24 August, 6 September 1949.

18. Small Address to Chamber of Commerce of Greater Philadelphia, 15 January 1952, Kniskern to Chief Bureau of Ships, 9 July 1951, Resume of Operations of Philadelphia Naval Shipyard for fiscal year 1951, File A3/NY4, box 8, Central Subject Files.

19. *Beacon*, 26 January, 21 December 1951, 4 April 1952.

20. Commander Philadelphia Naval Shipyard to Chief Bureau of Ships, 12 March 1951, F. S. Low, Deputy CNO, to Distribution List, 5 November 1951, Chief Bureau of Ships to Chief of Naval Operations via Chief Bureau of Yards & Docks, 22 June 1951, Chief Bureau of Ships to Commander Philadelphia Naval Shipyard, 7 March 1952, File A16–1, boxes 91–92, Central Subject Files.

21. Chief Office of Industrial Relations to All Naval and Marine Corps Activities, 31 January 1951, box 87, Acting Chief Bureau of Ships Wheelock to Commander Phila. Naval Shipyard, 14 September 1950, box 86, Wallin, Naval Shipyard Order No. 8-47 Revision 2, Local Disaster Plan Including Radiological, Bacteriological and Gas Defense, 10 January 1949, rev. 1 March 1948 Plan, File A16-1, box 90, Central Subject Files.

22. The Navy Yard reactivated *Capricornus, Enored, Niobrara, Adirondack*, and *Cascade* for Korean War in 1951, *Beacon*, 20 and 26 July, 3, 10, and 24 August 1951, 28 March and 18 April 1952.

23. *Evening Bulletin*, 17 July 1947, Philadelphia Navy Yard Aircraft Experimental Tests File, box 173A, Mounted Clippings; *Guppy* conversions, in John D. Alden, *The Fleet Submarine in the U.S. Navy: A Design and Construction History* (Annapolis, Md.: Naval Institute Press, 1979), 130–31, 137, 142, 152.

24. Kniskern to Chief Bureau of Ships, 9 July 1951, File A3/NY4, box 8, Central Subject Files; *Beacon*, 25 April 1947, 15 October 1948, 26 July and 3 August 1951; *New York Times*, 12 October 1947; Norman Friedman, *U.S. Small Combatants Including PT-Boats, Subchasers, and*

the *Brown-Water Navy* (Annapolis, Md.: Naval Institute Press, 1987), 396, 503; Alden, *Fleet Submarine*, 130ff.

25. *Beacon*, 18 May, 29 June, 19 October 1951.

26. *Evening Bulletin*, 24 August 1949; *Beacon*, 29 June, 13, 20, and 26 July 1951; K. Jack Bauer, "Dan Able Kimball, 31 July 1951–20 January 1953," in Paolo E. Coletta, ed., *American Secretaries of the Navy*, 2 vols. (Annapolis, Md. Naval Institute Press, 1980) 2: 829–39.

27. Kniskern to Chief Bureau of Ships, 23 April 1952, File A16-1, box 91, Commander Phila. Naval Shipyard to Commander Long Beach Naval Shipyard, 27 August 1951, File A16, box 86, Central Subject Files; *Beacon*, 25 July 1952, 25 September 1953; Bernard M. Kassell, "1,000 Submarines: Fact or Fiction?" *USNIP* 77 (March 1951): 267–75; Floyd D. Kennedy, Jr., "The Creation of the Cold War Navy, 1953–1962," in Hagan, *War and Peace*, 304–26; Paolo E. Coletta, "Francis P. Matthews, 25 May 1949–31 July 1951," in Coletta, *American Secretaries of the Navy*, 2: 783–827; J. J. Meyer, Jr., "Our Nation's Shipyards," *USNIP* 90 (November 1964): 34–35.

28. "Annual Report for 1954," Command Histories, (1954), page 10, typescript, box 1, PNSY History; *Beacon*, 6 March 1953.

29. Small Address, 15 January 1952, File A16-1, box 91, Central Subject Files; Alden, *Fleet Submarine*, 134–35, 155–59; Martin E. Holbrook, "Review of Post-War Construction," *USNIP* 76 (November 1950): 1230–35; W. J. O'Neill, "Our Zipper Fleet," *Philadelphia Inquirer Magazine*, 26 September 1954.

30. Command History (1958), typescript, 11, 29-30, 41, Organization Changes & Function, Chronology (1962), typescript, box 1, PNSY Histories; *Beacon*, 9 October 1959, 11 March 1966.

31. PNSY Chronology (1961), typescript, 7, LPH launchings, *Okinawa* (1962), *Guadalcanal* (1963), *Guam* (1965), *New Orleans* (1968) in Command Histories (1961–68), box 1, PNSY Histories; *Beacon*, 30 November 1959, 26 May 1960.

32. Command History (1960), typescript, 4, box 1, PNSY Histories; *Beacon*, 16 March, and 26 May 1960, 8 April and 29 June 1961, 7 December 1962; *Evening Bulletin*, 31 January 1967, Accidents File, box 173A, Mounted Clippings.

33. *Evening Bulletin*, 17 November 1959, Ships Scrapped and Sold File, box 173B, 11, 12, 17, and 18 December 1958 and 19 March 1958, Closing Rumors prior to 1964 File, box 173A, 20 December 1959, History Prior to 1960 File, box 173A, 29 May 1962, Housing 1960– File, box 173A, Mounted Clippings.

34. Dilworth, in *Evening Bulletin*, 3 July 1961, Housing 1960– File, box 173A, Van Zandt, in 9 May 1961, Closing Rumors prior to 1964 File, box 173A, 8 March 1958 and 13

November 1960, Aircraft Experimental Tests File, box 173A, 30 July 1961, Miscellaneous 1960–1969 File, box 173B, Mounted Clippings; George R. Kolbenschlag, "The Mothball Fleet," *USNIP* 91 (April l965): 88–101.

35. Command History (1960), typescript, 1, Command History (1961), typescript, 10, box 1, PNSY Histories; *Beacon*, 30 November 1962; *Evening Bulletin*, 12 April 1964, Private Ownership Proposed File, Box 173B, *Sunday Bulletin*, 19 April 1964, Closing Rumors, 1964–68 File, box 173A, Mounted Clippings; Clinton H. Whitehurst, Jr., "Is There a Future for Naval Shipyards?" *USNIP* 104 (April 1978): 31.

36. *Evening Bulletin*, 18 August 1961, 11 and 13 December 1963, Closing Rumors prior to 1964 File, box 173A, 20 February, 23 and 24 April and 9 May 1964, Closing Rumors 1964–68 File, box 173A, Mounted Clippings.

37. *Beacon*, 4 December 1964; *Time Magazine*, vol. 84, 27 November 1964: 30; *Evening Bulletin*, 19 February, 19 April, 9 and 23 May, 14 June, 14 and 15 September, 4, 8, 14, 15, 19, and 20 November 1964, Closing Rumors, 1964–68 File, box 173A, 12 November 1964, Editorials File, box 173A, Mounted Clippings.

38. *Beacon*, 11 March, 10 June, 8 July 1966, 20 (quote) and 27 January 1967; *Philadelphia Inquirer*, 25 and 27 February, 3 March 1968, Probe File, box 173B, Mounted Clippings; Paul R. Schratz, "Paul Henry Nitze, 29 November 1963–30 June 1967," in Coletta, *American Secretaries of the Navy*, 2: 950; Floyd D. Kennedy, Jr., "David Lamar McDonald, 1 August 1963–1 August 1967," Robert W. Love, ed., *The Chiefs of Naval Operations* (Annapolis, Md.: Naval Institute Press, 1980), 332–49.

39. Command Histories (1965), page 3 (1971), page 4 typescripts, box 1, PNSY Histories; *Beacon*, 14 January and 10 June 1966, 3 March and 16 June 1967, 6 and 14 November 1970.

40. *New Jersey*, in *Beacon*, 11 March and 29 April 1966, 16 June 1967, 12 April 1968; *Evening Bulletin*, 15 July 1967, Editorials File, box 173A, Mounted Clippings; *New York Times*, 20 January 1968; Paul Stillwell, *Battleship New Jersey: An Illustrated History* (Annapolis, Md.: Naval Institute Press, 1986); John Prados, *Hidden History of the Vietnam War* (Chicago: Ivan R. Dee, 1995), 30; Joseph-James Ahern, *Images of America: Philadelphia Naval Shipyard* (Dover, N.H.: Arcadia, 1967), 49.

41. Command History (1967), pp. 4–5, box 1, PNSY Histories; *Beacon*, 23 August 1963, 11 March 1966, 3 March, and 1 December 1967, 5 and 12 January 1969, 3 February 1969.

42. *New York Times*, 25 and 31 March 1967 on New York Ship; Lawrence J. Korb, "The Erosion of American Naval Preeminence, 1962–1978," in Hagan, *Peace and War*, 325–46.

43. Robert Andrews and Pat D'Amico Oral History transcripts, Oral History Project; *Evening Bulletin*, 15 July 1967, Editorials File, box 173A, Mounted Clippings; Whitehurst, "Is There a Future?"

CHAPTER 12. CULTURE OF CLOSURE

1. Peg Porter Oral History Transcript, p. 4, Bernard C. Moran Oral History Transcript, pp. 14–15, Oral History Project: *Beacon*, 13 February 1970, 24 March 1972, 8 August 1975; *Evening Bulletin*, 11 May 1972, [Philadelphia Navy Yard] Miscellaneous, 1970–79 File, Mounted Clippings; J. K. Holloway, Jr., "The Post Vietnam Navy: The Rhetoric and the Realities," *USNIP* 98 (August 1972): 52–59; Joseph F. Yurso, "Decline of the Seventies," in Randolph W. King, ed., *Naval Engineering and American Seapower* (Baltimore: Nautical & Aviation Publishing Company of America, 1989), 326–41; Lawrence J. Korb, "The Erosion of American Naval Preeminence, 1962–1978," in Kenneth J. Hagan, ed., *In Peace and War: Interpretations of American Naval History, 1775–1984*, 2nd ed. (Westport, Conn.: Greenwood Press, 1984), 329–30.

2. U.S. Congress, House Armed Service Committee, *Status of Shipyards Hearings by the Seapower Subcommittee of the Committee on Armed Service House of Representatives, 91st Congress, Second Session held April–July 1970* (Washington, D.C.: U.S. Government Printing Office, 1970), 10438–83, and *Current Status of Shipyards 1974, Hearings before the Seapower Subcommittee of the Committee on Armed Service, House of Representatives, 93rd Congress, 2nd Session*; Clinton H. Whitehurst, Jr., "Is There a Future for Naval Shipyards?" *USNIP* 104 (April 1978): 32–40; Korb, "Erosion of American Naval Pre-eminence," 328–31.

3. Robert Andrews Oral History Transcript, Oral History Project; Combat Camera Group, Atlantic Fleet, " 'Tidewater Navy': The Norfolk Naval Complex," *USNIP* 98 (August 1972): 71–85; "Tidewater Strategy," *Evening Bulletin*, 12 October 1977, [Philadelphia Navy Yard] Closing Rumors, 1969– File, box 173A, 6 November 1977, Features File, box 173A, Mounted Clippings.

4. Eilberg, in Philadelphia Naval Shipyard Chronology of Highlights—1973, typescript, page 5, also, Chronology–1974, for Kissinger, box 1, PNSY History.

5. *Beacon*, 27 April 1973; *Evening Bulletin*, 18 April 1973, [Philadelphia Navy Yard] Editorials File, also, see Closing Rumors 1969– File, box 173A, Mounted Clippings; William F. Trimble, *Wings for the Navy: A History of the Naval Aircraft Factory, 1917–1956* (Annapolis, Md.: Naval Institute Press, 1990), 327–31.

6. Chronology—1973, p. 6 for UNIVAC, box 1, PNSY Histories; *Beacon*, 13 July 1973, 28 February 1975, 16 July 1971; *Philadelphia Inquirer*, 28 June 1974, [Philadelphia Navy Yard] Housing, 1960– File, box 173A, Mounted Clippings; Commander William A. Cockell, Jr., "The DLG Modernization Program," *USNIP* 98 (January, 1972): 101–4.

7. Elmer Zellers Oral History Transcript, 21, Moran Oral History Transcript, 10, Oral History Project; *Beacon*, 23 October 1961, 18 January 1963, 10 June, 8 July 1966, 16 August 1968.

8. Command History—1973, p. 2, box 1, PNSY Histories; *Beacon*, 22 January and 7 May 1971, 24 March 1972, 3 August 1973; Elmo R. Zumwalt, Jr., *On Watch: A Memoir* (New York: Quadrangle/New York Times, 1976), 217–34; Norman Friedman, "Elmo Russel Zumwalt, Jr., 1 July 1970–1 July 1974," in Robert W. Love, Jr., ed., *The Chiefs of Naval Operations* (Annapolis, Md.: Naval Institute Press, 1980), 377.

9. Lorraine Daliessio Oral History Transcript, 5, Oral History Project; Command History—1977, page 6, PNSY Histories; *Beacon*, 3 August 1973, 11 January 1974, 9 January 1976, 16 February 1979.

10. Command History—1978, page 6, box 1, PNSY History; *Beacon*, 31 March and 21 April 1978; Floyd D. Kennedy, Jr., "From SLOC Protection to a National Maritime Strategy: The U.S. Navy Under Carter and Reagan, 1977–1984," in Hagan, *Peace and War*, 348; Norman Friedman, *U.S. Aircraft Carriers: An Illustrated Design History* (Annapolis, Md.: Naval Institute Press, 1983), 266–67.

11. *Beacon*, 21 April 1978; *Evening Bulletin*, 18 June 1975, Bicentennial File, box 173A, Mounted Clippings; Friedman, *Aircraft Carriers*, 266–67.

12. Naval Supply Systems Command Uniform Automated Data Processing System, in Chronology—1974, Quality Assurance Program in Chronology—1976, Planning & Engineering for Repair and Alteration (PERA) Office, in Chronology—1970–71, box 1, PNSY Histories; *Beacon*, 11 June 1971, 5 March 1976, 9 January and 5 March 1976, 13 April 1979; "Mothball Fleet Still Sleeping at Naval Base," *Philadelphia Inquirer*, 3 April 1973, Features File, box 173A, Mounted Clippings.

13. Command History—1973, page 4, box 1, PNSY Histories; *Beacon*, 27 April 1973, 21 April 1978, 11 May 1979; John D. Alden, "Tomorrow's Fleet," *USNIP* 105 (January 1979): 115; Robert R. Frump, "The Great Aircraft Carrier War," *Philadelphia Inquirer*, 20 July 1980; *Evening Bulletin*, 17 April 1978, [Philadelphia Navy Yard] Editorials File, box 173A, *Evening Bulletin*, 24 May 1980, Miscellaneous 1980–1982 File, box 173B, Mounted Clippings.

14. "Chronology of Highlights—1980," typescript, box 1,

PNSY Histories; *Beacon*, 21 April 1978; J. D. McCaffrey, "Navy Is Criticized over Yard in Philadelphia," *Evening Bulletin*, 14 April 1978, [Philadelphia Navy Yard] Employees, Employment 1965– File, box 173A, *Evening Bulletin*, 29 September 1980, Miscellaneous File, 1980–1982, box 173B, Mounted Clippings; *New York Times*, 11 June 1978.

15. Moran Oral History Transcript, 16, Oral History Project; Command History—1978, page 6, Chronology—1980, box 1, PNSY Histories; *Agreement Between Philadelphia Naval Shipyard and Philadelphia Metal Trades Council, 1965*, copy of contract, ISML; *Beacon*, 6 December 1963, 31 March and 21 April 1978; A. V. Guisini, *The Philadelphia Naval Shipyard Handbook* (Philadelphia: Philadelphia Employee Development Division, 1991), x–13b.

16. Moran Oral History Transcript, 19, Daliessio Oral History Transcript, 8–9, Oral History Project; Kennedy, "From SLOC Protection," 354–55; Charles Layton and Michael Mally, "The Navy Yard," *Oceans* 21 (January 1988): 32–37 (34).

17. *Beacon*, 10 and 24 August, 14 and 28 September, 26 October, and 30 November 1984.

18. *Beacon*, 16 July 1971, 30 November and 14 December 1984, 11 January, 29 March, 19 and 26 April, 17 May, and 4 October 1985.

19. Command History—1965, typescript, 1; box 1, PNSY History; *Beacon*, 1 November 1985, 11 December 1964.

20. *Beacon*, 1 November 1985, 7 March, 2 May and 3 October 1986; Layton and Mally, "The Navy Yard," 32–37; Kennedy, "From SLOC," 354–55.

21. *Beacon*, 7 April 8 and 22 December 1989 first base closure article, 4 November 1988; U.S. Department of the Navy, *Final Environmental Impact Statement: Disposal and Reuse of Naval Base Philadelphia, Pennsylvania* (Lester, Pa.: Northern Division Naval Facilities Engineering Command, 1996), S-2.

22. "How Do You Feel About the Yard Being on the 'Hit List' Again?" *Beacon*, 9 and 23 February 1990.

23. *Beacon*, 20 April and 4 May 1990; Guisini, *Philadelphia Naval Shipyard Handbook*, 1–4.

24. *Beacon*, 24 August 1990, 18 January and 22 June 1991; "Desert Shield, Getting There," *USNIP* 116 (October 1990): 104.

25. Pat D'Amico Oral History Transcript, Oral History Project; Wilmot biography, in *Baseline* (Newsletter of the Office of Defense Conversion, Commerce Department, City of Philadelphia) 1 (1994), copies in Free Library of Philadelphia; *Beacon*, 27 March, 22 May, 19 June 1992, 15 January 1993; *New York Times*, 17 October 1993.

26. Clark, in *Beacon*, 24 August 1990, 14 February 1992; *Final Environmental Impact Statement*, S1-1/2; U.S. Senate Committee on Armed Services, *Defense Base Clo-*

sure and Realignment Commission, [BRAC] Hearings before the Committee on Armed Services, United States Senate, 102nd Congress, 1st Session*; Brian Kelly, "Hogwild! Here's the Score on This Years's Biggest Pork-Barrel Hustle," *Regardie's* 11(June/July 1991): 47–48.

27. Claman to Chief of Naval Operations, 29 March 1991, in *Base Closure Hearings* (1992): 132–81 (156); *Beacon*, 31 January 1992; *Final Environmental Impact Statement*, S-1; Douglas C. McVarish, Thomas M. Johnson, John P. McCarthy, and Richard Meyer, *A Cultural Resources Survey of the Naval Complex Philadelphia, Philadelphia Pennsylvania*, 2 vols. (West Chester, Pa.: John Milner Associates, for Department of the Navy Northern Division Naval Facilities Engineering Command, Lester, Pennsylvania, 1994), 1.

28. *Base Closure Hearings* (1992): 178–80; *Beacon*, 12 April 1991.

29. *Base Closure Hearings* (1992): 132–50, 159–61, 180–81.

30. *Beacon*, 17 January, 28 February, 13 March, and 14 August 1992, 24 April, 7 May, 16 July, 14 September, and 8 October 1993, 4 November 1994, 11 August 1995; *New York Times*, 3 March and 24 May 1994.

31. Trash dump in Commandant's Order No. 18/40, 13 November 1940, river pollution in Commandant's Order No. 9/42, 28 February 1942, File A3-2, box 21, Commandants Orders, Central Subject Files; *Beacon*, 17 July and 11 August 1992; "Naval Yard Cleanup," *Philadelphia Inquirer*, 26 October 1997, E3; also see *Final Environmental Impact Statement*.

32. Pennsylvania Economy League and WEFA Group, *Economic Impact of the Philadelphia Naval Base and Shipyard on the Philadelphia Metropolitan Area* (Philadelphia: Pennsylvania Economy League, 1990); The Mayor's Commission of Defense Conversion, *Philadelphia Naval Base Conversion Initiative: A Concept Paper* (Philadelphia: Commission of Defense Conversion, 1993), 11, copy in Free Library of Philadelphia; *Baseline*, 1 (January 1994); *Final Environmental Impact Statement*, S-1.

33. Philadelphia Office of Defense Conversion, *League Island: An Environment of Innovation—Community Reuse Plan for the Philadelphia Naval Base and Shipyard, Philadelphia* (Philadelphia: Mayor's Commission of Defense Conversion, 1994); The Department of the Navy and the City of Philadelphia Master Lease Signing Ceremony, November 22, 1994, program, copy in ISML; *Baseline* 1 (1994), 2 (January/February; July/August, 1995); *Camden Courier-Post*, 22 September 1996, 17 July 1998; *Final Environmental Impact Statement*, 1–3.

34. *Beacon*, 17 December 1993, 25 March 1994, 27 January and 14 July 1995; *Baseline* 1 (1994); *Courier-Post*, 6 December 1997.

35. *Beacon*, 14 July 1995; *Baseline* 1 (1994); "Business Special Update on the Naval Base," *Philadelphia Inquirer*, 23 September 1996, 28 September 1997; *Courier-Post*, 25 July 1996, 20 January 1997; *New York Times*, 12 March 1994.

36. *Baseline* 1 (May/June, August/September, 1994): 2 (January/February, September/ October, 1995); *Courier-Post*, 22 September 1996, 24 October, 17 December 1997.

37. Ship's Deck Log Sheet, Philadelphia Naval Shipyard, 9/27/96, copy, ISML; Closure Ceremony Brochure, box 4, PNSY Histories; *Beacon*, 1 November 1985; *Courier-Post*, 10 September 1995, 28 September 1996; *Philadelphia Inquirer*, 23 and 28 September 1996.

INDEX